FOUNDATIONS OF

QUANTUM MECHANICS

From Photons to Quantum Computers

Reinhold Blümel

Wesleyan University

JONES AND BARTLETT PUBLISHERS
Sudbury, Massachusetts
BOSTON   TORONTO   LONDON   SINGAPORE

World Headquarters
Jones and Bartlett Publishers
40 Tall Pine Drive
Sudbury, MA 01776
978-443-5000
info@jbpub.com
www.jbpub.com

Jones and Bartlett Publishers
Canada
6339 Ormindale Way
Mississauga, Ontario L5V 1J2
Canada

Jones and Bartlett Publishers
International
Barb House, Barb Mews
London W6 7PA
United Kingdom

Jones and Bartlett's books and products are available through most bookstores and online booksellers. To contact Jones and Bartlett Publishers directly, call 800-832-0034, fax 978-443-8000, or visit our website, www.jbpub.com.

> Substantial discounts on bulk quantities of Jones and Bartlett's publications are available to corporations, professional associations, and other qualified organizations. For details and specific discount information, contact the special sales department at Jones and Bartlett via the above contact information or send an email to specialsales@jbpub.com.

Copyright © 2010 by Jones and Bartlett Publishers, LLC

All rights reserved. No part of the material protected by this copyright may be reproduced or utilized in any form, electronic or mechanical, including photocopying, recording, or by any information storage and retrieval system, without written permission from the copyright owner.

Production Credits:
Publisher: David Pallai
Editorial Assistant: Melissa Potter
Production Director: Amy Rose
Senior Marketing Manager: Andrea DeFronzo
V.P., Manufacturing and Inventory Control: Therese Connell
Composition: Northeast Compositors, Inc.
Cover and Title Page Design: Scott Moden
Cover and Title Page Image: © FloridaStock/ShutterStock, Inc.
Printing and Binding: Malloy, Inc.
Cover Printing: Malloy, Inc.

Library of Congress Cataloging-in-Publication Data
Blümel, R. (Reinhold)
 Foundations of quantum mechanics : from photons to quantum computers / Reinhold Blümel.
   p. cm.
 Includes bibliographical references and index.
 ISBN-13: 978-0-7637-7628-2 (hardcover)
 ISBN-10: 0-7637-7628-9 (hardcover)
 1. Quantum theory. 2. Wave-particle duality. 3. Einstein-Podolsky-Rosen experiment. 4. Photons. I. Title.
 QC174.12.B58 2010
 530.12—dc22
                                2009019915

6048
Printed in the United States of America
13 12 11 10 09   10 9 8 7 6 5 4 3 2 1

For Lynn

Contents

Preface ix

1 Photons 1
   1.1 Introduction 1
   1.2 Young's Double-Slit Experiment 2
   1.3 Photoelectric Effect: Einstein's Quanta 10
   1.4 Experiment of Aspect and Collaborators 13
   1.5 Properties of the Photon 18
   1.6 Summary 22
   Chapter Review Exercises 22

2 Wave-Particle Duality 25
   2.1 Introduction 25
   2.2 A Diffraction Experiment with Photons 26
   2.3 Young's Double-Slit Experiment Revisited 29
   2.4 Three Rules of Quantum Mechanics 34
   2.5 Two-Slit Which-Way Experiment 37
   2.6 Summary 40
   Chapter Review Exercises 42

3 The Machinery of Quantum Mechanics 43
   3.1 Introduction 43
   3.2 Schrödinger's Equation 44
   3.3 Observables 49
   3.4 Spectral Theory 62
   3.5 Dirac Notation 90
   3.6 Heisenberg Picture 101

Contents

    3.7   Two-Level Systems 105
    3.8   Summary 120
    Chapter Review Exercises 121

4 Measurement   125
    4.1   Introduction 125
    4.2   von Neumann Measurement 128
    4.3   Uncertainty Principle 135
    4.4   No-Cloning Theorem 138
    4.5   Quantum Zeno Effect 141
    4.6   Summary 145
    Chapter Review Exercises 146

5 Interaction-Free Measurements   149
    5.1   Introduction 149
    5.2   Seeing in the Dark: Conceptual Scheme 151
    5.3   Elitzur-Vaidman Scheme 154
    5.4   Optimal Interaction-Free Measurements 159
    5.5   Summary 165
    Chapter Review Exercises 166

6 EPR Paradox   169
    6.1   Introduction 169
    6.2   Hallmarks of Physical Theories 170
    6.3   EPR and Reality 174
    6.4   Bell's Theorem 182
    6.5   Mermin's Reality Machine 191
    6.6   Summary 194
    Chapter Review Exercises 196

7 Classical and Quantum Information   197
    7.1   Introduction 197
    7.2   Bits and Qubits 199
    7.3   Classical Gates 202
    7.4   Quantum Gates 205
    7.5   Classical and Quantum Circuits 209
    7.6   Teleportation 213
    7.7   Summary 220
    Chapter Review Exercises 221

8 Quantum Computing   223
    8.1   Introduction 223
    8.2   Our First Quantum Computer 224

	8.3 Deutsch's Algorithm	227
	8.4 Deutsch-Jozsa Algorithm	232
	8.5 Grover's Search Algorithm	238
	8.6 Summary	246
	Chapter Review Exercises	247

9 Classical Cryptology 251
    9.1 Introduction 251
    9.2 Private-Key Cryptosystems 252
    9.3 RSA Public-Key Cryptosystem 254
    9.4 How Does RSA Work? 260
    9.5 Why Is Integer Factorization So Difficult? 262
    9.6 Summary 265
    Chapter Review Exercises 266

10 Quantum Factoring 267
    10.1 Introduction 267
    10.2 Miller's Algorithm 268
    10.3 Quantum Fourier Transform 270
    10.4 Shor's Algorithm 274
    10.5 Summary 280
    Chapter Review Exercises 280

11 Ion-Trap Quantum Computers 281
    11.1 Introduction 281
    11.2 Linear Radio-Frequency Ion Trap 282
    11.3 Laser Cooling 287
    11.4 Cirac-Zoller Scheme 293
    11.5 Ca^+ Quantum Computer 297
    11.6 Summary 300
    Chapter Review Exercises 301

12 Outlook 303
    12.1 Introduction 303
    12.2 Quantum Internet 304
    12.3 Quantum Cryptography 306
    12.4 Quantum Computing 307
    12.5 Summary 309

Appendix 311

Index 313

Preface

Quantum mechanics is more than a physical theory of nature. It is a quantitative philosophy that provides us with a set of general, overarching principles that describe the innermost workings of our world at its most fundamental level. Due to its central importance, all physical and philosophical theories have to be consistent with quantum mechanics. There are two aspects of quantum mechanics: the "machinery" and the "spook." The machinery, epitomized by the Schrödinger equation and its various methods of solution, allows us to propagate the quantum state of a system deterministically forward in time. This aspect of quantum mechanics is not particularly "quantum;" we find it in similar form in all classical field theories. The "spook" are all those aspects of quantum mechanics that do not have a classical analogue, not even in principle. This part of quantum mechanics is connected with the theory of measurement and its implications. Since quantum mechanics is the most basic theory of nature, no wonder that quantum mechanics is capable of confronting basic philosophical questions head-on, prime among them the nature of reality.

For the longest time, the philosophical implications of quantum mechanics were far from main-stream physics. Only very few physicists studied them because it was felt that this is idle speculation, alien to a "hard," quantitative science such as physics. In particular, physicists thought that there are no consequences or applications. This point of view was fueled by the immense success of the quantum mechanical machinery, which, like a steam roller, flattened and solved any "practical" problem it was applied to. In such an atmosphere of success, questions probing the foundations of the theory appeared irrelevant and futile.

"Shut up and calculate" was the standard response of teachers and research advisers to students of quantum mechanics raising questions concerning the meaning and interpretation of quantum mechanics.

Surprisingly, starting in the mid 1970s and early 1980s, questions about the foundations of quantum mechanics led to direct technological advances and applications, culminating in the possibility of quantum computing, a qualitatively new way of data processing that promises to be orders of magnitudes faster and more efficient than any "classical" computer could ever hope to be. At this point in time, the promise of quantum information and quantum computation seems boundless and includes secure quantum communication using the quantum mechanical "no-cloning theorem," ultra-fast quantum computers using massive quantum parallelism, code breaking in polynomial time, and perhaps the exchange of secure and ultra-fast communications using a future quantum Internet. Even teleportation, the "beaming" of the quantum states of photons and matter from one point in space to another, has already been demonstrated in the lab. Thus, the current, thorough inspection of the foundations of quantum mechanics may lead to a new industrial revolution, that of quantum technology.

Quantum computing, teleportation, and quantum information may sound like science fiction, but are far from idle dreams: Proof-of-concept demonstrations of all components of the new quantum technology have already been provided in many laboratories around the world. Some, such as quantum cryptography, appealing for use in secure banking and government transactions, have already found commercial applications. Since the functionality of all components necessary for the new quantum information age have already been tested in the lab, there is no doubt that systems, such as quantum computers, built from these components, will work.

The conceptual basis for the new quantum technology consists of (a) superposition, (b) entanglement, and (c) quantum measurement (i.e., the collapse of the wave function). In connection with matter, these concepts simply do not exist in the classical description of our world and are not used in any of the "classical" machines and appliances around us. Thus, quantum technology is not just better, it is different. The new quantum concepts, implemented technologically, are expected to considerably enhance the possibilities, the power, and the repertoire of our technology-based civilization.

I designed this course in response to the currently unfolding quantum information revolution. While traditional courses in quantum mechanics emphasize the solution of Schrödinger's equation, the new quantum information technology emphasizes the "spooky" and "weird" parts of quantum mechanics. Therefore, in order to prepare students for the new quantum information age, this course reverses the paradigm and focuses on what is "quantum" about quantum mechanics. When researching the material for this book, I profited immensely from many excellent research articles and text books on the foundations of quantum mechanics, for instance the two excellent books "The Quantum Challenge" by George Greenstein and Arthur G. Zajonc, and "Introduction to Quantum Mechanics" by David J. Griffiths.

To follow this course, only a minimal set of prerequisites is required: Introductory courses in linear algebra (vectors and matrices) and (multivariable) calculus are sufficient. Familiarity with complex numbers is expected. An introductory course in traditional quantum mechanics, usually taught as "Quantum Mechanics I" in four-year colleges, is helpful but not required.

Directed primarily at undergraduate students in their Junior and Senior years, this course is also useful as an introductory text for graduate students and researchers. This course is not just directed at physicists, students and professionals alike; mathematicians and engineers will profit from it as well. Readers who have mastered the material of this course are well prepared to understand the current popular and research literature on the foundations of quantum mechanics, including its various technological applications.

True to its focus on the non-classical aspects of quantum mechanics, the course starts out in Chapters 1 and 2 by introducing photons, the quanta of light, and their surprising behavior in diffraction experiments that defies everyday logic. While Chapters 1 and 2 open our eyes to the strange world of the quantum, quantitative predictions require tools and equations. These are provided in Chapter 3, where we study the machinery of quantum mechanics such as the Schrödinger equation and its solution. Thus, although focusing on the foundations of quantum mechanics and what distinguishes quantum mechanics from a simple classical field theory of Schrödinger's ψ function, the course does not neglect the teaching of the more traditional material expected in a well-rounded course on quantum mechanics. In Chapters 4–10, ranging from the theory of measurement to code-breaking with quantum computers, we develop the core material of this course, i.e., quantum information and quantum computing. To demonstrate that quantum computing is more than wishful thinking, a real-life quantum computer is presented in Chapter 11. In Chapter 12, the course closes with a brief outline of currently existing quantum information technology and an outlook on things to come, including a quantum Internet that already exists in rudimentary form in several countries around the world.

I would like to express my thanks to the team at Jones and Bartlett Publishers for making this book a reality: First and foremost, thank you to Dave Pallai, who started this project by getting me interested in writing a text on quantum mechanics that includes some of the latest developments in the field. I am most grateful for his support, encouragement, and understanding in all phases of this project. Many thanks also to Amy Rose for expertly guiding this project through the production process, and to the marketing team for a superb job of extraordinary professional quality. I'd also like to thank the artist, compositor, designer, and copyeditor for a job well done.

Special thanks to my wife, Lynn Westling, for many substantial discussions during the writing phase of this book, and to my students at Wesleyan University for their useful suggestions in the trial runs of this course.

Middletown, Connecticut

chapter 1
Photons

In this chapter:

- Introduction
- Young's Double-Slit Experiment
- Photoelectric Effect: Einstein's Quanta
- Experiment of Aspect and Collaborators
- Properties of the Photon
- Summary

1.1 Introduction

What is the nature of light? Newton thought that light consists of corpuscles, tiny particles that explain the propagation of light in the form of rays and explain much of geometric optics. However, even Newton had to admit that not all light phenomena are easily explained by the picture of a grainy light. Huygens, a contemporary of Newton, had a different view. Huygens was convinced that light consists of waves. The struggle between the corpuscular theory of light and the wave theory of light seemed settled once and for all when Young, in 1801, showed that light produces interference fringes, something only waves can do. Based on the experiments of Young and other researchers, the wave theory of light was then accepted for more than a century when Einstein, in 1905, showed that the photoelectric effect is most naturally explained by assuming that light consists of packets of energy, later named *photons* (see Section 1.3). Einstein's theory of light appears to vindicate Newton's corpuscles. Einstein's photons, however, are much more complex. In Einstein's theory the energy E of a photon is related to its frequency ν, a characteristic of waves, via Planck's constant h, according to $E = h\nu$. With Einstein, the wave and particle pictures of light began to meld. And indeed, we now know that nature mixes wave and particle aspects for photons as well as any other objects whatsoever in a way that has become known as the *wave-particle duality*. We will study this aspect of nature in more detail in Chapter 2. While Einstein's photonic picture of light looked convincing at first, some scientists cast doubt on the existence of photons. An experiment conducted in 1956 by Hanbury-Brown and Twiss seemed to contradict the existence of photons. The photon puzzle was finally resolved by Aspect and collaborators who, in 1986, convincingly demonstrated the existence of photons with a textbook style experiment (see Section 1.4).

In this chapter we will follow the historic path of the evolution of our ideas about light. In Section 1.2 we start with Young's double-slit experiment and the wave theory of light that explains it. We continue with Einstein's photons in Section 1.3, and a demonstration of the existence of photons by Aspect and collaborators in Section 1.4. Some properties of the photon are discussed and summarized in Section 1.5.

1.2 Young's Double-Slit Experiment

For much of the latter half of the 18th century the corpuscular and the wave theories of light coexisted without a clear champion. There was perhaps a bit more evidence in favor of the wave theory of light, such as the phenomena of diffraction and birefringence, but the corpuscular theory, buoyed by Newton's authority, held its own. This changed rapidly when, in 1801, Young conducted his double-slit experiment with light. This experiment showed that light may produce interference patterns. Since interference is a typical wave property, incompatible with the behavior of particles, Newton's corpuscular theory was proved wrong, and the 19th century saw a vigorous development of wave optics. This development reached a high point in the latter half of the 19th century when Maxwell predicted that light is an electromagnetic wave, characterized by an electric field vector $\boldsymbol{E}(\boldsymbol{r};t)$, where $\boldsymbol{E}$ is the electric field strength of the light field at a point $\boldsymbol{r}$ at time t. Maxwell's theory was confirmed experimentally in 1888 by Hertz. Thus, toward the end of the 19th century, the wave theory of light was supported by numerous experiments and had a solid theoretical underpinning.

Consider the setup shown in Figure 1.1(a). Light with electric field $\boldsymbol{E}_i(x,z;t)$ is incident on a flat, opaque screen Σ, infinitely extended in the x and y directions, and positioned parallel to the x-y plane at $z = z_0$.

1.2 • YOUNG'S DOUBLE-SLIT EXPERIMENT 3

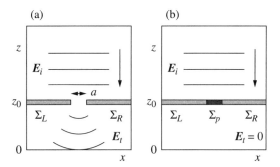

Figure 1.1 Diffraction of a light wave at a single slit in a screen Σ. (a) A slit of width a divides the screen Σ into two parts, Σ_L and Σ_R. (b) The slit is closed with a plug Σ_p.

A slit of width a, parallel to the y-axis, is cut into the screen, dividing Σ into two parts, Σ_L and Σ_R. Since the problem is translationally invariant in the y-direction, we consider it as a two-dimensional problem in the x-z plane. As shown in Figure 1.1(a), the light impinging on Σ is blocked by Σ_L and Σ_R, but is allowed to pass through the slit, slightly spreading out to the sides. The electric field of the light passing through the slit, the *transmitted light*, is denoted by $\boldsymbol{E}_t$. The spreading of $\boldsymbol{E}_t$ to the sides is the phenomenon of *diffraction*, which we are now going to analyze quantitatively. More specifically, our task is to compute the field strength of the light field on the x-axis (see Figure 1.1). To this end, imagine that we plug up the slit in Σ with a plug Σ_p as shown in Figure 1.1(b). In this case, since Σ is not transparent, no light is observed below the screen ($z < z_0$), i.e., the transmitted light field, $\boldsymbol{E}_t(x, z; t)$, is zero. Physically this is so since the incident light field $\boldsymbol{E}_i(x, z; t)$ excites the molecular dipoles in the screen to oscillate and radiate secondary fields $\boldsymbol{E}_L$, $\boldsymbol{E}_p$, and $\boldsymbol{E}_R$ produced by Σ_L, Σ_p, and Σ_R, respectively, such that the transmitted field in $z < z_0$ is exactly cancelled:

$$\boldsymbol{E}_t = \boldsymbol{E}_i + \boldsymbol{E}_L + \boldsymbol{E}_p + \boldsymbol{E}_R = 0. \tag{1.1}$$

Removing the plug, i.e., returning to the setup in Figure 1.1(a), molecular oscillators are present only in Σ_L and Σ_R. Therefore, with Equation 1.1,

$$\begin{aligned}\boldsymbol{E}_t &= \boldsymbol{E}_i + \boldsymbol{E}_L + \boldsymbol{E}_R \\ &= (\boldsymbol{E}_i + \boldsymbol{E}_L + \boldsymbol{E}_p + \boldsymbol{E}_R) - \boldsymbol{E}_p = -\boldsymbol{E}_p.\end{aligned} \tag{1.2}$$

Therefore, the transmitted field appears as if the slit region consists of radiation sources, formerly represented by the dipoles in the plug, generating the field $\boldsymbol{E}_p$. Consequently, we may represent the slit region as a collection of elementary radiation sources whose superposition yields the transmitted field $\boldsymbol{E}_t$. This is *Huygens' Principle*. It states that:

> Any point of the radiation field may be considered the source of elementary waves whose superposition accounts for the propagation of the radiation field. (1.3)

Let us apply Huygens' Principle to compute $\boldsymbol{E}_t$ for the single slit of Figure 1.1(a). According to Maxwell's theory the electromagnetic field in vacuum satisfies d'Alembert's equation

$$\left[\frac{\partial^2}{\partial x^2} + \frac{\partial^2}{\partial y^2} + \frac{\partial^2}{\partial z^2} - \frac{1}{c^2}\frac{\partial^2}{\partial t^2}\right] \boldsymbol{E}(\boldsymbol{r};t) = \boldsymbol{0}, \tag{1.4}$$

where $c = 2.9979\ldots \times 10^8$ m/s is the speed of light. The solution of Equation 1.4 can be written in the form

$$\boldsymbol{E}(\boldsymbol{r};t) = \boldsymbol{E}_0(r)\, e^{i(\boldsymbol{k}\cdot\boldsymbol{r}-\omega t)}, \tag{1.5}$$

where $\boldsymbol{E}_0(r)$, the amplitude of the electric field, is a slowly varying function of $r = |\boldsymbol{r}|$, $\boldsymbol{k}$ is the wave vector, ω is the angular frequency of the light, and the dot in the exponent indicates the scalar product, i.e.,

$$\boldsymbol{k}\cdot\boldsymbol{r} = k_x x + k_y y + k_z z. \tag{1.6}$$

In vacuum k and ω are related by the *dispersion relation*

$$k = |\boldsymbol{k}| = \frac{\omega}{c}. \tag{1.7}$$

Now consider Figure 1.2. It shows a magnification of the slit region of Figure 1.1(a). In the spirit of Huygens' Principle we placed N equidistant radiation sources into the gap with a spacing of

$$d = \frac{a}{N+1}. \tag{1.8}$$

Later, we will take N to infinity, thus representing a continuum of radiation sources spread out over the width of the slit. Assuming that the observation point P is very far from the gap (the *Fraunhofer limit*), the distances of the individual sources in the gap to the observation point P are

$$r_n = r_1 - (n-1)d\sin(\theta), \quad n = 1, 2, \ldots, N, \tag{1.9}$$

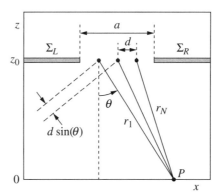

Figure 1.2 Magnification of the slit region of Figure 1.1(a). The dots in the slit region represent elementary radiation sources, implementing Huygens' Principle for the computation of the electric field strength in the observation point P on the x-axis.

where θ is the observation angle with respect to the vertical. We write the slowly varying part of the electric field in P as $\boldsymbol{E}_0(r)$, where

$$r = r_1 - \left(\frac{N-1}{2}\right) d \sin(\theta) \tag{1.10}$$

is the distance from the midpoint of the gap to the observation point P. The total field strength $\boldsymbol{E}_0(r)$ is the sum of the individual field strengths $\boldsymbol{E}_0^{(N)}(r)$ produced by each of the N sources in the gap. We assume that the electric field is practically constant over the width of the gap such that

$$\boldsymbol{E}_0^{(N)}(r) = \frac{1}{N} \boldsymbol{E}_0(r). \tag{1.11}$$

Now we sum the individual contributions of the N sources in the gap to obtain the field strength in the observation point P. Assuming that in the Fraunhofer limit $\boldsymbol{k}$ and $\boldsymbol{r}_n$, $n = 1, \cdots, N$, are practically parallel ($\boldsymbol{k} \cdot \boldsymbol{r}_n \approx k r_n$), we obtain:

$$\begin{aligned}
\boldsymbol{E}_t(r;t) &= \boldsymbol{E}_0^{(N)}(r) \sum_{n=1}^{N} e^{i(kr_n - \omega t)} \\
&= \boldsymbol{E}_0^{(N)}(r) e^{-i\omega t} \sum_{n=1}^{N} e^{ik[r_1 - (n-1)d\sin(\theta)]} \\
&= \boldsymbol{E}_0^{(N)}(r) e^{i(kr_1 - \omega t)} \sum_{n=0}^{N-1} \left[e^{-ikd\sin(\theta)}\right]^n .
\end{aligned} \tag{1.12}$$

With the help of the geometric-series summation formula

$$\sum_{m=0}^{M-1} q^m = \frac{q^M - 1}{q - 1}, \tag{1.13}$$

we sum Equation 1.12 analytically to obtain

$$\boldsymbol{E}_t(r;t) = \boldsymbol{E}_0^{(N)}(r)\, e^{i(kr_1 - \omega t)} \frac{e^{-ikNd\sin(\theta)} - 1}{e^{-ikd\sin(\theta)} - 1}. \tag{1.14}$$

With

$$\alpha = kd\sin(\theta), \tag{1.15}$$

Equation 1.14 may be written as

$$\begin{aligned}\boldsymbol{E}_t(r;t) &= \boldsymbol{E}_0^{(N)}(r)\, e^{i(kr_1 - \omega t)} \frac{e^{-iN\alpha/2}\left(e^{-iN\alpha/2} - e^{iN\alpha/2}\right)}{e^{-i\alpha/2}\left(e^{-i\alpha/2} - e^{i\alpha/2}\right)} \\ &= \boldsymbol{E}_0^{(N)}(r)\, e^{i(kr - \omega t)} \frac{\sin(N\alpha/2)}{\sin(\alpha/2)}. \end{aligned} \tag{1.16}$$

Since the intensity distribution of the light in the gap is continuous, we need to let the number N of sources go to infinity. Defining

$$\Delta = ka\sin(\theta), \tag{1.17}$$

we obtain in the limit of $N \to \infty$:

$$\frac{\sin(N\alpha/2)}{\sin(\alpha/2)} = \frac{\sin\left(\frac{N\Delta}{2(N+1)}\right)}{\sin\left(\frac{\Delta}{2(N+1)}\right)} \to N \frac{\sin(\Delta/2)}{(\Delta/2)}. \tag{1.18}$$

Inserting this result into Equation 1.16 and using Equation 1.11, we obtain:

$$\boldsymbol{E}_t(r;t) = \boldsymbol{E}_0(r)\, e^{i(kr - \omega t)} \frac{\sin(\Delta/2)}{(\Delta/2)}. \tag{1.19}$$

The physical field strength is the real part of the complex field strength of Equation 1.19:

$$\boldsymbol{E}_t^{(\text{phys})}(r;t) = \boldsymbol{E}_0(r) \frac{\sin(\Delta/2)}{(\Delta/2)} \cos(kr - \omega t). \tag{1.20}$$

According to Maxwell's theory, the time-averaged intensity $I(\boldsymbol{r})$ of light with electric field $\boldsymbol{E}(\boldsymbol{r}, t)$ in a point $\boldsymbol{r}$ is

$$I(\boldsymbol{r}) = \frac{1}{2} c\epsilon_0 |\boldsymbol{E}(\boldsymbol{r}, t)|^2, \tag{1.21}$$

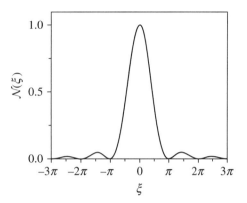

Figure 1.3   Normalized intensity distribution $\mathcal{N}(\xi)$.

where c is the speed of light and ϵ_0 is the permittivity of the vacuum. In our case, using the electric field specified in Equation 1.19, the time-averaged light intensity in P is:

$$I_t(r) = I_0(r) \frac{\sin^2(\Delta/2)}{(\Delta/2)^2}, \qquad (1.22)$$

where

$$I_0(r) = \frac{1}{2} c \epsilon_0 \boldsymbol{E}_0(r)^2. \qquad (1.23)$$

For small angles θ we have

$$\Delta \approx ka\theta = \frac{2\pi}{\lambda} a\theta, \qquad (1.24)$$

where λ is the wave length of the light. Therefore:

$$\mathcal{N}(\theta) = \frac{I_t(r)}{I_0(r)} \approx \frac{\sin^2(\pi a\theta/\lambda)}{(\pi a\theta/\lambda)^2}, \qquad (1.25)$$

where $\mathcal{N}(\theta)$ is the normalized intensity distribution. It is shown in Figure 1.3 as a function of

$$\xi = \pi a\theta/\lambda. \qquad (1.26)$$

According to Figure 1.3, the light intensity on the x-axis of Figure 1.2 (Figure 1.1(a), respectively) oscillates as a function of θ. It is maximal for $\theta = 0$, i.e., if the observation point P is located directly under the gap.

The intensity then drops and reaches a first minimum of zero intensity at

$$\xi_{\min}^{(1)} = \pm\pi, \tag{1.27}$$

i.e., for

$$\theta_{\min}^{(1)} = \pm\frac{\lambda}{a}. \tag{1.28}$$

Further minima are located at $\xi_{\min}^{(\nu)} = \pm\nu\pi$, $\nu = 2, 3, \ldots$. Therefore, in summary, the minima of $\mathcal{N}(\theta)$ are located at

$$\theta_{\min}^{(\nu)} = \pm\nu\frac{\lambda}{a}, \quad \nu = 1, 2, \ldots. \tag{1.29}$$

The maxima are located at

$$\theta_{\max}^{(0)} = 0, \quad \theta_{\max}^{(\nu)} = \pm\left(\nu + \frac{1}{2}\right)\frac{\lambda}{a}, \quad \nu = 1, 2, \ldots. \tag{1.30}$$

The normalized intensity distribution shown in Figure 1.3 is what is actually observed in the laboratory. If light were made of corpuscles, i.e., hard pellets, then what would be observed is an illuminated area of width a on the x-axis directly under the slit, and darkness outside of this illuminated strip, with a sharp border between the two. This is not what is observed. Instead, as shown in Figure 1.3, there is no sharply defined border between light and dark regions, an observational fact inconsistent with a corpuscular theory of light.

While Young might have performed experiments with single slits, his most famous, and most convincing experiment was performed with two slits. Suppose we cut two parallel slits into Σ, each of width a, and separated by a distance D from midpoint to midpoint, as shown in Figure 1.4. Again we work in the Fraunhofer limit, i.e., we assume that the observation point P is far from both slits. The distance of the midpoint of the left slit to P is denoted by r_L and the distance of the right slit to P is denoted by r_R. The total light field in P is the sum of the light fields in Equation 1.19 generated by the left and right slits, i.e.,

$$\boldsymbol{E}_t(r;t) = \boldsymbol{E}_0(r)\,e^{-i\omega t}\left[e^{ikr_L} + e^{ikr_R}\right]\frac{\sin(\Delta/2)}{(\Delta/2)}, \tag{1.31}$$

where r is now the distance of the midpoint between the two slits to P. In terms of r we may write:

$$r_L = r + \frac{D}{2}\sin(\theta),$$
$$r_R = r - \frac{D}{2}\sin(\theta). \tag{1.32}$$

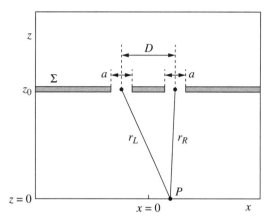

Figure 1.4 Young's double-slit experiment. The light field in point P on the x-axis is the sum of the light fields originating from the left and right slits.

The sum of the exponents in Equation 1.31 can now be written as:

$$e^{ikr_L} + e^{ikr_R} = e^{ikr}\left[e^{ikD\sin(\theta)/2} + e^{-ikD\sin(\theta)/2}\right]$$
$$= 2\,e^{ikr}\cos[kD\sin(\theta)/2]. \tag{1.33}$$

Therefore:

$$\boldsymbol{E}_t(r;t) = 2\,\boldsymbol{E}_0(r)\,e^{i(kr-\omega t)}\,\cos[kD\sin(\theta)/2]\,\frac{\sin(\Delta/2)}{(\Delta/2)}. \tag{1.34}$$

For the time-averaged intensity we obtain:

$$I_t(r) = 4\,I_0(r)\,\cos^2[kD\sin(\theta)/2]\,\frac{\sin^2[ka\sin(\theta)/2]}{[ka\sin(\theta)/2]^2}, \tag{1.35}$$

where $I_0(r)$ is defined in Equation 1.23. Equation 1.35 is the prediction for the intensity distribution of Young's double-slit experiment. As the observation point P moves along the x-axis, the intensity $I_t(r)$ at the observation point P displays fast oscillations produced by the term $\cos^2[kD\sin(\theta)/2]$ due to the finite separation D of the two slits, modulated by slow oscillations produced by the term $\sin^2[ka\sin(\theta)/2]/[ka\sin(\theta)/2]^2$ due to the finite widths a of the two slits.

In the course of the 19th century, Young's double-slit experiment had been performed many times and the observed intensity distributions were all found to be consistent with the prediction of Equation 1.35 computed on the basis of Maxwell's wave theory of light. Therefore, many researchers at the turn of the 19th to the 20th century were surprised when, in 1905, analyzing the physics of the photoelectric effect,

Einstein proposed a theory of "grainy light," seemingly reviving Newton's discredited corpuscular theory. The experimental evidence and the physical arguments that led Einstein to take a second look at particle theories of light are discussed in the following section.

Exercises:

1.2.1 For $D = a$ the double slit effectively becomes a single slit. Show that in this case the double-slit formula in Equation 1.35 predicts the same intensity distribution on the x-axis as the single-slit formula in Equation 1.22 for twice the slit width. Be careful about the meaning of $I_0(r)$ in both formulas.

1.2.2 On the basis of Equation 1.35, compute the angle $\theta_{\min}$ at which the first minimum of $I_t(r)$ occurs. Assume $D \gg a$.

1.3 Photoelectric Effect: Einstein's Quanta

In 1887 Hertz noticed that light had the capability to eject electrons from metal surfaces into the vacuum. This observation was subsequently called the *photoelectric effect*. A possible setup for demonstrating and investigating the photoelectric effect is shown in Figure 1.5. A metal surface $\mathcal{M}$ is illuminated with light of frequency ν. For large enough ν, electrons are observed to leave the metal surface. The ejected electrons are collected with an electrode $\mathcal{E}$ held at an electric potential $\mathcal{U}$. The collected electrons give rise to a photocurrent $\mathcal{I}$. In 1901 Lenard used a similar setup to measure the maximal kinetic energy $E_{\max}$ of the ejected electrons. To do this, as shown in Figure 1.5, he applied a negative voltage to the collection electrode $\mathcal{E}$. At a critical voltage $\mathcal{U}_c$ the ejected photoelectrons are no longer able to climb the potential hill due to the negative voltage and the photocurrent $\mathcal{I}$ vanishes. This method of determining $E_{\max}$ is known as the *counterfield method*. Lenard noticed that $E_{\max}$ is a function of the light's frequency ν, but is independent of the intensity of the incident light. This observation is hard to understand on the basis of the wave theory of light and motivated Einstein to postulate his quantum theory of light. According to Einstein, light consists of *quanta*, later called *photons*, i.e., point particles, each carrying an amount of energy $h\nu$, where

$$h = 6.626\ldots \times 10^{-34} \text{ Js} \qquad (1.36)$$

is Planck's constant. Einstein postulated that each of the light quanta may only interact with a single electron in the metal. In the process of interaction the photon is destroyed and transfers all of its energy to the electron it interacted with. The energy absorbed by the electron

1.3 • PHOTOELECTRIC EFFECT: EINSTEIN'S QUANTA

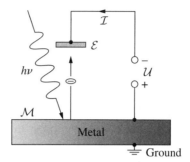

Figure 1.5 Photoelectric effect according to Einstein. A metal surface $\mathcal{M}$ illuminated with a stream of photons of frequency ν and energy $h\nu$ ejects electrons, if the frequency ν is large enough. The electrons, collected by the collector electrode $\mathcal{E}$, give rise to the photocurrent $\mathcal{I}$. A voltage $\mathcal{U}$ is applied between the metal surface $\mathcal{M}$ and the collector electrode $\mathcal{E}$. Applied with the polarity shown in the figure (Lenard's counterfield method), it allows one to determine the maximal kinetic energy $E_{\max}$ of the photoelectrons.

may now enable the electron to leave the metal surface, i.e., its total energy may now be enough to overcome the energy barrier W that, under normal conditions, confines the electron to the metal. Within this picture, Einstein predicted that the maximal kinetic energy $E_{\max}$ of the ejected electrons is

$$E_{\max}(\nu) = h\nu - W, \quad h\nu > W. \tag{1.37}$$

Millikan, who received the 1923 Nobel Prize for his work on the electron charge and the photoelectric effect, tested Einstein's formula, Equation 1.37, experimentally and found that it explained the data very well. If an electron escapes from the metal surface, its kinetic energy is positive, i.e., $E_{\max} > 0$. Using this observation in Equation 1.37, we obtain

$$\nu > \frac{W}{h} \tag{1.38}$$

as a necessary condition for ejecting photo electrons from the metal surface. This explains the existence of a threshold frequency below which no photoelectrons are observed.

The photon hypothesis explains another observation that is hard to explain with the wave theory of light: Even for extremely weak incident light, photoelectrons appear practically instantaneously. This observation cannot be explained within the framework of the wave theory of light since according to this theory the electron has to accumulate light from the weak beam over some time to gather enough energy so that the electron can overcome the energy barrier W and thus leave the metal surface. Within the photon picture, however, the sudden onset of photoelectrons, even for very weak light, is easily explained. Since the light consists of packets, the packets are instantly available and able to kick electrons out of the metal surface.

In summary, Einstein's photon hypothesis, for which he was awarded the 1921 Nobel prize in physics, explains all aspects of the photoelectric effect. Therefore, at this point, we may claim that the photon hypothesis is *consistent with* observations. It is, however, a different matter to claim that light actually *is* composed of photons. For this we need a more convincing experiment that directly shows the granular nature of light without invoking complicated intermediate processes or particles such as the photoelectrons in the photoelectric effect. Such an experiment was indeed performed by Aspect and collaborators in 1986. This experiment is both simple and direct, and leaves no doubt that light is indeed composed of photons. We describe this experiment in the following section.

Exercises:

1.3.1 A photovoltaic cell turns light into electricity. A possible setup is shown in Figure 1.6. Traversing a transparent, conducting collection electrode $\mathcal{E}$, a beam of light illuminates a metal surface $\mathcal{M}$. The collection electrode $\mathcal{E}$ and the metal surface $\mathcal{M}$ form a capacitor with capacitance C. A resistor R connects $\mathcal{E}$ with $\mathcal{M}$ and allows current to flow between $\mathcal{E}$ and $\mathcal{M}$. The charge on the collection electrode is denoted by Q; the voltage between $\mathcal{E}$ and $\mathcal{M}$ is denoted by $\mathcal{U}$. When the incident light hits $\mathcal{M}$, its photons cause electrons of maximal kinetic energy $E_{\max}$ to be ejected from the surface $\mathcal{M}$. Assume that, as long as $E_{\max} \geq e\mathcal{U}$,

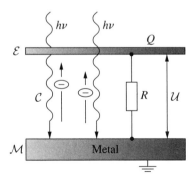

Figure 1.6 A photovoltaic cell.

where $e = 1.602\ldots \times 10^{-19}$ C is the elementary unit of charge, the photoelectrons flow to $\mathcal{E}$ at a constant rate α (= number of photoelectrons/s). Assume that $Q(t=0)=0$.

(i) Calculate $Q(t)$ and $\mathcal{U}(t)$ for $t>0$.

(ii) What is the steady-state current $\mathcal{I}_s$ through R?

(iii) Increasing the intensity of the incident light increases α, which in turn increases $\mathcal{U}_s$ and $\mathcal{I}_s$. Show that because of the condition $E_{\max} \geq e\mathcal{U}_s$, there is a maximal α corresponding to a maximal light intensity beyond which a further increase in the light intensity does not improve the performance of the device.

1.3.2 Lenard used an electric counterfield method to determine $E_{\max}$. One could think of using the Earth's gravitational field as a counterfield to determine $E_{\max}$. Why does this method fail?

1.3.3 A light beam with $\lambda = 300$ nm and intensity of $1\,\text{W/cm}^2$ illuminates a metal surface with an area of $5\,\text{cm}^2$. Calculate the photocurrent $\mathcal{I}$ under the assumption that the conversion of photons into photoelectrons is 100% effective.

1.4 Experiment of Aspect and Collaborators

Do photons exist? Einstein's photon hypothesis neatly explains all aspects of the photoelectric effect. But since the photoelectric effect measures electrons, and not photons, it is only indirect proof for the existence of photons. What we need is an experiment that is both simple and decisive, an experiment that works only with light, and has the power to prove unambiguously that photons exist. An experiment fulfilling these criteria is shown in Figure 1.7. A weak beam of light is incident on a *beam splitter*. A beam splitter is a half-silvered mirror that transmits

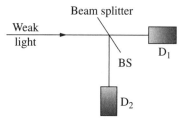

Figure 1.7 Naive version of a photon coincidence experiment. Weak light, incident from the left, is divided into two separate beams by a beam splitter and triggers detectors D_1 and D_2. If light is a wave, D_1 and D_2 are frequently triggered simultaneously (coincidence). If light consists of particles (photons), D_1 and D_2 are *never* triggered simultaneously (anticoincidence).

half of the incident light intensity and reflects the other half, as indicated in Figure 1.7. Detectors D_1 and D_2 terminate the transmitted and reflected parts of the beam, respectively. If the incident light is a continuous wave, both detectors D_1 and D_2, respectively, would continuously register light intensity at any instant t in time. This means that, in this case, D_1 and D_2 frequently click simultaneously. If, on the other hand, the incident beam of light consists of photons, the photon, an indivisible point particle, would either transmit, and thus trigger detector D_1, or it would reflect, and trigger D_2. Therefore, in this case, detectors D_1 and D_2 would never click simultaneously. Thus, the experiment consists of sending weak light toward the beam splitter and then looking for detection *coincidences*, i.e., instances in which both detectors are triggered at the same time. If we find coincidences, we have a strong argument for rejecting the photon hypothesis. If, on the other hand, D_1 and D_2 never respond in coincidence, i.e., we observe *anticoincidence*, we have a strong argument in favor of the existence of photons.

An experiment similar to the one illustrated in Figure 1.7 was actually performed by Hanbury-Brown and Twiss in 1956. This experiment showed a strong coincidence signal, thus favoring the wave theory of light. At the time this result was puzzling and seemed to contradict the photon theory of light. However, it was soon understood that the light source used by Hanbury-Brown and Twiss did not provide single photons, but provided mostly clusters of photons consisting of two or more photons very closely spaced in time. Any conventional light source, such as a spectroscopic lamp, produces this type of "clustered light." In this case the different photons cannot be temporally resolved and produce a coincidence signal within the temporal resolution window of the detectors. Therefore, in order to convincingly prove the existence of photons, the experiment of Hanbury-Brown and Twiss needs to be performed with a single-photon light source. This argument was understood very well by Aspect and collaborators, who in 1986 performed an experiment similar to the one of Hanbury-Brown and Twiss, but with an atomic transition serving as a single-photon light source instead of the clustered-light source used by Hanbury-Brown and Twiss.

A schematic sketch of the setup of the experiment of Aspect and collaborators is shown in Figure 1.8. A calcium atom emits two photons of frequencies ν_1 and ν_2, respectively, near simultaneously in time. The photon of frequency ν_1 triggers the detector D_1 indicating that a photon of frequency ν_2 is on its way to the beam splitter BS. As soon as D_1 is triggered, a gate pulse sensitizes the detectors D_2 and D_3 during a time interval Δt. As soon as the gate pulse is over, the two detectors, D_2 and D_3, are no longer accepting photons. While the gate is open, the photon of frequency ν_2 flies toward the beam splitter BS and then triggers either

1.4 • EXPERIMENT OF ASPECT AND COLLABORATORS

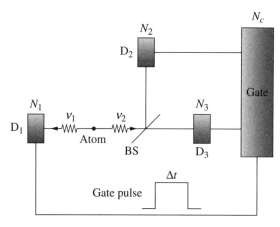

Figure 1.8 Schematic illustration of the single-photon experiment of Aspect and collaborators that decisively proved the existence of photons.

detector D_2 or detector D_3, corresponding to reflected and transmitted photons, respectively, or it triggers both D_2 and D_3, corresponding to the wave picture of light. The experiment will decide which of these two pictures is correct. If the photon picture of light is correct, D_2 and D_3 will never be triggered simultaneously during the time interval Δt during which the gate is open. This is so since a single photon, a point particle, either transmits or reflects, but cannot do both unless it is split in two by the beam splitter. This "splitting of the photon" can actually happen in certain nonlinear optical devices, but is not usually observed to happen at a beam splitter. In any event, since the splitting of the photon results in two (or more) photons, each with a lower frequency than the original photon (because of energy conservation), one easily guards against this possibility by blocking split photons with optical filters. With the filters in place (not shown in Figure 1.8) we ensure that only "whole" photons traverse the experimental apparatus. Therefore, a photon incident on a beam splitter has to "make up its mind" to either transmit or reflect; it cannot do both. Therefore, either D_2 clicks, or D_3 clicks, but never both.

We denote by N_1 the rate at which D_1 is triggered. Therefore, N_1 is also the rate of gate openings. We denote by N_2 the rate at which D_2 is triggered, and by N_3 the rate at which D_3 is triggered. We also define N_c, the rate at which both D_2 and D_3 are triggered simultaneously during the time the gate is open. If light consists of photons, we expect N_c to be zero. However, we have to admit the possibility of a nonzero N_c for several reasons: (1) The experiment of Aspect and collaborators is performed not with single calcium atoms (although this is now technologically possible with modern experimental equipment) but with an atomic vapor of calcium atoms. This means that despite the fact that Δt is chosen very small, there is still a possibility that a photon of frequency ν_2 is emitted by a second calcium atom within the time interval Δt and finds its way to the beam splitter. If this spurious photon chooses "reflection" when our original photon chooses "transmission," or vice versa, then we get a false coincidence that makes N_c different from zero. Aspect and collaborators chose their experimental conditions such that this contamination effect is negligibly small. (2) We need to allow for the possibility that light does not consist of photons. A classical electromagnetic wave will split evenly into a transmitted and reflected wave at the beam splitter BS and thus, with high probability, trigger both detectors D_2 and D_3 simultaneously. This would lead to a large rate N_c disproving the photon picture of light.

Having introduced the rates N_1, N_2, and N_3, we now define the probabilities p_r, p_t, and p_c for reflection, transmission, and coincidence, respectively, according to

$$p_r = \frac{N_2}{N_1}, \quad p_t = \frac{N_3}{N_1}, \quad p_c = \frac{N_c}{N_1}. \qquad (1.39)$$

In order to be able to discriminate between a classical wave theory and the photon theory of light, we have to compute what each of these two theories predicts for p_c. We already know that, ideally, the photon theory predicts $p_c = 0$. What remains is to compute the prediction of a classical wave theory.

We define the time-averaged intensity I_n during gate number n, open for a time interval Δt at time t_n according to:

$$I_n = \frac{1}{\Delta t} \int_{t_n}^{t_n + \Delta t} I(t)\, dt. \qquad (1.40)$$

We also define the ensemble averages $\langle I \rangle$ and $\langle I^2 \rangle$ of the time-averaged intensities I_n according to:

$$\langle I \rangle = \frac{1}{G} \sum_{n=1}^{G} I_n, \quad \langle I^2 \rangle = \frac{1}{G} \sum_{n=1}^{G} I_n^2, \qquad (1.41)$$

where G is the total number of gate openings for the duration of the experiment. During gate number n the probability of triggering D_2 is $p_r^{(n)} = \alpha_r I_n$, the probability of triggering D_3 is $p_t^{(n)} = \alpha_t I_n$, and the probability of a coincidence is $p_c^{(n)} = p_r^{(n)} p_t^{(n)} = \alpha_r \alpha_t I_n^2$, where α_r and α_t are proportionality constants that depend on the details of the detectors

1.4 • EXPERIMENT OF ASPECT AND COLLABORATORS

used in the experiments. Based on these local probabilities the total probabilities p_r, p_t, and p_c, averaged over G gates, are:

$$p_r = \frac{1}{G} \sum_{n=1}^{G} p_r^{(n)} = \alpha_r \langle I \rangle,$$

$$p_t = \frac{1}{G} \sum_{n=1}^{G} p_t^{(n)} = \alpha_t \langle I \rangle,$$

$$p_c = \frac{1}{G} \sum_{n=1}^{G} p_c^{(n)} = \alpha_r \alpha_t \langle I^2 \rangle. \tag{1.42}$$

Using the inequality

$$\left(\sum_{n=1}^{G} x_i \right)^2 \leq G \sum_{n=1}^{G} x_i^2, \tag{1.43}$$

which is generally valid for any set of real numbers x_i, $i = 1, \ldots, G$, we obtain:

$$\langle I^2 \rangle \geq \langle I \rangle^2. \tag{1.44}$$

Using this inequality in Equation 1.42, we get:

$$p_c = \alpha_r \alpha_t \langle I^2 \rangle \geq \alpha_r \alpha_t \langle I \rangle^2 = p_r p_t. \tag{1.45}$$

Therefore, the wave model of light predicts

$$p_c \geq p_r p_t. \tag{1.46}$$

Defining the ratio

$$\rho = \frac{p_c}{p_r p_t}, \tag{1.47}$$

Equation 1.46 yields

$$\rho \geq 1 \tag{1.48}$$

as a firm prediction of any classical wave theory of light. If this condition is violated, light cannot be a classical wave. According to Equation 1.39, the ratio ρ can be expressed with the help of the counting rates N_1, N_2, N_3, and N_c according to

$$\rho = \frac{N_c N_1}{N_2 N_3} \tag{1.49}$$

and is therefore directly accessible experimentally. When Aspect and collaborators performed the experiment illustrated in Figure 1.8 and evaluated ρ according to Equation 1.49, they found ρ values that ranged from about 0.2 to 0.8 in clear violation of the classical wave prediction in Equation 1.48. Thus it was established that light does not consist of waves that have finite spatial extent, but instead interacts with detectors and beam splitters in the form of point particles that have zero spatial extent. This answers the question posed at the beginning of this section: Yes, photons do exist!

Exercises:

1.4.1 Use the inequality of Cauchy and Schwarz

$$\left(\sum_{i=1}^{n} x_i y_i\right)^2 \leq \left(\sum_{j=1}^{n} x_j^2\right)\left(\sum_{k=1}^{n} y_k^2\right) \qquad (1.50)$$

to prove the inequality in Equation 1.43 we used to derive Equation 1.44.

1.4.2 We conduct a photon correlation experiment with a light source and a beam splitter in the spirit of Figure 1.7. The light source is $l = 1\,\text{cm}$ from the beam splitter and produces single, directed photons of wave length $\lambda = 500\,\text{nm}$ that travel straight from the source to the beam splitter. What is the maximally allowed power output of the light source if we require that, at any instant in time, there is only a single photon between the source and the beam splitter?

1.5 Properties of the Photon

Now that we have established the existence of photons, we will investigate their properties. At this point we know that photons are point particles and each photon carries an amount of energy

$$E = h\nu = \hbar\omega, \qquad (1.51)$$

where

$$\hbar = \frac{h}{2\pi} = 1.054\ldots \times 10^{-34}\,\text{Js} \qquad (1.52)$$

is also called "Planck's constant" (see Equation 1.36). Photons, as the elementary constituents of light, travel at the speed of light c. Therefore, according to Einstein's special theory of relativity, photons cannot have a rest mass. If they had a nonzero rest mass, traveling at the speed of light c would imply that they carried an infinite amount of energy, in contradiction to the fact that, according to Equation 1.51 they carry a finite amount of energy. Therefore, the photon rest mass is $m_{\text{ph}} = 0$.

On the basis of Einstein's energy-momentum formula

$$E^2 = p^2 c^2 + m_0^2 c^4, \qquad (1.53)$$

which is generally valid for any particle with rest mass m_0, we conclude that, since for photons $m_0 = m_{\text{ph}} = 0$, each photon carries momentum in the amount of

$$p = \frac{E}{c} = \frac{h\nu}{c} = \hbar \frac{2\pi}{\lambda} = \hbar k, \qquad (1.54)$$

where $k = 2\pi/\lambda$ is the wave number.

Photons are not charged. This is demonstrated easily by passing a light beam between two charged capacitor plates. Since light has no rest mass, we should see a substantial deflection of the light beam if photons carried charge. This, however, has never been observed experimentally. We conclude that the photon's charge is zero.

Pair production is a process that turns pure photonic energy into matter. For this to occur the photon has to collide with a second particle in order to conserve energy and momentum. For instance, a photon colliding with a heavy atomic nucleus may generate an electron-positron pair out of the vacuum. Since both the electron and the positron have rest mass m_e, where $m_e = 9.109\ldots \times 10^{-31}$ kg is the rest mass of the electron, this pair production process is possible only if the photon has a minimum energy of $E = h\nu = 2m_e c^2$. Momentum conservation may require a higher energy.

Pair annihilation is a process in which a massive particle and its antiparticle annihilate in a collision to produce two photons. Thus, this process turns mass into energy. According to Einstein, mass and energy are equivalent. Pair production and pair annihilation demonstrate this impressively.

Photons interact with, and are affected by, a gravitational field. This is immediately obvious when we consider the thought experiment shown in Figure 1.9. Two counter-propagating photons, each with energy $E_{\text{ph}} = m_e c^2$, one launched at point A, the other at point B, generate an electron-positron pair at a height H in Earth's gravitational field (gravitational acceleration $\boldsymbol{g}$). In addition to its rest-mass energy $2m_e c^2$, the electron-positron pair has gravitational energy $E_g = 2m_e g H$. We bring this pair to the Earth's surface. We gain an amount of energy E_g. At Earth-surface level we recombine the electron-positron pair (pair annihilation) and regain two photons of energy $m_e c^2$ each, which we can now use to

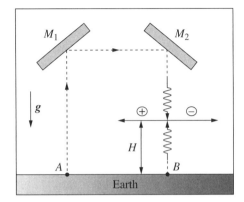

Figure 1.9 Thought experiment proving that photons interact with a gravitational field.

generate a second electron-positron pair at height H. Running this "machine" cyclically, we gain energy in the amount of E_g per cycle. Thus our machine violates the conservation of energy. The reason is that we assumed that the photons are unaffected by the Earth's gravitational field. Since this assumption leads to an impossible energy-generating machine, the converse must be true: photons *are* affected by the Earth's gravitational field. In order to balance the energies, we have to assume that in a homogeneous gravitational field photons have gravitational energy in the amount of

$$E_g^{(\text{ph})} = \frac{h\nu}{c^2} gH. \tag{1.55}$$

This formula for the gravitational energy of a photon looks like the equivalent formula for the gravitational potential energy of a massive particle of equivalent mass $h\nu/c^2$. We may use this analogy to state the gravitational energy of a photon in the gravitational field of a point mass M:

$$E_g^{(\text{ph})} = -G \frac{h\nu M}{c^2 r}, \tag{1.56}$$

where $G = 6.67 \ldots \times 10^{-11}\,\text{m}^3/(\text{kg}\,\text{s}^2)$ is the gravitational constant. According to Equations 1.55 and 1.56, photons moving upward in the gravitational field of the Earth lose energy, which, according to $E = h\nu$, has to manifest itself in a shift toward lower frequency. This effect, the *gravitational red shift*, was experimentally confirmed in 1959 by Pound and Rebka.

Photons have a property that classical point particles cannot have in principle: a constant, intrinsic amount of angular momentum of size $\hbar$.

This angular momentum is called the *spin* of the photon. It is called the spin, since we might imagine the photon rotating around its own "axis," just like the Earth rotates around its axis and has intrinsic angular momentum. The only problem with this classical picture is that photons are point particles. Point particles do not have a moment of inertia and can therefore, according to classical mechanics, not support a finite amount of spin angular momentum. Therefore, spin is a manifestly nonclassical property of the photon. Spin is usually measured in units of $\hbar$. Therefore, we say that photons have spin 1.

The spin of the photon leads to one of its most important characteristics: its *polarization*. Polarization is generated and detected with the help of polarization filters. Polarization is usually measured with respect to the photon's propagation direction. Imagine a plane orthogonal to the photon's propagation direction with an x-y coordinate system on it. We call the x-direction the horizontal polarization direction and the y-direction the vertical polarization direction. If a photon passes through a horizontal polarization filter, it is horizontally polarized from then on. A second horizontal polarization filter in its way transmits the horizontally polarized photon with 100% probability. A vertical polarization filter will stop a horizontally polarized photon. A photon may be linearly polarized in any direction. We characterize its polarization direction by the angle φ with respect to the horizontal. To determine the angle φ we rotate a horizontal polarization filter by an angle α counterclockwise until we reach an angle $\alpha_{\max}$ where 100% of photons are transmitted. Then, $\alpha_{\max}$ equals the polarization angle φ. The polarization degree of freedom of photons will play an important role in Section 5.4 in connection with interaction-free measurements.

In summary we obtain the following table for the properties of the photon

Table 1.1   Properties of the photon

property	value
size	0
charge	0
rest mass	0
energy	$h\nu$
momentum	$\hbar k$
spin	1

Exercises:

1.5.1 When a photon is retro-reflected from a mirror, it imparts momentum $2\hbar k$ to the mirror. Suppose light of intensity I is retro-reflected from a mirror. What is the pressure on the mirror? This effect is known as *radiation pressure*.

1.5.2 Since photons carry momentum, a strong light source can be used to propel a spaceship. Moreover, since photons travel at the speed of light, a spaceship equipped with such a "light drive" may, in principle, approach the speed of light arbitrarily closely, making the light drive a candidate for interstellar travel. Suppose a space probe of 100 kg mass is equipped with a laser that generates 100 GW of light power. How long does it take the probe to reach 10% of the speed of light?

1.6 Summary

Is light a wave or a stream of particles? Newton thought that light consists of point particles, called corpuscles. Huygens, a contemporary of Newton, thought that light is a wave. Not before Young did his diffraction experiments with single and double slits was this dispute temporarily settled in favor of the wave theory of light. In fact it took more than a century to realize that light is not really a continuous wave, but in fact sometimes behaves in ways better described with the help of point particles. A triumph for Newton at last? Not really, since the light particles, or photons, are unlike anything Newton ever dreamt of. Photons are massless, have spin, and, in a vacuum, move at the speed of light. In addition, photons may be polarized, a property that cannot be explained on the basis of classical point particles. But now that we are back to thinking of light in terms of particles, do we not encounter the same difficulties faced by the old corpuscular theories of light, such as the difficulty explaining diffraction and interference? Indeed we do. But there is a way out: quantum mechanics resolves this difficulty by assigning both wave-like and particle-like aspects to photons. This is called *wave-particle duality*. It is discussed in the following chapter.

Chapter Review Exercises:

1. The scattering of photons on other particles is called the *Compton effect*. Consider the situation shown in Figure 1.10. A photon of wave length λ is incident from the left and collides with an electron of mass m_e at rest. The photon's direction of incidence coincides with the x-axis. After the collision the photon flies off with an angle θ with respect to the x-axis; the electron is scattered at an angle φ

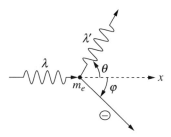

Figure 1.10 Collision of a photon of wave length λ with an electron of mass m_e at rest. The photon is scattered into a new direction characterized by the angle θ, and acquires a new wavelength λ'. The electron starts moving into a direction characterized by the angle φ. This process is called *Compton scattering*.

with respect to the x-axis. This process is called *Compton scattering*. Use energy and momentum conservation as well as Einstein's energy-momentum formula in Equation 1.53 to derive the Compton scattering formula

$$\lambda' - \lambda = \lambda_e \left[1 - \cos(\theta)\right], \tag{1.57}$$

where

$$\lambda_e = \frac{h}{m_e c} \approx 2.426 \times 10^{-12}\,\text{m} \tag{1.58}$$

is the Compton wavelength of the electron, λ is the wavelength of the incident photon, and λ' is the wavelength of the scattered photon.

2. On a bright, sunny day the surface of the Earth receives about 500 W of sunlight per square meter in the visible range of the electromagnetic spectrum. Assume an average wavelength of visible photons of 550 nm and compute the photon flux [= number of photons / (m$^2 \times$ s)] in the visible range.

3. In 1959 Pound and Rebka conducted an experiment to measure the gravitational red shift of photons. In this experiment 14 keV photons, emitted by a radioactive ^{57}Fe source in the basement of Harvard's Jefferson Lab were directed upward, against Earth's gravitational field, and were detected 22.5 m above the source in the attic of Jefferson Lab. Compute the fractional gravitational red shift $\delta\nu/\nu$ of the photons.

4. Suppose the screen Σ, illuminated with light of wavelength λ, contains two pin holes, one at $P = (-a, 0, z_0)$ and one at $Q = (a, 0, z_0)$. Compute the loci of the intensity maxima in the x-y plane and characterize their geometric shapes.

chapter 2

Wave–Particle Duality

In this chapter:

- ◆ Introduction
- ◆ A Diffraction Experiment with Photons
- ◆ Young's Double-Slit Experiment Revisited
- ◆ Three Rules of Quantum Mechanics
- ◆ Two-Slit Which-Way Experiment
- ◆ Summary

2.1 Introduction

Our everyday, macroscopic, classical world can be described naturally with the help of two concepts: particles and waves. Particles are the "solid stuff" our world is made of. Usually, we think of particles as tiny objects that are sharply localized in space and time. In the ideal limit of vanishing diameter they become *point particles*. Point particles occupy no space and cannot be at two points in space at the same time. In our classical, macroscopic world, particles follow the laws of Newtonian mechanics. In particular, particles travel in straight lines (rays) unless a force acts on them.

A wave, on the other hand, cannot be touched. It is an abstract concept, a spatiotemporal phenomenon, a "ripple in space-time" that frequently carries momentum and energy from one point in space-time to another. Sometimes waves manifest themselves as a collective phenomenon of particle motion, an *emergent phenomenon* of many particles cooperatively acting together. In this case, while individual particles move little, the ripple produced by temporarily displaced particles may move with great speed over large distances. Examples of this type of wave include a kink traveling on a taught rope, a surface wave on a pond, or the wave of temporarily moving cars in stop-and-go traffic. There are also wave phenomena that do not need carrier particles, i.e., a *medium*, to spread. Examples are light waves and gravitational waves.

Waves and particles, so clearly distinct in the macro-world, are not present as separate phenomena in the micro-world. Ever since the early years of the 20th century, scientists have been confounded by a kind of dualism exhibited by the entities of the micro-world. This ambiguous nature of physical objects has been called *wave-particle duality*. Wave-particle duality does not mean that some given phenomenon "really is" a wave, and it is just more convenient, in some circumstances, to describe this wave, approximately, as a collection of particles, like we frequently do in classical optics when we make the transition from wave optics to ray optics, which often is an excellent approximation. No, wave-particle duality is more than "choose the most convenient description." Wave-particle duality means that physical entities are neither pure waves, nor pure particles, but depending on the circumstances, can act as one or the other. But if all physical entities have this dual nature, what about "manifestly material" particles, such as electrons? And yes, indeed, individual electrons have been stored in electromagnetic traps, and their presence in these traps has been continuously monitored and verified for up to nine months! Surely, therefore, far from spreading and oozing out of the trap, like a wave would do, these electrons have a grainy, indivisible, localized, material nature that could not be further from a wave! But as De Broglie conjectured, and subsequent experiments confirmed in detail, even electrons have a wave nature that is revealed in electron diffraction experiments and most convincingly in the electron microscope, a technical application of the wave nature of the electron. Exploring the fundamental nature of wave-particle duality and some of its ramifications is the topic of this chapter.

2.2 A Diffraction Experiment with Photons

In Chapter 1 we showed that photons exist. Does this make the wave picture of light superfluous? Does it perhaps even prove it "wrong"? To answer this question, consider the setup shown in Figure 2.1. A light source S, far away from a screen Σ, produces single photons, one at a time, and emits them radially outward with an isotropic angular dis-

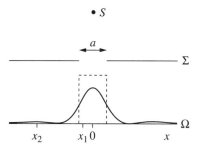

Figure 2.1 Photon diffraction experiment. The source S, assumed to be located far from Σ, emits single photons that pass through a slit of width a in the screen Σ and are recorded on the observation screen Ω as single dots of blackened emulsion. If photons are point particles, as corroborated by the tiny dots they make on Ω, their probability distribution of arrival on Ω should look like the dashed rectangular curve. Instead, what is observed, is an intensity distribution resembling the smooth line. Dots on Ω (particles), whose density is described by a continuous diffraction curve (waves), illustrate wave-particle duality for photons.

tribution. Some of the photons will fly through a slit of width a in the screen Σ. The photons will then proceed to the observation screen Ω, which is covered with a photographic emulsion that turns black precisely at the spot where the photon hits Ω. We imagine that the source S is so far from the screen Σ that photons in the vicinity of the slit hit Σ practically at vertical incidence. If photons are point particles, we predict that the photons will hit Ω in the interval $-a/2 < x < a/2$, and no photon will ever be observed outside of this interval ($|x| > a/2$). Thus, within the framework of the particle picture, the probability of hitting Ω in a point x is a flat rectangular distribution (represented as a dashed line in Figure 2.1), which is nonzero only in the interval $-a/2 < x < a/2$.

To test the particle hypothesis of photons and its ramifications, we switch on the photon source S and wait. After a while a photon is emitted by S that passes through the slit in Σ and lands at the point x_1 on Ω (see Figure 2.1). We examine the black spot and are pleased: (1) The small size of the black spot indicates that photons are indeed point particles, and (2) the position $-a/2 < x_1 < a/2$ of the black spot, located inside of the interval $-a/2 < x < a/2$, is precisely what we predicted should happen for point particles. We wait for the next photon .... It lands at x_2, squarely outside the interval $-a/2 < x < a/2$. This position is completely inconsistent with the picture of photons as classical point particles. To investigate this phenomenon further we wait for many more photons to arrive at Ω and examine the resulting density of dots on Ω.

We characterize the density of dots on Ω with the help of the function $D(x)$, which is the number of dots ΔN in a vicinity Δx of a point x on

Ω, i.e., $D(x) = \Delta N/\Delta x$ for $\Delta x \to 0$. If we normalize $D(x)$ by the total number N of dots on Ω, we get the normalized function $\rho(x) = D(x)/N$ with

$$\int_{-\infty}^{\infty} \rho(x)\, dx = 1. \tag{2.1}$$

Since we also observe that photons appear at random positions on Ω with no correlation between the locations of successive photons on Ω, we may interpret $\rho(x)$ as the *probability density* of a photon to land at x on Ω. Experimentally we find that the probability density $\rho(x)$ is well described by the smooth function shown in Figure 2.1. This function reminds us of the intensity distribution function that a classical electromagnetic wave produces on Ω when diffracting at a slit (see Figure 1.3). This result is baffling. On the one hand, we know that we are dealing with point particles: The source S produces single photons. Correspondingly, only tiny black spots, consistent with the picture of photons as point particles, are ever produced on Ω. On the other hand, photons, one-by-one, and independent of each other (there is a long waiting period between the arrival of two photons on Ω), produce a diffraction pattern! The result of this experiment is what we mean by wave-particle duality. Photons have the capability of exhibiting, individually, both particle and wave properties. This also answers our question of whether the existence of photons has made the wave theory of light superfluous. Not at all! In fact, as our experiment shows, for a proper description of light we need both pictures. Any single picture alone, i.e., wave *or* particle, is not sufficient to account for the observed phenomena: Waves do not produce tiny spots on Ω and particles do not produce diffraction patterns!

Exercises:

2.2.1 For the single-photon experiment shown in Figure 2.1, compute the probability distribution $\rho(x)$ of photons hitting Ω in the point x. We assume that the source S is very far from Σ, the distance between the screens Σ and Ω is $L \gg a$, and that $\rho(x)$ is proportional to the light intensity, Equation 1.22, such that in the small-angle approximation ($\theta \ll 1$, $\theta \approx x/L$, I_0 approximately constant):

$$\rho(x) = \rho_0 \frac{\sin^2[\pi ax/(\lambda L)]}{[\pi ax/(\lambda L)]^2}, \tag{2.2}$$

where ρ_0 is a constant. Using

$$\int_{-\infty}^{\infty} \frac{\sin^2(\xi)}{\xi^2}\, d\xi = \pi,$$

and the normalization condition expressed in Equation 2.1, show that

$$\rho_0 = \frac{a}{\lambda L}. \qquad (2.3)$$

2.2.2 Referring to Figure 2.1, compute the degree of violation of the particle picture of photons by computing the probability $P_>$ of finding the photon with $|x| > a/2$ on the observation screen Ω. Use the probability distribution in Equation 2.2 with ρ_0 given in Equation 2.3.

(a) Express your result for $P_>$ in terms of the sine integral function

$$\mathrm{Si}(z) = \int_0^z \frac{\sin(\eta)}{\eta}\, d\eta.$$

(b) Obtain a numerical value for $P_>$ using $a = \lambda$, $L = 10\lambda$, $\sin(z) \approx z$, and $\mathrm{Si}(z) \approx z$ for small z. You will find that $P_>$ is rather large, demonstrating that in this case the violation of the particle picture of photons is a large effect.

2.2.3 For the case of single-slit diffraction (see Figure 2.1) show that for a, L fixed, but $\lambda \to 0$, the probability density $\rho(x) \to 0$ for $x \neq 0$. Within our approximations this means that in the $\lambda \to 0$ limit we observe particles only with $\theta = 0$, but not with $\theta \neq 0$, where θ is the incident angle of photons on Ω. Within the framework of our approximations, in particular $a/L << 1$, this is the expected result if the particle picture of photons were correct. Physically this means that for fixed experimental conditions the particle picture of photons improves as the photon energy $\to \infty$. This is the reason why it is often a good approximation to represent high-energy γ photons as particles, justifying the term "γ-rays."

2.3 Young's Double-Slit Experiment Revisited

In the previous section we learned that although photons are created by a source S as point particles, they somehow manage to produce a diffraction pattern when passing through a single slit. Investigating this unexpected result further, consider the double-slit experiment shown in Figure 2.2. It reminds us of Young's double-slit experiment except that instead of a strong light source that generates plane waves incident on the screen Σ, we now work with a single-photon source S, just like we did in the single-slit experiment shown in Figure 2.1. Although, at this

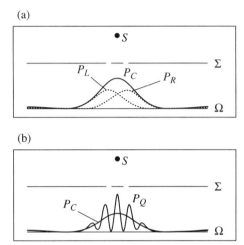

Figure 2.2 Double-slit experiment with single photons. The source S, assumed to be located far from Σ, emits single photons with an isotropic distribution. (a) If photons are point particles, they will either go through the left slit with probability distribution $P_L(x)$ (dashed curve peaked on the left) or through the right slit with probability distribution $P_R(x)$ (dashed curve peaked on the right), resulting in the total probability distribution $P_C(x) = P_L(x) + P_R(x)$ (full line) for photon hits on Ω. (b) What is actually observed is the probability distribution $P_Q(x)$ (heavy full line).

point, we do not know the reason why, we do know that if we block off the right-hand slit in Σ, we obtain the probability distribution $P_L(x)$ (dashed line with a maximum on the left in Figure 2.2(a)) due to the photons passing through the left slit. This much we already established in the previous section. In complete analogy, blocking off the left slit, we obtain the probability distribution $P_R(x)$ (dashed line with a maximum on the right in Figure 2.2(a)) for photons passing through the right slit and recorded on Ω. We normalize $P_L(x)$ and $P_R(x)$ to $1/2$, i.e.,

$$\int_{-\infty}^{\infty} P_L(x)\,dx = \int_{-\infty}^{\infty} P_R(x)\,dx = \frac{1}{2}, \qquad (2.4)$$

since only $1/2$ of the photons with both slits open are recorded in the case where only a single slit is open. All this is correct, and an actual experiment with photons would confirm our conclusions.

Now we open both slits .... A point particle has no choice but to go either through the left slit or through the right slit. It cannot do both since going through both slits simultaneously would contradict the basic property of a point particle of being indivisible and localized in space and time, which means it cannot be at two different places at the same time. When it goes through the left slit, it produces the diffraction pattern $P_L(x)$, i.e., the probability distribution with a maximum on the left in Figure 2.2(a). When it goes through the right slit, it produces $P_R(x)$, i.e., the probability distribution with a maximum on the right in Figure 2.2(a). Since these two processes are independent of each other, we can group all hits on Ω into two classes corresponding to photons that went through the left slit, producing $P_L(x)$, and photons that went through the right slit, producing $P_R(x)$. Because of the independence of the individual events, the two classes are not correlated. Therefore, if photons are indivisible, classical point particles, the total probability distribution of photons recorded on Ω is predicted to be the sum of the events that go through the left slit and those that go through the right slit, i.e.,

$$P_C(x) = P_L(x) + P_R(x). \qquad (2.5)$$

The probability distribution $P_C(x)$, our prediction based on the classical point-particle picture of photons, is shown as the full line in Figure 2.2(a).

What we observe instead, when we actually perform the experiment, is the probability distribution $P_Q(x)$, shown in Figure 2.2(b). It strongly resembles the intensity distribution of a continuous light wave passing through a double slit. In this case we understand the intricate oscillations in Figure 2.2(b) as the result of interference between the light fields produced by the left and right slits. But how can point particles produce an interference pattern on Ω? The observed probability distribution $P_Q(x)$ for a single-photon experiment completely contradicts our classical logic that led to the probability distribution $P_C(x)$, which, it must be emphasized, is the *correct* distribution if photons were indeed classical, indivisible point particles.

The unexpected result $P_Q(x)$ for the double-slit experiment is what Feynman once called the central mystery of quantum mechanics. It is a result that cannot be explained using everyday reasoning. We have entered the quantum world in which particles somehow manage to produce diffraction and interference patterns. Nobody has ever been able to explain, in macroscopic, classical terms what is going on in the single-particle two-slit experiment illustrated in Figure 2.2. This does not mean that we wouldn't be able to formally, mathematically describe, with perfect accuracy, the outcome of quantum experiments of the type shown in Figure 2.2. And indeed, in Chapter 3, we will review the "quantum machinery" that allows us to make these perfect predictions of experimental outcomes. What we are saying is that although we may be able to develop a mathematical machinery that is perfectly able to handle any kind of quantum situations (see Chapter 3), a fundamental *understanding* of "what's going on?" during the experiment may forever be beyond human intellectual reach. A first taste of this is provided by the outcome of the double-slit experiment for single photons illustrated in Figure 2.2(b), and many more quantum mysteries lie in store for us to be discovered and explored in this and the following chapters.

Perhaps we accept that photons behave in the strange way described previously. Photons, after all, are "special" in that they have no rest mass and always go at the speed of light. But what about electrons? Unlike photons, electrons do have a rest mass and therefore, intuitively, should be excellent examples of classical, Newtonian point particles. Surely an electron will be able to "make up its mind" and go either through the left slit or through the right slit, producing the probability distribution $P_C(x)$ of Figure 2.2(a)? However, contrary to our expectations, when we perform the double-slit experiment of Figure 2.2 with single electrons, we again obtain the interference pattern $P_Q(x)$ shown in Figure 2.2(b), corresponding to the interference pattern of a wave; not the naive probability distribution $P_C(x) = P_L(x) + P_R(x)$ in Figure 2.2, corresponding to individual electrons going either through the left slit or through the right slit. Apparently, just like photons, electrons, too, have a dual wave-particle nature, despite the fact that they have a rest mass and are as close an embodiment of classical point particles as we might hope to find in nature.

This dual nature of material particles was first conjectured by the French physicist Louis de Broglie in the early 1920s. His argument is both ingenious and simple. De Broglie argued that, since according to Einstein's special theory of relativity mass and energy are equivalent, there is no reason why we should treat electrons any differently than photons. Since we assign a wavelength λ to photons with momentum p according to

$$\lambda = \frac{h}{p}, \tag{2.6}$$

where h is Planck's constant, based on Einstein's mass-energy equivalence principle, the same relationship should hold for material particles. In fact, if we use Equation 2.6 to assign a wave length to electrons of momentum p, we can quantitatively explain the interference pattern produced by single electrons passing through two slits by assigning these electrons a wavelength according to Equation 2.6 and treating them like photons. This "explains" why electrons produce the same interference pattern $P_Q(x)$ as photons when we arrange for both to have the same de Broglie wavelength λ.

It is straightforward to generalize de Broglie's theory to three dimensions. According to de Broglie, the matter wave $\psi(\boldsymbol{r}, t)$ associated with a particle traveling with momentum $\boldsymbol{p}$ in free space is a plane wave of

the form
$$\psi(\mathbf{r},t) = \exp\left[i(\mathbf{k}\cdot\mathbf{r}-\omega t)\right], \qquad (2.7)$$
where
$$\mathbf{k} = \frac{\mathbf{p}}{\hbar} \qquad (2.8)$$
is the wave number of the matter wave and
$$\omega = \frac{E}{\hbar} \qquad (2.9)$$
is its angular frequency. Note that for material particles the energy E in Equation 2.9 is the particle's total energy, including its rest-mass energy, i.e.,
$$E = mc^2 = \gamma m_0 c^2, \quad \gamma = \frac{1}{\sqrt{1-(v/c)^2}}. \qquad (2.10)$$
With Equations 2.8 and 2.9 we may write de Broglie's matter wave, Equation 2.7, in the form
$$\psi(\mathbf{r},t) = \exp\left[\frac{i}{\hbar}(\mathbf{p}\cdot\mathbf{r} - Et)\right]. \qquad (2.11)$$

Exercises:

2.3.1 A baseball of mass 145 g is thrown with a speed of 30 m/s.

  (a) Compute its de Broglie wavelength.

  (b) The ball flies through an opening with a width of 10 cm. Approximating the baseball as a point particle, where does the first diffraction minimum occur on an observation screen Ω located 1 m from the slit? Any chance of observing the diffraction pattern of a baseball on Ω?

2.3.2 The resolution of a microscope is on the order of a wave length. For light microscopes this is about 500 nm, corresponding to a typical wave length of visible photons. Estimate the resolution of an electron microscope in which electrons are accelerated to a kinetic energy of 10^4 eV. Your result will show that the resolution of an electron microscope vastly exceeds the resolution of a light microscope. Note: Due to imperfections in the electron optics this theoretically possible resolution is not usually achieved in practice. Currently, practically achievable resolutions are on the order of 0.5 nm, still about 1000 times better than the resolution of a light microscope.

2.3.3 In an attempt to produce a narrow, collimated beam of electrons, the electrons are passed through a slit of width 1 mm. However, due to wave-particle duality, the electrons diffract at the slit. Estimate the width of the beam 1 m downstream from the slit by computing the distance between the first diffraction minima to the left and to the right of the center of the beam. Assume that the velocity of the electrons in the beam is 1 m/s.

2.4 Three Rules of Quantum Mechanics

In the previous section we saw that Young's double-slit experiment, operated in single-particle mode, produced some truly puzzling results. On the basis of these results we were forced to accept that no entity in nature is strictly a particle or strictly a wave. All things in nature exhibit wave-particle duality. The purpose of this section is to make some sense of this puzzling situation. Because of wave-particle duality we need a description that involves characteristics of both waves and particles. Waves are characterized by amplitudes, whereas particles are characterized by positions. To combine these two aspects of nature we associate *amplitudes* with *events*. Amplitudes represent the wave aspect of natural entities, whereas events represent their particle aspect. Referring to the double-slit experiment shown in Figure 2.2, the following is an example for an event. A photon is generated by the light source S and arrives at position x on the observation screen Ω, where it is measured. An amplitude is associated with this event. Dirac invented a short-hand notation for the amplitude of this event:

$$\langle x|S\rangle. \tag{2.12}$$

We read this symbol in the following way: $\langle x|S\rangle$ is the amplitude for a photon generated at S to arrive at x. Notice that this symbol is read backward. The *initial condition* is S, on the right of the vertical bar; the *final condition* is x, on the left of the vertical bar. The amplitude $\langle x|S\rangle$ is, in general, a complex number. Notice, too, that the symbol displayed in Equation 2.12 represents both wave and particle aspects. S and x in the symbol displayed in Equation 2.12 refer to clearly defined locations; they represent the particle aspect. The symbol itself, the amplitude, represents the wave aspect.

Now, $\langle x|S\rangle$ is the complete amplitude with both slits open. We can break it up into an amplitude where we imagine that the photon goes through the left slit and an amplitude where we imagine that the photon goes through the right slit:

$$\langle x|S\rangle = \langle x|S\rangle_L + \langle x|S\rangle_R. \tag{2.13}$$

Just like in classical electrodynamics, a wave theory, we added the amplitudes for both individual processes, $\langle x|S\rangle_L$ and $\langle x|S\rangle_R$, to obtain the amplitude $\langle x|S\rangle$ for the combined process. This is the *superposition principle* that we assume to be rigorously valid, without exception, in the quantum world.

The determination of the total amplitude $\langle x|S\rangle$ as the sum of the two partial amplitudes $\langle x|S\rangle_L$ and $\langle x|S\rangle_R$ is an example of Feynman's Rule:

$$\boxed{\text{If a particle can produce a certain result by two different routes, the total amplitude for the process is the } sum\ of\ the\ amplitudes \text{ for the two routes considered separately.}} \qquad (2.14)$$

Of course Feynman's Rule can be extended immediately to N different routes, or infinitely many different routes. In this case Feynman's Rule reads: If a particle can produce a certain outcome by N different routes, the total amplitude for the process is the *sum of the amplitudes* for each of the N routes considered separately. Of course, long before Feynman, many researchers had explicitly, or implicitly, made use of Feynman's Rule whenever they computed quantum amplitudes. However, Feynman was the first to formulate this quantum rule concisely (see Equation 2.14) and use it as a conceptual foundation for his path integral formulation of quantum mechanics.

Although quantum amplitudes are a very useful mathematical construct in quantum mechanics, our macroscopic measurement devices do not measure quantum amplitudes directly. Our devices measure the probability of the occurrence of certain events. This is different from classical electrodynamics, e.g., where we can directly measure the amplitude of an electromagnetic wave. The question, therefore, is how we make the transition from the amplitude of an event to the probability of an event. The solution was provided by Max Born in 1926 for which he received the 1954 Nobel Prize in physics. According to Born we obtain the *probability* of finding a particle at position x by squaring its amplitude. The following is Born's Rule:

$$\boxed{P(x) \;=\; |\langle x|S\rangle|^2.} \qquad (2.15)$$

If a path can be broken up into stages, we may compute the total amplitude using the Composition Rule:

> If a path can be broken up into stages, the total amplitude for traversal of the path is the product of the amplitudes for each of the individual stages. (2.16)

As an example consider the path for going from S to x via the left slit in Figure 2.2. This path can be broken up into two stages:

Stage 1: Go from S to the left slit.
Stage 2: Go from the left slit to x.

Therefore:

$$\langle x|S\rangle_L = \langle x|\text{left slit}\rangle \langle \text{left slit}|S\rangle. \tag{2.17}$$

Again, note that the expression on the right is to be read from right to left.

At this point we have all the rules in place that allow us to compute the probability distribution $P_Q(x)$ in Figure 2.2(b). Using Feynman's Rule 2.14 and the Composition Rule 2.16 we obtain:

$$\langle x|S\rangle = \langle x|\text{left slit}\rangle \langle \text{left slit}|S\rangle + \langle x|\text{right slit}\rangle \langle \text{right slit}|S\rangle. \tag{2.18}$$

Applying Born's Rule 2.15 we obtain

$$\begin{aligned} P_Q(x) &= |\langle x|S\rangle|^2 \\ &= |\langle x|\text{left slit}\rangle \langle \text{left slit}|S\rangle + \langle x|\text{right slit}\rangle \langle \text{right slit}|S\rangle|^2 \\ &= |\langle x|\text{left slit}\rangle \langle \text{left slit}|S\rangle|^2 + |\langle x|\text{right slit}\rangle \langle \text{right slit}|S\rangle|^2 \\ &\quad + 2\Re \langle x|\text{left slit}\rangle^* \langle \text{left slit}|S\rangle^* \langle x|\text{right slit}\rangle \langle \text{right slit}|S\rangle, \end{aligned} \tag{2.19}$$

where $\Re$ stands for taking the real part. We can write this result as

$$P_Q(x) = P_L(x) + P_R(x) + \mathcal{I}(x), \tag{2.20}$$

where

$$P_L(x) = |\langle x|\text{left slit}\rangle \langle \text{left slit}|S\rangle|^2 \tag{2.21}$$

is the probability of the photon to go through the left slit,

$$P_R(x) = |\langle x|\text{right slit}\rangle \langle \text{right slit}|S\rangle|^2 \tag{2.22}$$

is the probability of the photon to go through the right slit, and

$$\mathcal{I}(x) = 2\Re\,\langle x|\text{left slit}\rangle^*\,\langle\text{left slit}|S\rangle^*\langle x|\text{right slit}\rangle\,\langle\text{right slit}|S\rangle \quad (2.23)$$

is an *interference term*. The form of $P_Q(x)$ in Equation 2.20 is illuminating. The sum of the first two terms in Equation 2.20 is our "naive," classical expectation $P_C(x)$. The third term in Equation 2.20 is a nonintuitive correction term, the *interference term*, which is responsible for the fine structure we see on the screen Ω when both slits are open. This term cannot be explained on the basis of a pure particle picture. It is of quantum origin, reflecting the dual wave-particle nature of photons and electrons. Thus, the application of the three basic quantum rules provides a complete description of $P_Q(x)$.

Exercises:

2.4.1 Use Dirac notation to formally state the amplitude of the following event: An electron is generated by an electron gun G, flies through the left hole of a screen with two pin holes in it, proceeds to fly toward a screen with three slits in it, passes through the middle of the three slits, and is recorded in an electron counter C.

2.4.2 Consider Figure 2.2, but assume that the two slits are different. The difference in the slits manifests itself by different amplitudes for traversal of the left or right slit. As an example, suppose that the amplitude for an electron to go from S through the left slit and land at x on Ω is

$$\langle x|\text{left slit}\rangle\,\langle\text{left slit}|S\rangle = \frac{1}{\sqrt{2}},$$

and that the amplitude for an electron to go from S through the right slit and land at x on Ω is

$$\langle x|\text{right slit}\rangle\,\langle\text{right slit}|S\rangle = \frac{i}{\sqrt{3}}.$$

Compute (a) the amplitude $\langle x|S\rangle$ and (b) the total probability of an electron starting at S to be recorded at x.

2.5 Two-Slit Which-Way Experiment

We still don't buy it. Especially in the case of a single electron, it is hard to imagine that it would not go through one slit or the other. After all, an electron is an indivisible point particle! Nobody has ever seen fractional electron charges, or fractional electron masses. Electrons always appear in one piece, as a single, indivisible entity. So, we argue, if the electron cannot "split," it has to go through either the left slit or the right slit. Surely the question of which slit the electron went through can be decided experimentally, by measurement! Experiments that try to detect the path of quantum particles in a single-particle double-slit experiment are known as *which-way experiments*.

With the help of two electron detectors, D_L and D_R, mounted under the left and right slits, respectively, we turn the double-slit experiment of Figure 2.2 into a which-way experiment, as shown in Figure 2.3. For our version of a which-way experiment it is important that we have control over the detector efficiencies. Therefore, we denote by t_L the amplitude that an electron transmits through the left detector without being detected and by t_R the amplitude that an electron transmits through the right detector without being detected. If an electron is detected in one of the detectors, it is absorbed with 100% efficiency and will not continue its journey to the detection screen Ω.

Using the Composition Rule 2.16, the amplitude to go undetected from S via the left slit to x on Ω is

$$A_L = \langle x|\text{left slit}\rangle \, t_L \, \langle \text{left slit}|S\rangle. \qquad (2.24)$$

Similarly, the amplitude to go undetected from S via the right slit to x on Ω is

$$A_R = \langle x|\text{right slit}\rangle \, t_R \, \langle \text{right slit}|S\rangle. \qquad (2.25)$$

Therefore, the total amplitude to go undetected from S to x is

$$\begin{aligned} A(x) &= A_L(x) + A_R(x) \\ &= \langle x|\text{left slit}\rangle \, t_L \, \langle \text{left slit}|S\rangle + \langle x|\text{right slit}\rangle \, t_R \, \langle \text{right slit}|S\rangle. \end{aligned} \qquad (2.26)$$

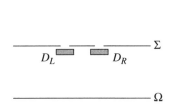

Figure 2.3   Two slits equipped with photodetectors.

2.5 • TWO-SLIT WHICH-WAY EXPERIMENT

The probability to go undetected from S to x is:

$$P_u(x) = |A(x)|^2$$
$$= |\langle x|\text{left slit}\rangle\, t_L\, \langle \text{left slit}|S\rangle + \langle x|\text{right slit}\rangle\, t_R\, \langle \text{right slit}|S\rangle|^2. \quad (2.27)$$

The probability of detection is

$$P_d(x) = 1 - P_u(x). \quad (2.28)$$

We now discuss two cases.

(A) $t_L = t_R = t$ (symmetric case). In this case we can take the amplitude t out of the absolute square in Equation 2.27 and obtain:

$$P_u(x) = |t|^2 \left| \langle x|\text{left slit}\rangle \langle \text{left slit}|S\rangle \right.$$
$$\left. + \langle x|\text{right slit}\rangle \langle \text{right slit}|S\rangle \right|^2 = |t|^2 P_Q(x). \quad (2.29)$$

Although reduced in height by the factor $|t|^2 < 1$, the shape of the interference pattern stays the same. This is good news. But do we get any path information? Unfortunately, we don't. If we record an electron on Ω, we know that it was *not* detected by either one of the electron detectors. This is so, since, if it had been detected, it could not possibly have arrived on Ω because of our assumption of 100% absorption efficiency in case of detection. But since both detectors have the same transmission amplitude, i.e., the same transmission probability, it is impossible for us to tell through which slit the electron passed to arrive on Ω. As a result, we have the interference pattern, but no path information.

(B) Now we choose $|t_L| \gg |t_R|$ (asymmetric case). In this case, if an electron arrives at x, we know that it most likely took the path through the left slit, since most electrons that take the path through the right slit get absorbed with probability $1 - |t_R|^2 \approx 1$. Therefore, in this case, we obtain path information. But what happens to the interference pattern? With $|t_R| \ll |t_L| \approx 1$, we obtain for the probability distribution of the electrons that arrive on Ω:

$$P_u(x) \approx |\langle x|\text{left slit}\rangle\, t_L\, \langle \text{left slit}|S\rangle|^2 \approx P_L(x), \quad (2.30)$$

i.e., the interference pattern is gone! We obtain the probability distribution of a point particle taking the path through the left slit.

If we now increase the detection efficiency of t_R, the interference pattern slowly returns, but we start to lose path information. When we are back to the symmetric case, the full interference pattern is restored, but all path information is lost.

We obtain the following important result:

$$\boxed{\begin{array}{l}\text{Whenever we obtain path information,}\\ \text{the interference pattern vanishes.}\\ \text{Whenever we observe interference,}\\ \text{we lose path information.}\end{array}} \quad (2.31)$$

There have been many discussions and attempts to find a way to observe the interference pattern and still be able to tell precisely which slit each electron went through. To date no convincing experiments have been reported that accomplish this task.

Exercises:

2.5.1 In order to obtain path information, we fill the left half of the two-slit setup shown in Figure 2.2 with a fluorescent dye solution and observe the intensity pattern in the right half of the setup shown in Figure 2.2. If the photon takes the path through the left hole, the dye may light up and betray the presence of the photon in the left half of the experiment. If the photon takes the path through the right hole, the photon does not come in contact with the dye and the dye will not light up. The absence of fluorescence will betray the presence of the photon in the right half of the experiment. The hope is that the addition of the dye solution will reveal the path of the photon without destroying the interference pattern. Unfortunately this idea does not work. Why?

2.6 Summary

Because of their mutually exclusive properties all phenomena we encounter in our macroscopic, classical world can be grouped naturally into two categories: particles and waves. Particles have a localized nature in space and time whereas waves are extended. This natural division of the classical world does not hold in the microscopic quantum world: The objects of the micro-world exhibit both wave and particle aspects. This is called *wave-particle duality*. Even the simplest quantum experiments, such as the single-particle double-slit experiment, cannot be understood without invoking this duality. Therefore, in this chapter, we chose the quantum version of Young's double-slit experiment to take our first steps into the quantum world. In fact, Feynman once remarked that everything one needs to know about quantum mechanics is contained in the double-slit experiment.

Analyzing the double-slit experiment in detail we learned about the curious coexistence between wave and particle pictures. Both are needed for a proper description of light and matter. What is this duality? Does it mean that photons, electrons, and other fundamental objects of our world are "really" particles if we look closely enough, and that the wave description is only a convenient approximation whenever many particles are involved simultaneously? In the case of light this idea seems natural, since macroscopic light, consisting of many photons, behaves like a wave to an excellent approximation, whereas individual photons make tiny spots on photographic film and clicks in detectors, behavior more consistent with the particle picture. However, when interpreting the photon as a pure point particle, we saw that we ran into irreconcilable logical difficulties when interpreting single- and double-slit experiments with photons. Therefore, we were forced to assign both wave and particle aspects to photons. Even electrons, the best examples of point-particles in nature, sometimes behave as waves to such an extent that even scientific instruments, such as the electron microscope, would not work, were it not for wave-particle duality. In fact, with the electron microscope we encounter a first example that "quantum philosophy" is more than idle speculation and leads to concrete technical applications that would not work on the basis of classical physics alone.

So, is there a summary statement, a take-home message, that concisely characterizes what we learned by studying the quantum version of Young's double-slit experiment? There is. There is no doubt that the quantum objects begin and end their journeys as particles. The source S is defined as a device that generates single *particles*, one by one, and the observation screen Ω records tiny dots of blackened photographic emulsion, characteristic of particle hits, and not "grey areas" characteristic of the effects of a wave. While the objects are in flight, and in particular when they encounter the two slits, the particles manage to generate an interference phenomenon characteristic of a wave. So, in summary, the message is this:

$$\boxed{\begin{array}{l}\text{A quantum object is}\\ \text{produced as a } \mathbf{particle},\\ \text{propagates like a } \mathbf{wave},\\ \text{is detected as a } \mathbf{particle}\\ \text{with a probability distribution}\\ \text{that corresponds to a } \mathbf{wave}.\end{array}} \quad (2.32)$$

This message, concisely expressed in Equation 2.32, is what we mean by wave-particle duality.

Chapter Review Exercises:

1. The Large Electron Positron collider (LEP) at the European Center for Nuclear Research (CERN) was able to accelerate electrons to $100\,\text{GeV}$ ($1\,\text{GeV} = 10^9\,\text{eV}$). Estimate the smallest length scales that LEP was able to resolve. Compare your result with the diameter of a proton ($\approx 10^{-15}\,\text{m}$).

2. We refer to Figure 2.1. According to our assumptions S is very far from Σ such that photons in the vicinity of the slit, to a good approximation, have no momentum in x direction ($p_x = 0$). The photon landing at x_2, however, has a large negative momentum $p_x \neq 0$ in x direction.

    (a) Where does this momentum come from?

    (b) Is momentum conserved in this experiment?

3. Explain why $P_Q = 2P_C$ at the center of the interference pattern in Figure 2.2(b).

4. Use Einstein's energy-momentum formula, Equation 1.53, to show that the phase velocity $v = \omega/k$ of de Broglie's matter waves is:

    (a) $v = c$ for photons and

    (b) $v > c$ for any particle with nonzero rest mass m_0.

    (c) Apparently, for material particles ($m_0 \neq 0$), the phase velocity of de Broglie's matter waves exceeds the speed of light. Is this a problem? Show that for any material particle the group velocity $d\omega/dk$ is smaller than c.

chapter 3

The Machinery of Quantum Mechanics

In this chapter:

- ◆ Introduction
- ◆ Schrödinger's Equation
- ◆ Observables
- ◆ Spectral Theory
- ◆ Dirac Notation
- ◆ Heisenberg Picture
- ◆ Two-Level Systems
- ◆ Summary

3.1 Introduction

In Chapters 1 and 2 we encountered many surprising quantum phenomena whose analysis did not require a large amount of calculation. However, if we want to predict in detail the outcome of quantum mechanical experiments, we need a mathematical framework that allows us to do so. This framework is provided by the "machinery of quantum mechanics." We start in Section 3.2 with an introduction to Schrödinger's equation that allows us to compute the precise values of quantum amplitudes for the prediction of quantum measurement results. In Section 3.3 we connect the mathematical formalism of quantum mechanics to the actual physical observables that are measured in laboratory experiments.

In Section 3.4 we study several model systems of great practical importance in quantum mechanics: The infinite square-well potential, the harmonic oscillator potential, the finite square-well potential, and the delta-function potential. These potentials may be used to model many of the most important features of atoms, molecules, and atomic nuclei, such as their discrete energy levels that reveal themselves, e.g., in the line spectra of atoms and molecules. In Section 3.5 we provide a more detailed foundation for the use of Dirac's notation first introduced in Chapter 2. An alternative formulation of quantum mechanics, "Heisenberg's Picture," is discussed in Section 3.6. In Section 3.7 we take a close look at two-level systems, the most fundamental quantum systems in atomic spectroscopy and quantum computing.

3.2 Schrödinger's Equation

In 1801 Young showed that light behaves like a wave. But what was the wave equation? This question remained unanswered until, about 60 years later, Maxwell provided the answer. In 1924 quantum mechanics was in a similar situation. De Broglie had proposed matter waves, but what was the wave equation? This problem was solved considerably faster than finding the wave equation of light. Only two years later, in 1926, Schrödinger provided the answer. He showed that de Broglie's matter wave, $\psi(\boldsymbol{r},t)$, associated with each quantum particle, in its non-relativistic form, satisfies the partial differential equation

$$i\hbar \frac{\partial \psi(\boldsymbol{r},t)}{\partial t} = -\frac{\hbar^2}{2m} \nabla^2 \psi(\boldsymbol{r},t) + V(\boldsymbol{r},t)\psi(\boldsymbol{r},t), \qquad (3.1)$$

where $\boldsymbol{r}$ is the position of the particle, t is the time variable, $V(\boldsymbol{r},t)$ is the potential energy of the particle at position $\boldsymbol{r}$ at time t, and

$$\nabla^2 = \frac{\partial^2}{\partial x^2} + \frac{\partial^2}{\partial y^2} + \frac{\partial^2}{\partial z^2} \qquad (3.2)$$

is the Laplacian operator. Equation 3.1 for the matter wave $\psi(\boldsymbol{r},t)$ is known as *Schrödinger's equation*. If we know the matter wave $\psi(\boldsymbol{r},t)$ of a particle at time $t = t_0$, we can use Schrödinger's Equation 3.1 to compute the time evolution of the matter wave $\psi(\boldsymbol{r},t)$ for $t > t_0$. Because the matter wave $\psi(\boldsymbol{r},t)$ is a solution of Schrödinger's wave equation, we call $\psi(\boldsymbol{r},t)$ the *wave function* of the particle. Interpreted physically, the wave function $\psi(\boldsymbol{r},t)$ is the *amplitude* for finding the quantum particle at position $\boldsymbol{r}$ at time t. In Section 2.4 we had already introduced amplitudes for quantum events, but, apart from ad hoc procedures based on reasonable similarities with light waves, did not provide a method for computing these amplitudes. With Schrödinger's Equation 3.1, the wave equation for quantum amplitudes, this gap is now filled.

The importance of Schrödinger's equation cannot be overstated. Any nonrelativistic, quantitative quantum computation starts with Schrödinger's Equation 3.1 or an equivalent formulation. Acknowledging the importance of his Equation 3.1, Schrödinger received the 1933 Nobel Prize for the application of his equation to atomic structure calculations.

According to quantum mechanics, all we can possibly know about a quantum object is encoded in its wave function $\psi(\boldsymbol{r}, t)$. Once $\psi(\boldsymbol{r}, t)$ is determined, for instance by solving Schrödinger's Equation 3.1, we would like to extract physical information about the object from its wave function $\psi(\boldsymbol{r}, t)$. The most basic information of interest is the location of the particle in space. Since, physically, the wave function $\psi(\boldsymbol{r}, t)$ is an amplitude, as introduced and discussed in Section 2.4, we answer this question by applying Born's Rule. According to Born's Rule 2.15, the probability density for finding a quantum particle at position $\boldsymbol{r}$ at time t is

$$\rho(\boldsymbol{r}, t) = |\psi(\boldsymbol{r}, t)|^2. \tag{3.3}$$

This result is significant. Since, in general, $\psi(\boldsymbol{r}, t)$ is a smooth function that extends over all space, the associated quantum object, in general, is not sharply located at any one position $\boldsymbol{r}$ at a given time t. Instead, the quantum object may be found anywhere in space with a probability density $\rho(\boldsymbol{r}, t)$ computed via Equation 3.3 from its wave function $\psi(\boldsymbol{r}, t)$. In real experiments we cannot probe individual space points $\boldsymbol{r}$ to determine, point-by-point, the probability density $\rho(\boldsymbol{r}, t)$. The physical question to ask is this: What is the probability $P(\mathcal{V}, t)$ of finding the quantum object in a volume $\mathcal{V}$ at time t? The answer is:

$$P(\mathcal{V}, t) = \int_{\mathcal{V}} \rho(\boldsymbol{r}, t) \, d\mathcal{V} = \int_{\mathcal{V}} |\psi(\boldsymbol{r}, t)|^2 \, d\mathcal{V}, \tag{3.4}$$

where the integration is over the specified volume $\mathcal{V}$. Since the wave function describes a particle that is actually present somewhere in space, the probability of finding the particle anywhere in space is 1. Therefore, if we choose the volume $\mathcal{V}$ in Equation 3.4 to be all space, i.e., we integrate $|\psi(\boldsymbol{r}, t)|^2$ over all space, we obtain

$$\int_{\text{all space}} |\psi(\boldsymbol{r}, t)|^2 \, d\mathcal{V} = \int_{\text{all space}} \psi(\boldsymbol{r}, t)^* \, \psi(\boldsymbol{r}, t) \, d\mathcal{V} = 1. \tag{3.5}$$

We call this result the *normalization of the wave function*.

We mention that $\psi(\boldsymbol{r}, t)$ stays normalized to 1 over the course of time only if $V(\boldsymbol{r}, t)$ is a real function, i.e., $V(\boldsymbol{r}, t)^* = V(\boldsymbol{r}, t)$. Absorption of quantum particles may be modeled by choosing a complex potential with $V(r,t)^* \neq V(r,t)$. If the imaginary part of $V(r,t)$ is chosen negative, the normalization integral, Equation 3.5, decreases over time corresponding to the vanishing of particles due to absorption. This method of modeling absorption is borrowed from optics where a complex index of refraction is capable of modeling absorption of electromagnetic waves in a lossy optical medium.

In the introduction to Chapter 2 we mentioned that waves may carry energy and momentum from one point in space to another. An example are electromagnetic waves that carry energy from the Sun to Earth. Since $\psi(r,t)$ is a wave, it, too, may carry energy and momentum. We discuss this aspect of $\psi(r,t)$ further in Section 3.3. In addition, since $\psi(r,t)$ is a probability amplitude, it may carry probability from one region in space to another. The transport of probability is described physically with the help of a probability current density defined as

$$j(r,t) = \frac{\hbar}{2mi}\left[\psi(r,t)^* \nabla \psi(r,t) - \psi(r,t) \nabla \psi(r,t)^*\right]. \quad (3.6)$$

In the absence of absorption, i.e., $V(r,t)^* = V(r,t)$, Equation 3.5 holds, i.e., the total probability of finding a quantum particle anywhere in space is normalized to 1 and does not change in time. Therefore, the loss of probability from one region in space has to be compensated by a gain of probability in another. For an arbitrary volume $\mathcal{V}$ this means that the absolute value of the time change of probability in $\mathcal{V}$,

$$\frac{d}{dt}P(\mathcal{V}) = \frac{d}{dt}\int_\mathcal{V} \rho(r,t)\,d\mathcal{V} = \int_\mathcal{V} \frac{\partial \rho(r,t)}{\partial t}\,d\mathcal{V}, \quad (3.7)$$

is equal to the absolute value of the total probability

$$Q(\mathcal{V}) = \int_{\mathcal{S}(\mathcal{V})} j(r,t) \cdot d\mathcal{S} \quad (3.8)$$

streaming through the surface $\mathcal{S}(\mathcal{V})$ of the volume $\mathcal{V}$. In order to conserve the balance of probability, a decrease in probability inside of $\mathcal{V}$, $dP(\mathcal{V})/dt < 0$, is offset by an increase in probability, $Q(\mathcal{V}) > 0$, outside of $\mathcal{V}$. Therefore we have

$$\int_\mathcal{V} \frac{\partial \rho(r,t)}{\partial t}\,d\mathcal{V} = -\int_{\mathcal{S}(\mathcal{V})} j(r,t) \cdot d\mathcal{S}. \quad (3.9)$$

With the help of Gauss' Theorem, well known from electrostatics, we transform the surface integral in Equation 3.9 into a volume integral according to

$$\int_{\mathcal{S}(\mathcal{V})} j(r,t) \cdot d\mathcal{S} = \int_\mathcal{V} \nabla \cdot j(r,t)\,d\mathcal{V} \quad (3.10)$$

3.2 • SCHRÖDINGER'S EQUATION

to obtain

$$\int_{\mathcal{V}} \frac{\partial \rho(\boldsymbol{r},t)}{\partial t} d\mathcal{V} = -\int_{\mathcal{V}} \nabla \cdot \boldsymbol{j}(\boldsymbol{r},t) d\mathcal{V}. \tag{3.11}$$

This is the continuity equation of quantum mechanics in integral form. Since Equation 3.11 holds for all volumes $\mathcal{V}$, the integrands in Equation 3.11 have to be the same. Thus, we obtain the continuity equation of quantum mechanics,

$$\frac{\partial \rho(\boldsymbol{r},t)}{\partial t} + \nabla \cdot \boldsymbol{j}(\boldsymbol{r},t) = 0, \tag{3.12}$$

in differential form.

According to the normalization condition, Equation 3.5, the volume integral over $|\psi(\boldsymbol{r},t)|^2$ exists and is finite ($= 1$). Therefore, $\psi(\boldsymbol{r},t)$ is a *square-integrable function*. If we perform the integral in Equation 3.5 in spherical coordinates, we obtain

$$\int_{\text{all space}} |\psi(\boldsymbol{r},t)|^2 d\mathcal{V} = \int_0^{\pi} \sin(\theta) d\theta \int_0^{2\pi} d\varphi \int_0^{\infty} dr\, r^2 |\psi(\boldsymbol{r},t)|^2, \tag{3.13}$$

where $r = |\boldsymbol{r}|$ is the radial coordinate, θ is the polar angle, and φ is the azimuthal angle. The r integral in Equation 3.13 converges only if $r^2|\psi(\boldsymbol{r},t)|^2$ decreases faster than $1/r$, the function at the border between integrable functions (decreasing faster than $1/r$) and nonintegrable functions (decreasing at $1/r$ or slower). Therefore, we need:

$$r^2 |\psi(\boldsymbol{r},t)|^2 \sim \frac{1}{r^{1+\epsilon}}, \quad \epsilon > 0 \tag{3.14}$$

for large r, independent of t. Equation 3.14 implies

$$|\psi(\boldsymbol{r},t)| \sim \frac{1}{r^{\frac{3}{2}+\delta}}, \quad \delta = \frac{\epsilon}{2} > 0, \quad \text{for large } r \text{ and all } t. \tag{3.15}$$

This is a powerful result that shows that if $\psi(\boldsymbol{r},t)$ represents a single particle, we necessarily have

$$\psi(\boldsymbol{r},t) \to 0 \quad \text{for } |\boldsymbol{r}| \to \infty. \tag{3.16}$$

In physical language we say that $\psi(\boldsymbol{r},t)$ has to satisfy the *boundary condition* $\psi(\boldsymbol{r},t) = 0$ at $|\boldsymbol{r}| = \infty$ for all times t. Boundary conditions, such as the one stated in Equation 3.16, play an important role in the solution of Schrödinger's Equation 3.1, as we will see in Section 3.4.

For smooth, physical potentials $V(\mathbf{r},t)$ the solutions $\psi(\mathbf{r},t)$ of Schrödinger's Equation 3.1 are smooth functions with continuous first and second derivatives. Sometimes, as we will do in Section 3.4, model potentials are used that have discontinuities.

In order to find out how to proceed in this case, let us imagine a surface $\mathcal{S}$ that divides space into two regions, Region I and Region II. We assume that the potential $V(\mathbf{r},t)$ is smooth in Regions I and II, but jumps discontinuously across the surface $\mathcal{S}$. Solving Schrödinger's equation, we obtain the wave function $\psi_I(\mathbf{r},t)$ in Region I and $\psi_{II}(\mathbf{r},t)$ in Region II. Since $V(\mathbf{r},t)$ is smooth in Regions I and II, both $\psi_I(\mathbf{r},t)$ and $\psi_{II}(\mathbf{r},t)$ are smooth in their respective regions and have a continuous derivative. This means that we can define the probability currents

$$\mathbf{j}_I(\mathbf{r},t) = \frac{\hbar}{2mi}\left[\psi_I(\mathbf{r},t)^*\nabla\psi_I(\mathbf{r},t) - \psi_I(\mathbf{r},t)\nabla\psi_I(\mathbf{r},t)^*\right],$$
$$\mathbf{j}_{II}(\mathbf{r},t) = \frac{\hbar}{2mi}\left[\psi_{II}(\mathbf{r},t)^*\nabla\psi_{II}(\mathbf{r},t) - \psi_{II}(\mathbf{r},t)\nabla\psi_{II}(\mathbf{r},t)^*\right] \quad (3.17)$$

in Regions I and II, respectively. Since $\mathbf{j}$, the probability current, measures the probability per unit time and surface area that traverses the surface $\mathcal{S}$, a discontinuous jump in $\mathbf{j}$ would correspond to the creation or destruction of particles. Since, because of the Continuity Equation 3.12, this cannot happen in nonrelativistic quantum theory, we have to require that $\mathbf{j}$ is continuous in the vicinity of the surface $\mathcal{S}$. Because of the structure of $\mathbf{j}$, involving the product of the wave function and its derivative (see Equation 3.6), continuity of $\mathbf{j}$ requires continuity of $\psi(\mathbf{r},t)$ and its derivative $\nabla\psi(\mathbf{r},t)$ in the vicinity of $\mathcal{S}$. Since $\psi_I(\mathbf{r},t)$ and $\psi_{II}(\mathbf{r},t)$ are already separately continuous in Regions I and II, respectively, all we have to require to achieve global continuity of the wave function is

$$\psi_I(\mathbf{r},t) = \psi_{II}(\mathbf{r},t) \quad \text{for } \mathbf{r} \text{ on } \mathcal{S}, \text{ and all times } t. \quad (3.18)$$

We call Equation 3.18 a *matching condition*. Next, we have to achieve continuity of the derivative $\nabla\psi(\mathbf{r},t)$ of the wave function in the vicinity of $\mathcal{S}$. Since both $\psi_I(\mathbf{r},t)$ and $\psi_{II}(\mathbf{r},t)$ have a continuous derivative in their respective Regions I and II, and since, according to Equation 3.18, they are matched on the surface $\mathcal{S}$, the derivative of the wave function tangentially to the surface $\mathcal{S}$ is already continuous. Therefore, defining the unit vector $\mathbf{n}$, orthogonal to $\mathcal{S}$ in the point $\mathbf{r}$, all we have to require to obtain a globally continuous derivative of the wave function is for the derivative of the wave function to be continuous in the direction orthogonal (normal) to the surface, i.e., we require the matching condition

$$\frac{\partial \psi_I(\mathbf{r},t)}{\partial \mathbf{n}} = \frac{\partial \psi_{II}(\mathbf{r},t)}{\partial \mathbf{n}} \quad \text{for } \mathbf{r} \text{ on } \mathcal{S}, \text{ and all times } t, \quad (3.19)$$

where

$$\frac{\partial}{\partial \boldsymbol{n}} = \boldsymbol{n} \cdot \nabla \qquad (3.20)$$

is the derivative in the direction orthogonal (normal) to $\mathcal{S}$. In summary, by fulfilling both matching conditions of Equations 3.18 and 3.19, we may allow discontinuous model potentials in our discussions of quantum mechanics.

Exercises:

3.2.1 A quantum particle is described by the wave function

$$\psi(\boldsymbol{r},t) = \alpha \exp(-r/\beta) \exp(-iEt/\hbar),$$

where α, β, and E are constants, and $r = |\boldsymbol{r}|$. What is the probability of finding the particle inside of a sphere of radius β centered at the origin?

3.2.2 Using Schrödinger's Equation 3.1, check by explicit calculation that for real potentials, $V(\boldsymbol{r},t)^* = V(\boldsymbol{r},t)$, the probability density $\rho(\boldsymbol{r},t)$ and the probability current density $\boldsymbol{j}(\boldsymbol{r},t)$, defined in Equations 3.3 and 3.6, respectively, satisfy the Continuity Equation 3.12.

3.2.3 Assume that the potential $V(\boldsymbol{r},t)$ of a quantum particle governed by Schrödinger's Equation 3.1 can be written as

$$V(\boldsymbol{r},t) = \alpha(\boldsymbol{r},t) - i\beta(\boldsymbol{r},t), \qquad (3.21)$$

where $\beta(\boldsymbol{r},t) > 0$ for all $\boldsymbol{r}$ and t. Show that

$$\frac{d}{dt} \int_{\text{all space}} \rho(\boldsymbol{r},t) \, d\mathcal{V} < 0,$$

i.e., the total probability of finding the quantum particle anywhere in space decreases monotonically for increasing t. Thus, the potential in Equation 3.21 may be used to model absorption of particles.

3.3 Observables

Since $\psi(\boldsymbol{r},t)$ is the amplitude for finding a particle at position $\boldsymbol{r}$ and since, in general, $\psi(\boldsymbol{r},t)$ is nonzero in large regions of space, the location of a quantum particle is not usually sharply defined. In practice this manifests itself in the following way. Suppose we have N identical copies of our particle, each described by $\psi(\boldsymbol{r}, t)$, and we measure the position of each of these N particles. As a result of our measurements, we obtain a set of random positions $\boldsymbol{r}_1, \boldsymbol{r}_2, \ldots, \boldsymbol{r}_N$, which, in general, are all different, and distributed according to $\rho(\boldsymbol{r}, t) = |\psi(\boldsymbol{r}, t)|^2$ as defined in Equation 3.3. In order to characterize the sequence of measurements statistically, the mean value

$$\bar{\boldsymbol{r}} = \frac{1}{N} \sum_{j=1}^{N} \boldsymbol{r}_j \qquad (3.22)$$

and the variance

$$(\Delta r)^2 = \frac{1}{N} \sum_{j=1}^{N} (\boldsymbol{r}_j - \bar{\boldsymbol{r}})^2 \qquad (3.23)$$

are of interest. In quantum mechanics $\bar{\boldsymbol{r}}$ is also known as the *expectation value* of $\boldsymbol{r}$ and

$$\Delta r = \sqrt{(\Delta r)^2} \qquad (3.24)$$

is known as its *uncertainty*. The expectation value of a physical quantity is the average value we obtain by measuring this quantity repeatedly on quantum objects prepared in the same wave function $\psi(\boldsymbol{r}, t)$. Clearly, both quantities are of great practical value, since $\bar{\boldsymbol{r}}$ tells us where to look and Δr tells us how far away from $\bar{\boldsymbol{r}}$ we have to look to observe a significant number of the particles, i.e., how sharply the expectation value $\bar{\boldsymbol{r}}$ is defined. In the limit of many measurements, i.e., $N \to \infty$, we compute the expectation value according to

$$\bar{\boldsymbol{r}} = \int_{\text{all space}} \rho(\boldsymbol{r}, t) \, \boldsymbol{r} \, d\mathcal{V} = \int_{\text{all space}} \psi(\boldsymbol{r}, t)^* \boldsymbol{r} \psi(\boldsymbol{r}, t) \, d\mathcal{V} \qquad (3.25)$$

and the variance according to

$$(\Delta r)^2 = \int_{\text{all space}} \rho(\boldsymbol{r}, t) \, (\boldsymbol{r} - \bar{\boldsymbol{r}})^2 \, d\mathcal{V}$$

$$= \int_{\text{all space}} \psi(\boldsymbol{r}, t)^* (\boldsymbol{r} - \bar{\boldsymbol{r}})^2 \psi(\boldsymbol{r}, t) \, d\mathcal{V}, \qquad (3.26)$$

where we have written the second integrals in Equations 3.25 and 3.26 in a suggestive form that later allows for a generalization to the expectation values and variances of other physical observables.

3.3 • OBSERVABLES

The position is not the only observable of physical significance. We would also like to know, for instance, the momentum, the angular momentum, and the energy of our particle. But all we have is the wave function $\psi(\boldsymbol{r},t)$ that does not directly refer to momentum, angular momentum, or energy. On the other hand, implicit in our use of Schrödinger's equation as the central equation of quantum mechanics is the assertion that all we can possibly know about a quantum particle must be contained in $\psi(\boldsymbol{r},t)$. Therefore, although not immediately obvious, all information on additional physical observables, such as momentum, angular momentum, and energy must be contained in $\psi(\boldsymbol{r},t)$. We extract the relevant information from $\psi(\boldsymbol{r},t)$ with the help of operators.

Operators are mathematical objects that take $\psi(\boldsymbol{r},t)$ as input and, as output, assign either another function or a number to $\psi(\boldsymbol{r},t)$. Examples of operators are ∇ and $\int_{\text{all space}} \ldots d\mathcal{V}$. The operator ∇ assigns the function $\varphi(\boldsymbol{r},t) = \nabla \psi(\boldsymbol{r},t)$ to the wave function $\psi(\boldsymbol{r},t)$, whereas the operator $\int_{\text{all space}} \ldots d\mathcal{V}$ assigns the number $\int_{\text{all space}} \psi(\boldsymbol{r},t) d\mathcal{V}$ to the wave function $\psi(\boldsymbol{r},t)$. The obvious question to ask is this: How do we assign operators to observables? To make a long story short, there is no hard and fast mathematical formula for assigning operators to observables. This has to be done in a physical way by choosing operators that "make physical sense." We illustrate the types of reasoning that may be used in the construction of operators with the example of linear momentum.

According to de Broglie (see Equation 2.11), the (non-normalized) wave function of a particle traveling with momentum $\boldsymbol{p}$ in free space is given by

$$\psi(\boldsymbol{r},t) = \exp\left[\frac{i}{\hbar}(\boldsymbol{p}\cdot\boldsymbol{r} - Et)\right]. \tag{3.27}$$

Applying the operator

$$\hat{\boldsymbol{p}} = -i\hbar\nabla \tag{3.28}$$

to the wave function in Equation 3.27, we obtain

$$\hat{\boldsymbol{p}}\psi(\boldsymbol{r},t) = -i\hbar\nabla \exp\left[\frac{i}{\hbar}(\boldsymbol{p}\cdot\boldsymbol{r} - Et)\right]$$
$$= \boldsymbol{p} \exp\left[\frac{i}{\hbar}(\boldsymbol{p}\cdot\boldsymbol{r} - Et)\right] = \boldsymbol{p}\,\psi(\boldsymbol{r},t). \tag{3.29}$$

We see that applying the operator $\hat{\boldsymbol{p}}$ to the wave function $\psi(\boldsymbol{r},t)$ reproduces the wave function $\psi(\boldsymbol{r},t)$ with the momentum of the particle appearing as a multiplicative factor in front of $\psi(\boldsymbol{r},t)$. This way we extract the momentum $\boldsymbol{p}$ from the wave function $\psi(\boldsymbol{r},t)$. In case a potential $V(\boldsymbol{r},t)$ is switched on, the form of the wave function, Equation 3.27, stays approximately the same, but only if $V(\boldsymbol{r},t)$ varies slowly in space and time. In addition, $\psi(\boldsymbol{r},t)$ acquires an amplitude factor $A(\boldsymbol{r},t)$ such that in the presence of $V(\boldsymbol{r},t)$ the wave function is given by

$$\psi(\boldsymbol{r},t) = A(\boldsymbol{r},t)\exp\left[\frac{i}{\hbar}(\boldsymbol{p}\cdot\boldsymbol{r} - Et)\right], \quad (3.30)$$

where $\boldsymbol{p}$ and E are only defined for short intervals of space and time where $V(\boldsymbol{r},t)$ may be considered approximately constant. If we now apply the operator $\hat{\boldsymbol{p}}$ to the wave function of Equation 3.30, we no longer reproduce the entire wave function with a factor in front. Still, if we look only locally, we approximately reproduce the wave function $\psi(\boldsymbol{r},t)$ in this interval multiplied with the local momentum $\boldsymbol{p}(\boldsymbol{r},t)$ of the particle, where $|\boldsymbol{p}| = \sqrt{2m(E-V)}$. Since $\hat{\boldsymbol{p}}$, defined in Equation 3.28, extracts the exact momentum in the free-particle case and does a good job extracting the local momentum in the case with a potential present, we adopt $\hat{\boldsymbol{p}}$ defined in Equation 3.28 as the *operator of momentum*.

In connection with the construction of the operator $\hat{\boldsymbol{p}}$ of momentum we encountered two situations. (1) $\hat{\boldsymbol{p}}$ applied to the wave function reproduces the wave function exactly with a multiplicative factor $\boldsymbol{p}$, i.e.,

$$\hat{\boldsymbol{p}}\psi(\boldsymbol{r},t) = \boldsymbol{p}\psi(\boldsymbol{r},t), \quad (3.31)$$

where $\boldsymbol{p}$ is a constant. In this case the momentum of the wave function is $\boldsymbol{p}$ and it is sharply defined. We call Equation 3.31 an *eigenvalue equation* for the operator $\hat{\boldsymbol{p}}$ with *eigenvalue* $\boldsymbol{p}$ and *eigenfunction* $\psi(\boldsymbol{r},t)$. (2) $\hat{\boldsymbol{p}}$ applied to the wave function does not reproduce the wave function globally, but only locally with local eigenvalue $\boldsymbol{p}(\boldsymbol{r},t)$. In this case the momentum of the wave function is not sharp, but we may still define the expectation value of momentum. We do this in analogy to Equation 3.25 in the following way:

$$\bar{\boldsymbol{p}} = \int_{\text{all space}} \rho(\boldsymbol{r},t)\,\boldsymbol{p}(\boldsymbol{r},t)\,d\mathcal{V} = \int_{\text{all space}} \psi(\boldsymbol{r},t)^*\boldsymbol{p}(\boldsymbol{r},t)\psi(\boldsymbol{r},t)\,d\mathcal{V}$$

$$= \int_{\text{all space}} \psi(\boldsymbol{r},t)^*\hat{\boldsymbol{p}}\psi(\boldsymbol{r},t)\,d\mathcal{V}. \quad (3.32)$$

In analogy to Equation 3.26 we compute the variance of momentum according to

$$(\Delta p)^2 = \int_{\text{all space}} \rho(\boldsymbol{r},t)\,[\boldsymbol{p}(\boldsymbol{r},t) - \bar{\boldsymbol{p}}]^2\,d\mathcal{V}$$

$$= \int_{\text{all space}} \psi(\boldsymbol{r},t)^*[\boldsymbol{p}(\boldsymbol{r},t) - \bar{\boldsymbol{p}}]^2\psi(\boldsymbol{r},t)\,d\mathcal{V}$$

$$= \int_{\text{all space}} \psi(\boldsymbol{r},t)^*(\hat{\boldsymbol{p}} - \bar{\boldsymbol{p}})^2\,\psi(\boldsymbol{r},t)\,d\mathcal{V}. \quad (3.33)$$

3.3 • OBSERVABLES

We are now in a position to generalize the theory to any observable Θ with associated operator $\hat{\Theta}$. If

$$\hat{\Theta}\psi(\boldsymbol{r},t) = \vartheta\psi(\boldsymbol{r},t), \qquad (3.34)$$

where ϑ is a constant, then Equation 3.34 is an eigenvalue equation for $\hat{\Theta}$ with eigenvalue ϑ and eigenfunction $\psi(\boldsymbol{r},t)$. In this case the associated observable Θ has a sharp value ϑ for the wave function $\psi(\boldsymbol{r},t)$. If $\psi(\boldsymbol{r},t)$ does not satisfy an eigenvalue equation such as Equation 3.34, the observable Θ does not have a sharp value for the wave function $\psi(\boldsymbol{r},t)$. However, in analogy to Equations 3.32 and 3.33, we can still define the expectation value $\bar{\Theta}$ and the variance $(\Delta\Theta)^2$ according to

$$\bar{\Theta} = \int_{\text{all space}} \psi(\boldsymbol{r},t)^* \hat{\Theta} \psi(\boldsymbol{r},t)\, d\mathcal{V} \qquad (3.35)$$

and

$$(\Delta\Theta)^2 = \int_{\text{all space}} \psi(\boldsymbol{r},t)^* (\hat{\Theta} - \bar{\Theta})^2 \psi(\boldsymbol{r},t)\, d\mathcal{V}. \qquad (3.36)$$

In order to put the position into this general operator framework, we define the position operator $\hat{\boldsymbol{r}}$ according to

$$\hat{\boldsymbol{r}}\,\psi(\boldsymbol{r},t) = \boldsymbol{r}\,\psi(\boldsymbol{r},t). \qquad (3.37)$$

This looks like an eigenvalue equation of the form of Equation 3.34, but it is not, since $\boldsymbol{r}$, on the right-hand side of Equation 3.37, is not a constant. Therefore, a quantum particle described by a smooth wave function $\psi(\boldsymbol{r},t)$ does not have a sharply defined position, which is consistent with the nature of $\psi(\boldsymbol{r},t)$ as describing the amplitude of finding the particle at position $\boldsymbol{r}$, which is nonzero for many possible $\boldsymbol{r}$ if $\psi(\boldsymbol{r},t)$ is nonzero in an extended region of space. Only if $\psi(\boldsymbol{r},t)$ is sharply peaked at some $\boldsymbol{r}_0$, somewhere in space, and nowhere else, would $\psi(\boldsymbol{r},t)$ be an (approximate) eigenfunction of $\hat{\boldsymbol{r}}$ with eigenvalue $\boldsymbol{r}_0$, and therefore describe a particle with a sharp position. We will discuss this case further in Section 3.5. With the position operator $\hat{\boldsymbol{r}}$ as defined in Equation 3.37, we may now write $\bar{\boldsymbol{r}}$ and $(\Delta r)^2$ in the form

$$\bar{\boldsymbol{r}} = \int_{\text{all space}} \psi(\boldsymbol{r},t)^* \hat{\boldsymbol{r}} \psi(\boldsymbol{r},t)\, d\mathcal{V} \qquad (3.38)$$

and

$$(\Delta r)^2 = \int_{\text{all space}} \psi(\boldsymbol{r},t)^* (\hat{\boldsymbol{r}} - \bar{\boldsymbol{r}})^2 \psi(\boldsymbol{r},t)\, d\mathcal{V}. \qquad (3.39)$$

Equations 3.38 and 3.39, the operator analogues of Equations 3.25 and 3.26, incorporate the position into the general operator framework established in Equations 3.34–3.36.

Now that we defined the operator analogues of the observables position and momentum, we may define the operator analogue of any function $f(\boldsymbol{p},\boldsymbol{r},t)$ of classical arguments $\boldsymbol{p}$ and $\boldsymbol{r}$ in the following way:

$$f(\boldsymbol{p},\boldsymbol{r},t) \;\rightarrow\; f(\hat{\boldsymbol{p}},\hat{\boldsymbol{r}},t). \tag{3.40}$$

This procedure is known as *canonical quantization*. We apply this procedure immediately to compute the operators of angular momentum and energy. In classical mechanics angular momentum is defined as

$$\boldsymbol{L} \;=\; \boldsymbol{r}\times\boldsymbol{p}. \tag{3.41}$$

Applying canonical quantization we obtain the following operator analogue of angular momentum:

$$\hat{\boldsymbol{L}} \;=\; \hat{\boldsymbol{r}}\times\hat{\boldsymbol{p}}. \tag{3.42}$$

In classical, analytical mechanics the energy is associated with Hamilton's function

$$\mathcal{H}(\boldsymbol{p},\boldsymbol{r},t) \;=\; \frac{\boldsymbol{p}^2}{2m} + V(\boldsymbol{r}). \tag{3.43}$$

Applying the canonical quantization rule, Equation 3.40, we obtain the Hamiltonian operator, the operator analogue of Hamilton's function, according to:

$$\hat{\mathcal{H}}(\hat{\boldsymbol{p}},\hat{\boldsymbol{r}},t) \;=\; \frac{\hat{\boldsymbol{p}}^2}{2m} + V(\hat{\boldsymbol{r}},t). \tag{3.44}$$

Using the definition of the momentum operator, Equation 3.28, this can also be written as:

$$\hat{\mathcal{H}}(\hat{\boldsymbol{p}},\hat{\boldsymbol{r}},t) \;=\; -\frac{\hbar^2}{2m}\nabla^2 + V(\hat{\boldsymbol{r}},t), \tag{3.45}$$

where the operator ∇^2 is defined in Equation 3.2. This form of the Hamiltonian operator allows us to write Schrödinger's Equation 3.1 in the following, more concise way as:

$$i\hbar\,\frac{\partial \psi(\boldsymbol{r},t)}{\partial t} \;=\; \hat{\mathcal{H}}(\hat{\boldsymbol{p}},\hat{\boldsymbol{r}},t)\,\psi(\boldsymbol{r},t). \tag{3.46}$$

This is the "standard form" of the Schrödinger equation frequently found in the literature.

For time-dependent potentials $V(\mathbf{r}, t)$ the energy E of the quantum particle is not a constant. However, as we know from classical mechanics, for time-independent potentials $V(\mathbf{r})$, it is. Therefore, classically,

$$\mathcal{H}(\mathbf{p}, \mathbf{r}) = E. \tag{3.47}$$

This means, quantum mechanically, that we can find wave functions $\psi(\mathbf{r}, t)$ of the quantum particle that satisfy the eigenvalue equation

$$\hat{\mathcal{H}}(\hat{\mathbf{p}}, \hat{\mathbf{r}})\, \psi(\mathbf{r}, t) = E\, \psi(\mathbf{r}, t), \tag{3.48}$$

which means that the energy E is sharp in the wave function $\psi(\mathbf{r}, t)$. In the general case, where $\psi(\mathbf{r}, t)$ is not an eigenfunction of the Hamiltonian $\hat{\mathcal{H}}(\hat{\mathbf{p}}, \hat{\mathbf{r}})$, we can still compute the expectation value $\bar{E}$ and the variance of the energy, $(\Delta E)^2$ according to our general rules, Equations 3.35 and 3.36, as

$$\bar{E} = \int_{\text{all space}} \psi(\mathbf{r}, t)^* \hat{\mathcal{H}}(\hat{\mathbf{p}}, \hat{\mathbf{r}}) \psi(\mathbf{r}, t) \, d\mathcal{V} \tag{3.49}$$

and

$$(\Delta E)^2 = \int_{\text{all space}} \psi(\mathbf{r}, t)^* [\hat{\mathcal{H}}(\hat{\mathbf{p}}, \hat{\mathbf{r}}) - \bar{E}]^2\, \psi(\mathbf{r}, t) \, d\mathcal{V}. \tag{3.50}$$

There is a peculiar difficulty with the canonical quantization procedure that we have not yet talked about. Suppose we have the classical function

$$f(\mathbf{p}, \mathbf{r}) = \mathbf{p}r, \quad r = |\mathbf{r}|, \tag{3.51}$$

which we would like to turn into an operator using canonical quantization. There are two possibilities:

$$\hat{f}_1 = \hat{\mathbf{p}}\hat{r}, \quad \text{and} \quad \hat{f}_2 = \hat{r}\,\hat{\mathbf{p}}. \tag{3.52}$$

Since $\hat{\mathbf{p}}$ and $\hat{r}$ are operators, it matters whether we write $\hat{\mathbf{p}}\hat{r}$ or $\hat{r}\hat{\mathbf{p}}$. The two are not the same! Therefore, a single classical function, $f = \mathbf{p}r$ gives rise to (at least) two different quantum operators, $\hat{f}_1$ and $\hat{f}_2$. To investigate this further, we need to define more accurately what it means that two operators are the same. We define: Two operators, $\hat{\theta}_1$ and $\hat{\theta}_2$, are the same if

$$\hat{\theta}_1\, \psi(\mathbf{r}, t) = \hat{\theta}_2\, \psi(\mathbf{r}, t) \tag{3.53}$$

for any wave function $\psi(\mathbf{r}, t)$. We call the wave functions $\psi(\mathbf{r}, t)$ in Equation 3.53 test functions. If we can find even one single test function $\varphi(\mathbf{r}, t)$

for which Equation 3.53 is not satisfied, $\hat{\theta}_1$ and $\hat{\theta}_2$ are different. Let us use this definition to show that $\hat{f}_1 \neq \hat{f}_2$. Our task is to find a test function $\varphi(\boldsymbol{r},t)$ for which $\hat{f}_1\varphi(\boldsymbol{r},t) \neq \hat{f}_2\varphi(\boldsymbol{r},t)$. We choose

$$\varphi(\boldsymbol{r},t) = \sqrt{\beta^3/\pi} \exp(-\beta r), \qquad (3.54)$$

where β is a positive constant. Our test function $\varphi(\boldsymbol{r},t)$ is normalized and depends only on $r = |\boldsymbol{r}|$; it is not dependent on the time t. For this wave function we have:

$$\begin{aligned}
\hat{f}_1\,\varphi(\boldsymbol{r},t) &= \hat{\boldsymbol{p}}\hat{r}\,\varphi(\boldsymbol{r},t) = -i\hbar\nabla[r\sqrt{\beta^3/\pi}\exp(-\beta r)] \\
&= -i\hbar\sqrt{\beta^3/\pi}[\exp(-\beta r) - r\beta\exp(-\beta r)]\boldsymbol{r}/r \\
&= i\hbar\sqrt{\beta^3/\pi}(\beta r - 1)\exp(-\beta r)\boldsymbol{r}/r, \qquad (3.55)
\end{aligned}$$

whereas

$$\begin{aligned}
\hat{f}_2\,\varphi(\boldsymbol{r},t) &= \hat{r}\hat{\boldsymbol{p}}\,\varphi(\boldsymbol{r},t) = -i\hbar \boldsymbol{r}\sqrt{\beta^3/\pi}\nabla\exp(-\beta r) \\
&= i\hbar\beta\boldsymbol{r}\sqrt{\beta^3/\pi}\exp(-\beta r). \qquad (3.56)
\end{aligned}$$

Clearly $\hat{f}_1\,\varphi(\boldsymbol{r},t) \neq \hat{f}_2\,\varphi(\boldsymbol{r},t)$, since the term $-i\hbar\sqrt{\beta^3/\pi}\exp(-\beta r)\boldsymbol{r}/r$ is clearly missing in Equation 3.56. Therefore, since we are able to produce a counterexample to Equation 3.53 (for $\hat{\theta}_1 = \hat{f}_1$ and $\hat{\theta}_2 = \hat{f}_2$), and one single counterexample is enough (!), we showed that $\hat{f}_1 \neq \hat{f}_2$, i.e., $\hat{\boldsymbol{p}}\hat{r} \neq \hat{r}\hat{\boldsymbol{p}}$.

There is a more formal way to investigate whether the ordering of two operators $\hat{\Omega}_1$ or $\hat{\Omega}_2$ matters, or not. We define the *commutator*

$$[\hat{\Omega}_1, \hat{\Omega}_2] = \hat{\Omega}_1\hat{\Omega}_2 - \hat{\Omega}_2\hat{\Omega}_1 \qquad (3.57)$$

of two operators. Then, if $[\hat{\Omega}_1, \hat{\Omega}_2] = 0$, we have $\hat{\Omega}_1\hat{\Omega}_2 = \hat{\Omega}_2\hat{\Omega}_1$ and the ordering of the two operators does not matter. In this case, i.e., if $[\hat{\Omega}_1, \hat{\Omega}_2] = 0$, we say that $\hat{\Omega}_1$ and $\hat{\Omega}_2$ *commute*. If $[\hat{\Omega}_1, \hat{\Omega}_2] \neq 0$ we say that $\hat{\Omega}_1$ and $\hat{\Omega}_2$ *do not commute*, i.e., the ordering of the operators matters. But how do we actually compute commutators? "Computation" of a commutator means that we establish an *operator identity*

$$[\hat{\Omega}_1, \hat{\Omega}_2] = \hat{C}, \qquad (3.58)$$

where the commutator, an operator, is on the left-hand side and the result, $\hat{C}$, also an operator, is on the right-hand side. But we already have a criterion to decide whether Equation 3.58 is true: According to Equation 3.53, the operator identity, Equation 3.58, is true if it holds for all wave functions $\psi(\boldsymbol{r},t)$.

3.3 • OBSERVABLES   57

As an example, let us evaluate the commutator of $\hat{\boldsymbol{p}}$ with $\hat{r}$. We have:

$$\begin{aligned}
\hat{\boldsymbol{p}}\hat{r}\,\psi(\boldsymbol{r},t) &= -i\hbar\nabla[r\psi(\boldsymbol{r},t)] \\
&= -i\hbar\left[(\nabla r)\psi(\boldsymbol{r},t) + r\nabla\psi(\boldsymbol{r},t)\right] \\
&= -i\hbar\boldsymbol{r}^0\,\psi(\boldsymbol{r},t) + r\hat{\boldsymbol{p}}\psi(\boldsymbol{r},t),
\end{aligned} \quad (3.59)$$

where

$$\boldsymbol{r}^0 = \frac{\boldsymbol{r}}{r} \quad (3.60)$$

is the unit vector in $\boldsymbol{r}$ direction. From Equation 3.59 we obtain:

$$\begin{aligned}
(\hat{\boldsymbol{p}}\hat{r} - \hat{r}\hat{\boldsymbol{p}})\,\psi(\boldsymbol{r},t) &= [\hat{\boldsymbol{p}},\hat{r}]\,\psi(\boldsymbol{r},t) = -i\hbar\boldsymbol{r}^0\,\psi(\boldsymbol{r},t) \\
&= -i\hbar\hat{\boldsymbol{r}}^0\,\psi(\boldsymbol{r},t).
\end{aligned} \quad (3.61)$$

Since Equation 3.61 holds for any wave function $\psi(\boldsymbol{r},t)$ (we made no assumptions about $\psi(\boldsymbol{r},t)$ when deriving Equations 3.59 and 3.61), we may omit the wave function in Equation 3.61 and establish the operator identity:

$$[\hat{\boldsymbol{p}},\hat{r}] = -i\hbar\hat{\boldsymbol{r}}^0. \quad (3.62)$$

It was first noticed by Born that quantum mechanics is different from classical mechanics precisely because the commutators of quantum observables, represented by operators, are, in general, nonzero, whereas the commutators of classical variables, since they are ordinary numbers, are always trivially zero.

For later applications we need to establish the commutator relations of the components $\hat{L}_x$, $\hat{L}_y$, and $\hat{L}_z$ of the orbital angular momentum operator $\hat{\boldsymbol{L}}$ defined in Equation 3.42. According to Equation 3.42, we have:

$$\hat{\boldsymbol{L}} = \hat{\boldsymbol{r}}\times\hat{\boldsymbol{p}} = \hat{\boldsymbol{r}}\times(-i\hbar\boldsymbol{\nabla}) = -i\hbar\begin{pmatrix} y\frac{\partial}{\partial z} - z\frac{\partial}{\partial y} \\ z\frac{\partial}{\partial x} - x\frac{\partial}{\partial z} \\ x\frac{\partial}{\partial y} - y\frac{\partial}{\partial x} \end{pmatrix}. \quad (3.63)$$

Therefore:

$$\begin{aligned}
[\hat{L}_x, \hat{L}_y]\psi(\mathbf{r},t) &= (\hat{L}_x\hat{L}_y - \hat{L}_y\hat{L}_x)\psi(\mathbf{r},t) \\
&= -\hbar^2\left[\left(y\frac{\partial}{\partial z} - z\frac{\partial}{\partial y}\right)\left(z\frac{\partial}{\partial x} - x\frac{\partial}{\partial z}\right)\right. \\
&\quad \left. - \left(z\frac{\partial}{\partial x} - x\frac{\partial}{\partial z}\right)\left(y\frac{\partial}{\partial z} - z\frac{\partial}{\partial y}\right)\right]\psi(\mathbf{r},t) \\
&= -\hbar^2\left[y\frac{\partial}{\partial z}\left(z\frac{\partial}{\partial x}\right) - yx\frac{\partial^2}{\partial z^2} - z^2\frac{\partial^2}{\partial y\partial x} + zx\frac{\partial^2}{\partial y\partial z}\right. \\
&\quad \left. - zy\frac{\partial^2}{\partial x\partial z} + z^2\frac{\partial^2}{\partial x\partial y} + xy\frac{\partial^2}{\partial z^2} - x\frac{\partial}{\partial z}\left(z\frac{\partial}{\partial y}\right)\right]\psi(\mathbf{r},t). \quad (3.64)
\end{aligned}$$

For smooth wave functions with a smooth first derivative the order of differentiation may be exchanged so that, e.g.,

$$\frac{\partial^2}{\partial x\partial y} = \frac{\partial^2}{\partial y\partial x}. \qquad (3.65)$$

Therefore:

$$\begin{aligned}
[\hat{L}_x, \hat{L}_y]\psi(\mathbf{r},t) &= -\hbar^2\left[y\frac{\partial}{\partial z}\left(z\frac{\partial}{\partial x}\right) + zx\frac{\partial^2}{\partial y\partial z}\right. \\
&\quad \left. - zy\frac{\partial^2}{\partial x\partial z} - x\frac{\partial}{\partial z}\left(z\frac{\partial}{\partial y}\right)\right]\psi(\mathbf{r},t) \\
&= -\hbar^2\left[y\frac{\partial}{\partial x} + yz\frac{\partial^2}{\partial z\partial x} + zx\frac{\partial^2}{\partial y\partial z}\right. \\
&\quad \left. - zy\frac{\partial^2}{\partial x\partial z} - x\frac{\partial}{\partial y} - xz\frac{\partial^2}{\partial z\partial y}\right]\psi(\mathbf{r},t) \\
&= \hbar^2\left(x\frac{\partial}{\partial y} - y\frac{\partial}{\partial x}\right)\psi(\mathbf{r},t) = i\hbar\hat{L}_z\psi(\mathbf{r},t). \quad (3.66)
\end{aligned}$$

Since this result is valid for all $\psi(\mathbf{r},t)$, we established the operator identity

$$[\hat{L}_x, \hat{L}_y] = i\hbar\hat{L}_z. \qquad (3.67)$$

In the same way we obtain:

$$[\hat{L}_y, \hat{L}_z] = i\hbar\hat{L}_x, \quad [\hat{L}_z, \hat{L}_x] = i\hbar\hat{L}_y. \qquad (3.68)$$

If we take into account that any operator $\hat{\Omega}$ commutes with itself, i.e., $[\hat{\Omega}, \hat{\Omega}] = 0$, and that for any two operators $\hat{\theta}$ and $\hat{\Omega}$ we have $[\hat{\theta}, \hat{\Omega}] = -[\hat{\Omega}, \hat{\theta}]$, we may generalize Equations 3.67 and 3.68 to:

$$[\hat{L}_m, \hat{L}_n] = i\hbar \sum_{j=1}^{3} \epsilon_{mnj} \hat{L}_j, \tag{3.69}$$

where $m, n, j \in \{x, y, z\}$, and

$$\epsilon_{mnj} = \begin{cases} 1, & \text{if } m, n, j = x, y, z \text{ cyclic,} \\ -1, & \text{if } m, n, j = y, x, z \text{ cyclic,} \\ 0, & \text{otherwise.} \end{cases} \tag{3.70}$$

Any type of angular momentum in quantum mechanics satisfies the commutator identity of Equation 3.69. It is customary in quantum mechanics to turn this result around and call any set $\{\hat{\Omega}_x, \hat{\Omega}_y, \hat{\Omega}_z\}$ of three operators *angular momentum operators* if they fulfill Equation 3.69.

As a generalization of the normalization integral, Equation 3.5, we introduce the scalar product

$$\int_{\text{all space}} \psi(\boldsymbol{r}, t)^* \varphi(\boldsymbol{r}, t) \, d\mathcal{V} \tag{3.71}$$

of two wave functions $\psi(\boldsymbol{r}, t)$ and $\varphi(\boldsymbol{r}, t)$. Then, the expectation value

$$\bar{\theta} = \int_{\text{all space}} \psi(\boldsymbol{r}, t)^* \hat{\theta} \varphi(\boldsymbol{r}, t) \, d\mathcal{V} = \int_{\text{all space}} \psi(\boldsymbol{r}, t)^* [\hat{\theta} \varphi(\boldsymbol{r}, t)] \, d\mathcal{V} \tag{3.72}$$

may be interpreted as the scalar product of the wave function $\psi(\boldsymbol{r}, t)$ with the wave function $\varphi(\boldsymbol{r}, t) = \hat{\theta} \psi(\boldsymbol{r}, t)$. If for two operators $\hat{\theta}$ and $\hat{\Omega}$ we have

$$\int_{\text{all space}} \psi(\boldsymbol{r}, t)^* \hat{\theta} \varphi(\boldsymbol{r}, t) \, d\mathcal{V} = \int_{\text{all space}} [\hat{\Omega} \psi(\boldsymbol{r}, t)]^* \varphi(\boldsymbol{r}, t) \, d\mathcal{V} \tag{3.73}$$

for any pair of wave functions $\psi(\boldsymbol{r}, t)$ and $\varphi(\boldsymbol{r}, t)$, we call $\hat{\Omega}$ the *Hermite conjugate* of $\hat{\theta}$ and write

$$\hat{\Omega} = \hat{\theta}^\dagger. \tag{3.74}$$

Of utmost importance in quantum mechanics are operators $\hat{A}$ that are their own Hermite conjugates, i.e.,

$$\hat{A}^\dagger = \hat{A}. \tag{3.75}$$

If an operator $\hat{A}$ fulfills Equation 3.75, we call $\hat{A}$ a *Hermitian operator*. Using the additivity of integrals it is straightforward to show that if $\hat{A}$ and $\hat{B}$ are Hermitian operators, so is their sum, i.e.,

$$\hat{A}^\dagger = \hat{A}, \quad \hat{B}^\dagger = \hat{B}, \quad \hat{C} = \hat{A} + \hat{B} \quad \to \quad \hat{C}^\dagger = \hat{C}. \qquad (3.76)$$

We show now that $\hat{r}$, $\hat{p}$, and $\hat{\mathcal{H}}$ are all Hermitian operators. In the case of $\hat{r}$ we have

$$\int_{\text{all space}} \psi(r,t)^* \, \hat{r} \, \varphi(r,t) \, d\mathcal{V} = \int_{\text{all space}} \psi(r,t)^* \, r \, \varphi(r,t) \, d\mathcal{V}$$

$$= \int_{\text{all space}} r \, \psi(r,t)^* \, \varphi(r,t) \, d\mathcal{V} = \int_{\text{all space}} [r \, \psi(r,t)]^* \, \varphi(r,t) \, d\mathcal{V}$$

$$= \int_{\text{all space}} [\hat{r} \, \psi(r,t)]^* \, \varphi(r,t) \, d\mathcal{V}, \qquad (3.77)$$

where we used the definition in Equation 3.37 of $\hat{r}$. Therefore, according to Equations 3.73, 3.74, and 3.75, the position operator $\hat{r}$ is Hermitian with

$$\hat{r}^\dagger = \hat{r}. \qquad (3.78)$$

In the case of the momentum, we use the fact that, according to Equation 3.16, the wave function vanishes at infinity. Then, in Cartesian coordinates and with the definition in Equation 3.28, we have for the x component of the momentum operator:

$$\int_{\text{all space}} \psi(r,t)^* \, \hat{p}_x \, \varphi(r,t) \, d\mathcal{V}$$

$$= -i\hbar \int_{-\infty}^{\infty} dz \int_{-\infty}^{\infty} dy \int_{-\infty}^{\infty} \psi(r,t)^* \, \frac{\partial}{\partial x} \varphi(r,t) \, dx$$

$$= -i\hbar \int_{-\infty}^{\infty} dz \int_{-\infty}^{\infty} dy \Bigg\{ [\psi(r,t)^* \varphi(r,t)]_{x=-\infty}^{x=\infty}$$

$$- \int_{-\infty}^{\infty} \left[\frac{\partial}{\partial x} \psi(r,t) \right]^* \varphi(r,t) \, dx \Bigg\}$$

$$= \int_{-\infty}^{\infty} dz \int_{-\infty}^{\infty} dy \int_{-\infty}^{\infty} \left[-i\hbar \frac{\partial}{\partial x} \psi(r,t) \right]^* \varphi(r,t) \, dx$$

$$= \int_{\text{all space}} [\hat{p}_x \psi(r,t)]^* \, \varphi(r,t) \, d\mathcal{V}, \qquad (3.79)$$

which shows that $\hat{p}_x$ is a Hermitian operator. Similarly we show that $\hat{p}_y$ and $\hat{p}_z$ are Hermitian operators. Therefore,

$$\hat{p}^\dagger = \hat{p}. \qquad (3.80)$$

If $\hat{\Omega}$ is a Hermitian operator, then $\hat{\Omega}^2$ is Hermitian, too. We show this in the following way:

$$\int_{\text{all space}} \psi(\boldsymbol{r},t)^* \, \hat{\Omega}^2 \, \varphi(\boldsymbol{r},t) \, d\mathcal{V}$$

$$= \int_{\text{all space}} [\hat{\Omega}\psi(\boldsymbol{r},t)]^* \, \hat{\Omega} \varphi(\boldsymbol{r},t) \, d\mathcal{V}$$

$$= \int_{\text{all space}} \left\{ \hat{\Omega} \left[\hat{\Omega}\psi(\boldsymbol{r},t) \right] \right\}^* \varphi(\boldsymbol{r},t) \, d\mathcal{V}$$

$$= \int_{\text{all space}} \left[\hat{\Omega}^2 \psi(\boldsymbol{r},t) \right]^* \varphi(\boldsymbol{r},t) \, d\mathcal{V}. \tag{3.81}$$

This shows that

$$\hat{\Omega}^\dagger = \hat{\Omega} \ \to \ (\hat{\Omega}^2)^\dagger = \hat{\Omega}^2. \tag{3.82}$$

It is straightforward to generalize the proof in Equation 3.81 for Equation 3.82 to any power n of $\hat{\Omega}$, i.e.,

$$\hat{\Omega}^\dagger = \hat{\Omega} \ \to \ \left(\hat{\Omega}^n\right)^\dagger = \hat{\Omega}^n. \tag{3.83}$$

This implies that if a real function $f(x)$ has a Taylor series expansion

$$f(x) = \sum_{n=0}^{\infty} \omega_n \, x^n, \quad \omega_n \text{ real}, \tag{3.84}$$

the operator $\hat{W} = f(\hat{\Omega})$ is Hermitian, $\hat{W}^\dagger = \hat{W}$, as soon as $\hat{\Omega}$ is Hermitian. Since $\hat{\boldsymbol{r}}$ is a Hermitian operator, we apply this reasoning to the potential operator $V(\hat{\boldsymbol{r}},t)$ of the Hamiltonian operator $\hat{\mathcal{H}}(\hat{\boldsymbol{p}},\hat{\boldsymbol{r}},t)$ in Equation 3.44, which shows that $V(\hat{\boldsymbol{r}},t)$ is a Hermitian operator if $V(\boldsymbol{r},t)$ is sufficiently smooth to possess a Taylor series expansion. Since according to Equations 3.80 and 3.82 $\hat{\boldsymbol{p}}^2$ is a Hermitian operator, so is the operator $\hat{K} = \hat{\boldsymbol{p}}^2/(2m)$, since m is real. Since, according to Equation 3.44, the Hamiltonian operator $\hat{\mathcal{H}}(\hat{\boldsymbol{p}},\hat{\boldsymbol{r}},t)$ is the sum of the two Hermitian operators $\hat{K}$ and $V(\hat{\boldsymbol{r}},t)$, the Hamiltonian operator $\hat{\mathcal{H}}(\hat{\boldsymbol{p}},\hat{\boldsymbol{r}},t)$ is Hermitian, i.e.,

$$\hat{\mathcal{H}}(\hat{\boldsymbol{p}},\hat{\boldsymbol{r}},t)^\dagger = \hat{\mathcal{H}}(\hat{\boldsymbol{p}},\hat{\boldsymbol{r}},t). \tag{3.85}$$

Similarly we may show that the angular momentum operator $\hat{\boldsymbol{L}}$ is Hermitian.

In summary we have the result that all observables we studied so far, i.e., position, momentum, angular momentum, and energy have associated quantum operators that are Hermitian operators. This is generally true. We have the important result that

> In quantum mechanics observables are represented by Hermitian operators. (3.86)

Exercises:

3.3.1 Three Hermitian operators, $\hat{A}$, $\hat{B}$, and $\hat{C}$, satisfy the commutator relation
$$[\hat{A}, \hat{B}] = \gamma \hat{C},$$
where γ is a complex constant. Show that γ is purely imaginary.

3.3.2 Show that the test function $\varphi(r, t)$, defined in Equation 3.54, is properly normalized.

3.3.3 Show that the operators
$$\hat{\theta}_1 = \hat{\boldsymbol{p}}^2 \hat{r}, \quad \hat{\theta}_2 = \hat{r}\hat{\boldsymbol{p}}^2, \quad \hat{\theta}_3 = \hat{\boldsymbol{p}}\hat{r}\hat{\boldsymbol{p}} = \hat{\boldsymbol{p}} \cdot (\hat{r}\hat{\boldsymbol{p}})$$
are pairwise different.

3.3.4 Using the definitions in Equations 3.73 and 3.74 of the Hermitian conjugate of an operator $\hat{\theta}$, show that
$$(\hat{\theta}^\dagger)^\dagger = \hat{\theta}.$$

3.3.5 Show that for two operators $\hat{A}$ and $\hat{B}$:
$$(\hat{A}\hat{B})^\dagger = \hat{B}^\dagger \hat{A}^\dagger.$$

3.3.6 Prove the following useful commutator identities:
(a) $[\hat{A}\hat{B}, \hat{C}] = \hat{A}[\hat{B}, \hat{C}] + [\hat{A}, \hat{C}]\hat{B}$,
(b) $[\hat{A}, \hat{B}\hat{C}] = [\hat{A}, \hat{B}]\hat{C} + \hat{B}[\hat{A}, \hat{C}]$,

where $\hat{A}$, $\hat{B}$, and $\hat{C}$ are arbitrary operators.

3.4 Spectral Theory

Fully time-dependent potentials $V(r, t)$ need to be considered only in rare circumstances, for instance if we place a quantum system in a time-dependent electric or magnetic field. In the context of our discussion of the foundations of quantum mechanics, we do not need to consider such systems. Therefore, from now on, we will focus on time-independent potentials only.

Although quantum objects exist and move in three-dimensional space, it is sufficient to consider the Schrödinger equation in one dimension

3.4 • SPECTRAL THEORY 63

for studying its structure and its physical implications. In addition, it is often possible to reduce the full three-dimensional Schrödinger Equation 3.1 (Equation 3.46, respectively) to three decoupled one-dimensional Schrödinger equations, one for each spatial dimension. In one dimension, for instance on the x-axis, and in the time-independent case, the potential takes the form $V(x)$ and Schrödinger's Equation 3.46 simplifies to

$$i\hbar \frac{\partial \psi(x,t)}{\partial t} = \hat{\mathcal{H}}(\hat{p},\hat{x})\,\psi(x,t), \qquad (3.87)$$

where

$$\hat{\mathcal{H}}(\hat{p},\hat{x}) = \frac{\hat{p}^2}{2m} + V(x) = -\left(\frac{\hbar^2}{2m}\right)\frac{\partial^2}{\partial x^2} + V(x) \qquad (3.88)$$

is the Hamiltonian operator in one dimension. In one dimension the analogue of the normalization integral, Equation 3.5, is:

$$\int_{-\infty}^{\infty} \psi(x,t)^*\,\psi(x,t)\,dx = 1. \qquad (3.89)$$

It is possible to solve the one-dimensional Schrödinger Equation 3.87 for many physically significant potentials $V(x)$. The following is a solution technique that is generally applicable for any time-independent potential. In Section 3.3 (see Equation 3.48) we learned that for time-independent Hamiltonians we may find wave functions whose energy E is sharp, i.e., E is a constant. In this case we try solutions of the form

$$\psi(x,t) = e^{-iEt/\hbar}\,\varphi(x), \qquad (3.90)$$

where $\psi(x,t)$ is written as a product of two factors, the first of which depends only on the time variable t, while the second factor depends only on the space variable x. If $\psi(x,t)$ is normalized according to Equation 3.89, $\varphi(x)$ is normalized according to

$$\int_{-\infty}^{\infty} \varphi(x)^*\,\varphi(x)\,dx = 1. \qquad (3.91)$$

Inserting Equation 3.90 into the Schrödinger Equation 3.87,

$$\hat{\mathcal{H}}(\hat{p},\hat{x})\,\psi(x,t) = i\hbar\frac{\partial \psi(x,t)}{\partial t} = E\,\psi(x,t), \qquad (3.92)$$

we verify that E is indeed the energy eigenvalue of the Hamiltonian operator $\hat{\mathcal{H}}(\hat{p},\hat{x})$. Using the explicit form of $\hat{\mathcal{H}}(\hat{p},\hat{x})$ given in Equation 3.88, and the product form in Equation 3.90 of the wave function $\psi(x,t)$, we may write Equation 3.92 more explicitly as

$$-\frac{\hbar^2}{2m}e^{-iEt/\hbar}\frac{\partial^2 \varphi(x)}{\partial x^2} + V(x)e^{-iEt/\hbar}\varphi(x) = E\,e^{-iEt/\hbar}\,\varphi(x). \qquad (3.93)$$

Since the exponent factor $\exp(-iEt/\hbar)$ is never zero, we may cancel it on both sides of Equation 3.93 and obtain:

$$-\frac{\hbar^2}{2m}\varphi''(x) + V(x)\varphi(x) = E\,\varphi(x), \qquad (3.94)$$

where the double primes indicate the second derivative with respect to x. Since the time variable does not occur in Equation 3.94, Equation 3.94 is called the *stationary Schrödinger equation*. Once the stationary Schrödinger Equation 3.94 is solved and $\varphi(x)$ is determined, we use $\varphi(x)$ in Equation 3.90 to construct the complete time-dependent wave function $\psi(x,t)$.

To gain some insight into the nature of the solutions of the stationary Schrödinger Equation 3.94 and to familiarize ourselves with some common solution techniques, we will study three important model potentials: The infinite square-well potential, the harmonic oscillator potential, and the finite square-well potential. Each of these potentials has important insights to offer on the nature of the solutions of the stationary Schrödinger Equation 3.94.

We start with the infinite square-well potential

$$V(x) = \begin{cases} \infty, & \text{for } x \leq 0, \\ 0, & \text{for } 0 < x < a, \\ \infty, & \text{for } x \geq a, \end{cases} \qquad (3.95)$$

where $a > 0$ is the width of the potential. A sketch of the infinite square-well potential is shown in Figure 3.1. A particle placed inside ($0 < x < a$) the infinite square-well potential will forever be trapped inside the potential, since the potential walls ($x \leq 0$, $x \geq a$) are infinitely high and infinitely thick. Since $|\psi(x,t)|^2 = |\varphi(x)|^2$ is the probability of finding the particle at position x, and since the particle will never be found inside the potential walls, this implies that $\varphi(x) = 0$ for $x \leq 0$ and $x \geq a$. In

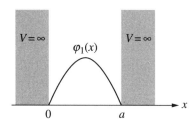

Figure 3.1 Infinite square-well potential. The ground-state wave function $\varphi_1(x)$ is also shown.

particular, this implies that

$$\varphi(0) = 0, \quad \text{and} \quad \varphi(a) = 0. \tag{3.96}$$

Equation 3.96 imposes a *boundary condition* on the solutions of the stationary Schrödinger Equation 3.94. In our case the boundary condition demands that the wave function $\varphi(x)$, the solution of the stationary Schrödinger Equation 3.94, has to vanish at $x = 0$ and at $x = a$. Since $V(x) = 0$ inside the infinite square-well potential, the stationary Schrödinger Equation 3.94 becomes:

$$-\frac{\hbar^2}{2m} \varphi''(x) = E\varphi(x). \tag{3.97}$$

Defining the wave number

$$k = \sqrt{\frac{2mE}{\hbar^2}}, \tag{3.98}$$

Equation 3.97 can be written as:

$$\varphi''(x) + k^2 \varphi(x) = 0. \tag{3.99}$$

This is the well-known harmonic oscillator equation (written in x instead of t) whose general solution is

$$\varphi(x) = A \sin(kx) + B \cos(kx), \tag{3.100}$$

where A and B are constants. The boundary condition $\varphi(0) = 0$ (see Equation 3.96) requires $B = 0$. Therefore, $\varphi(x) = A \sin(kx)$. We notice that A cannot be zero, since otherwise the wave function $\varphi(x)$ would be identically zero inside the potential well, representing no particle at all. Next, we consider the boundary condition $\varphi(a) = 0$ (see Equation 3.96). This boundary condition requires that $A \sin(ka) = 0$. But since $A \neq 0$, as we argued previously, we may divide this equation by A to obtain:

$$\sin(ka) = 0. \tag{3.101}$$

This equation is fulfilled only if k has one of the special values

$$k_n = \frac{\pi}{a} n, \quad n = 1, 2, 3, \ldots. \tag{3.102}$$

Since the allowed values of k are discrete and can be counted with a counting index n, we say that the wave number k is *quantized*. Notice that we do not include $n = 0$ as a possibility, since in this case, again, the wave function $\varphi(x)$ would vanish identically inside the potential, corresponding to the *absence*, not the *presence* of a quantum particle inside the well. Notice, too, that $k_n = -\pi n/a$ also satisfies Equation 3.101. However, these solutions do not produce qualitatively new wave functions since $\sin(kx)$ is an odd function and the minus sign may thus be taken out of the argument of the sine function and be absorbed in the as yet undetermined constant A. Also, in order to obtain a one-to-one relationship between E and k, we chose the positive branch of the square root in Equation 3.98, which excludes negative k_n according to our chosen convention.

Quantization of k implies quantization of the energy E of a particle in the infinite square well. Solving Equation 3.98 for E, and using the allowed k values 3.102, we obtain

$$E_n = \frac{\hbar^2 k_n^2}{2m} = \frac{\hbar^2 \pi^2}{2ma^2} n^2, \quad n = 1, 2, 3, \ldots . \tag{3.103}$$

Apparently, due to the quantization of k, the energy of a particle in the infinite square well is quantized, too. This reminds us of the quantized energy levels of atoms. Therefore we call the discrete energies in Equation 3.103 the *energy levels* of the infinite square-well potential. In general, the complete set of all the allowed energy levels of a quantum system is known as the *spectrum* of this system. Therefore, the complete set $\{E_n\}_{n=1}^{\infty}$ of the energy levels in Equation 3.103 of the infinite square well is the spectrum of the infinite square-well potential. Since the spectrum of the infinite square-well potential is determined by Equation 3.101, Equation 3.101 is known as the *spectral equation* of the infinite square-well potential. This, again, may be generalized to any quantum system. An equation, such as Equation 3.101, that determines the energy levels E_n of a quantum system is a spectral equation of this quantum system.

For each energy level E_n we have an associated time-independent wave function

$$\varphi_n(x) = \begin{cases} 0, & \text{for } x \leq 0, \\ A_n \sin(k_n x), & \text{for } 0 < x < a, \\ 0, & \text{for } x \geq a, \end{cases} \tag{3.104}$$

where A_n is a *normalization constant*. The normalization condition in Equation 3.91 requires

$$\int_{-\infty}^{\infty} \varphi_n(x)^* \varphi_n(x)\, dx = |A_n|^2 \int_0^a \sin^2(k_n x)\, dx = \frac{a}{2}|A_n|^2 = 1, \tag{3.105}$$

from which we obtain immediately that $|A_n| = \sqrt{2/a}$. In principle this equation determines only the modulus $|A_n|$ of A_n. Therefore, A_n itself could still be a complex number. However, since only the absolute square of the wave function, the probability of finding the particle at x, is measurable, we may choose A_n to be real, since any complex phase will be eliminated when taking $|\psi(x,t)|^2 = |\varphi(x)|^2$ to determine the probability. Therefore, in summary, we obtain the stationary wave functions

$$\varphi_n(x) = \sqrt{\frac{2}{a}} \sin(k_n x), \quad n = 1, 2, \ldots, \tag{3.106}$$

and the time-dependent wave functions

$$\psi_n(x,t) = \sqrt{\frac{2}{a}} \exp\left(-\frac{i}{\hbar} E_n t\right) \sin(k_n x), \quad n = 1, 2, \ldots, \tag{3.107}$$

for the infinite square-well potential, where k_n and E_n are given in Equations 3.102 and 3.103, respectively.

From linear algebra we remember an important concept about scalar products: If the scalar product of two vectors is zero, the vectors are *orthogonal*. In Equation 3.71 we defined a scalar product for wave functions. Applying the scalar product in Equation 3.71 to the stationary wave functions in Equation 3.106, we obtain:

$$\int_{\text{all space}} \varphi_n(x)^* \varphi_m(x)\, dx = \int_0^a \varphi_n(x)^* \varphi_m(x)\, dx$$

$$= \frac{2}{a} \int_0^a \sin(n\pi x/a) \sin(m\pi x/a)\, dx = \delta_{nm}, \tag{3.108}$$

where

$$\delta_{nm} = \begin{cases} 1, & \text{for } n = m, \\ 0, & \text{for } n \neq m \end{cases} \tag{3.109}$$

is the *Kronecker symbol*. Applying the scalar product in Equation 3.71 to the time-dependent wave functions in Equation 3.107, we obtain:

$$\int_{\text{all space}} \psi_n(x,t)^* \psi_m(x,t)\, dx$$

$$= \exp\left[\frac{i}{\hbar}(E_n - E_m)t\right] \int_0^a \varphi_n(x,t)^* \varphi_m(x,t)\, dx$$

$$= \exp\left[\frac{i}{\hbar}(E_n - E_m)t\right] \delta_{nm} = \delta_{nm}, \tag{3.110}$$

since δ_{nm} is nonzero only for $n = m$, in which case $E_n = E_m$ and the exponential function is identically 1 for all times t. This means that the stationary and time-dependent wave functions of the infinite square-well potential are orthogonal for $n \neq m$, and they are normalized to 1 for $n = m$. We say that the wave functions in Equations 3.106 and 3.107 are *orthonormal* with respect to the scalar product defined in Equation 3.71.

Next, we study the harmonic oscillator potential

$$V(x) = \frac{1}{2}m\omega^2 x^2, \qquad (3.111)$$

shown in Figure 3.2. Here, m is the mass of the quantum particle and ω is the oscillator frequency. The stationary Schrödinger equation with the oscillator potential $V(x)$ stated in Equation 3.111 is

$$-\varphi''(x) + \left(\frac{x^2}{b^4}\right)\varphi(x) = k^2 \varphi(x), \qquad (3.112)$$

where we introduced the *oscillator length*

$$b = \sqrt{\frac{\hbar}{m\omega}}, \qquad (3.113)$$

and k is defined in Equation 3.98. To get a first idea of the overall behavior of the wave function, we investigate the Schrödinger Equation 3.112 for large $|x|$. In this case we may neglect the constant k^2 with respect to x^2/b^4 and arrive at the equation

$$\varphi''(x) = \left(\frac{x^2}{b^4}\right)\varphi(x), \qquad (3.114)$$

approximately valid for large $|x|$. An approximate solution of Equation 3.114 is

$$\varphi(x) \approx \exp\left[\left(\frac{\sigma}{2b^2}\right)x^2\right], \qquad (3.115)$$

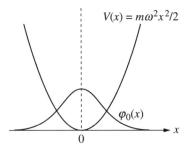

Figure 3.2 Harmonic oscillator potential.

where $\sigma = \pm 1$. Indeed, taking the second derivative of the exponential function in Equation 3.115 we obtain

$$\left\{\exp\left[\left(\frac{\sigma}{2b^2}\right)x^2\right]\right\}'' = \left(\frac{\sigma}{b^2} + \frac{\sigma^2 x^2}{b^4}\right)\exp\left[\left(\frac{\sigma}{2b^2}\right)x^2\right]$$

$$\approx \frac{x^2}{b^4}\exp\left[\left(\frac{\sigma}{2b^2}\right)x^2\right] \qquad (3.116)$$

for both values of σ. While, formally, Equation 3.115 allows for both signs, $\sigma = \pm 1$, only $\sigma = -1$ is consistent with a normalizable wave function. For $\sigma = +1$, the exponential term in Equation 3.115 diverges and the normalization integral in Equation 3.91 does not exist. As discussed in Section 3.2, the existence of the normalization integral requires that the wave function vanishes at infinity. In one dimension, this implies the boundary conditions $\varphi(\pm\infty) = 0$, consistent with our choice of $\sigma = -1$.

Based on the asymptotic form of $\varphi(x)$ for $x \to \pm\infty$, we write $\varphi(x)$ in the form

$$\varphi(x) = N f\left(\frac{x}{b}\right)\exp\left(-\frac{x^2}{2b^2}\right), \qquad (3.117)$$

where $f(z)$ is an, as yet, undetermined function of a single variable $z = x/b$, and N is a normalization constant. Inserting Equation 3.117 into Equation 3.112 we obtain

$$\left\{f''\left(\frac{x}{b}\right) - 2\left(\frac{x}{b}\right)f'\left(\frac{x}{b}\right) + [(kb)^2 - 1]f\left(\frac{x}{b}\right)\right\}\exp\left(-\frac{x^2}{2b^2}\right) = 0. \qquad (3.118)$$

The exponential function in Equation 3.118 is never zero. Therefore, canceling the exponential function, we obtain the following differential equation for $f(z)$:

$$f''(z) - 2zf'(z) + [(kb)^2 - 1]f(z) = 0. \qquad (3.119)$$

There are many functions that fulfill the differential Equation 3.119. But not all are allowed. Most of the solutions of Equation 3.119 are strongly divergent and ruin the boundary condition $\varphi(\pm\infty) = 0$. There is, however, a class of solutions that works. The Hermite polynomials $H_n(z)$, $n = 0, 1, 2, \ldots$, a class of polynomials studied intensively more than a century ago, fulfills the differential equation

$$H_n''(z) - 2zH_n'(z) + 2nH_n(z) = 0, \quad n = 0, 1, 2, \ldots, \qquad (3.120)$$

where n is the degree (i.e., the highest power of z) of the polynomial $H_n(z)$. Since the differential equation for the Hermite polynomials in

Equation 3.120 is linear, it determines the Hermite polynomials only up to multiplicative constants. We choose these constants such that the coefficient of the highest power z^n of the Hermite polynomial $H_n(z)$ is 2^n. Then, the first three Hermite polynomials are given by:

$$\begin{aligned} H_0(z) &= 1, \\ H_1(z) &= 2z, \\ H_2(z) &= 4z^2 - 2. \end{aligned} \quad (3.121)$$

The Hermite polynomials satisfy the important *orthogonality relation*:

$$\int_{-\infty}^{\infty} H_n(z) H_m(z) e^{-z^2} dz = \delta_{nm} 2^n n! \sqrt{\pi}, \quad (3.122)$$

where δ_{nm} is the *Kronecker symbol* defined in Equation 3.109. Equation 3.119 looks precisely like Equation 3.120 if we set

$$(kb)^2 - 1 = 2n. \quad (3.123)$$

Equation 3.123 is the spectral equation of the harmonic oscillator. Solved for E, we obtain the energy levels

$$E_n = \hbar\omega \left(n + \frac{1}{2} \right), \quad n = 0, 1, 2, \ldots. \quad (3.124)$$

Apparently, the spectrum of the harmonic oscillator is discrete and equispaced. The harmonic oscillator and its spectrum play an important role in Chapter 11, where we discuss a real-life implementation of a quantum computer.

According to Equation 3.117 the wave functions $\varphi_n(x)$ associated with the energy eigenvalues E_n have the form

$$\varphi_n(x) = N_n H_n \left(\frac{x}{b} \right) \exp\left(-\frac{x^2}{2b^2} \right), \quad n = 0, 1, 2, \ldots, \quad (3.125)$$

where H_n are the Hermite polynomials and N_n are normalization constants. Using the orthogonality relation stated in Equation 3.122, the properly normalized oscillator wave functions are given explicitly by

$$\varphi_n(x) = \frac{1}{\sqrt{2^n n! b \sqrt{\pi}}} H_n \left(\frac{x}{b} \right) \exp\left(-\frac{x^2}{2b^2} \right), \quad n = 0, 1, 2, \ldots. \quad (3.126)$$

The harmonic-oscillator wave functions defined in Equation 3.126 are orthonormal according to

$$\int_{-\infty}^{\infty} \varphi_n(x) \varphi_m(x) \, dx = \delta_{nm}. \quad (3.127)$$

3.4 • SPECTRAL THEORY

Infinitely high potentials, such as the infinite square-well potential or the harmonic oscillator potential are excellent examples for taking our first steps in spectral theory, but they do not occur in nature: Physical potentials remain finite as $x \to \pm\infty$. As an example of a more physical potential we study the finite square-well potential

$$V(x) = \begin{cases} 0, & \text{for } x \leq 0, \\ V_0, & \text{for } 0 < x < a, \\ 0, & \text{for } x \geq a, \end{cases} \qquad (3.128)$$

shown for $V_0 < 0$ in Figure 3.3.

Let us first look at the special case $V_0 = 0$. This corresponds to the case where no potential is switched on, i.e., the case of *free motion*. In this case the stationary Schrödinger equation is

$$-\frac{\hbar^2}{2m}\varphi''(x) = E\varphi(x), \qquad (3.129)$$

which can also be written as

$$\varphi''(x) + k^2\varphi(x) = 0, \qquad (3.130)$$

where we introduced the wave number k defined in Equation 3.98. There are two types of functions that solve Equation 3.130:

$$\varphi_k^{(+)}(x) = N_k^{(+)} \exp(ikx),$$
$$\varphi_k^{(-)}(x) = N_k^{(-)} \exp(-ikx), \qquad (3.131)$$

where $N_k^{(\pm)} \neq 0$ are normalization constants. The solutions $\varphi_k^{(+)}(x)$ correspond to quantum particles moving from left to right, while the solutions $\varphi_k^{(-)}(x)$ correspond to quantum particles moving from right to left. This is so since

$$\hat{p}\,\varphi_k^{(\pm)}(x) = -i\hbar\frac{d}{dx}\varphi_k^{(\pm)}(x) = \pm\hbar k\,\varphi_k^{(\pm)}(x), \qquad (3.132)$$

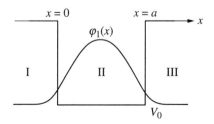

Figure 3.3   The finite square-well potential.

i.e., $\varphi_k^{(\pm)}(x)$ are eigenfunctions of the momentum operator $\hat{p}$ with eigenvalues $p = \pm \hbar k$, respectively, i.e., positive momentum (moving left to right) in the case of a particle represented by $\varphi_k^{(+)}(x)$ and negative momentum (moving right to left) in the case of a particle represented by $\varphi_k^{(-)}(x)$.

Compared with the infinite square-well potential and the harmonic-oscillator potential there is one profound novelty: There is no k for which either $\varphi_k^{(+)}(x)$ or $\varphi_k^{(-)}(x)$ are square-normalizable! This is so since $\varphi_k^{(\pm)}(x)^* \varphi_k^{(\pm)}(x) = |N_k^{(\pm)}|^2$, i.e., a nonzero constant, for all x. Therefore, there is no k for which the integral defined in Equation 3.91 exists. As a consequence, there is no quantization condition, no spectral equation, and there are no discrete energy levels. We say that the solutions defined in Equation 3.131 form a *continuum*, where all $k \geq 0$, i.e., all energies $E \geq 0$, are allowed.

To motivate what has to be done in this case, we introduce a *convergence generating factor* into the wave functions defined in Equation 3.131 according to

$$\varphi_k^{(\pm)}(x) \to \tilde{\varphi}_k^{(\pm)}(x) = \tilde{N}_k^{(\pm)} \exp(\pm i k x - \epsilon |x|/2), \qquad (3.133)$$

where $\epsilon > 0$ is very small. We may imagine ϵ to be so small that over large intervals of x around $x = 0$ the deviation of the wave functions defined in Equation 3.133 from the original wave functions in Equation 3.131 is vanishingly small. Still, as long as $\epsilon > 0$, the wave functions defined in Equation 3.133 are square-normalizable. Let us now consider the scalar product of $\tilde{\varphi}_k^{(\pm)}(x)$ with $\tilde{\varphi}_{k'}^{(\pm)}(x)$:

$$\int_{-\infty}^{\infty} \tilde{\varphi}_k^{(\pm)}(x)^* \tilde{\varphi}_{k'}^{(\pm)}(x) \, dx$$

$$= \int_{-\infty}^{\infty} \tilde{N}_k^{(\pm)*} \tilde{N}_{k'}^{(\pm)} \exp[\pm i(k' - k)x - \epsilon|x|] \, dx$$

$$= \tilde{N}_k^{(\pm)*} \tilde{N}_{k'}^{(\pm)} \left\{ \int_0^{\infty} \exp[\pm i(k' - k)x - \epsilon x] \, dx \right.$$

$$\left. + \int_{-\infty}^{0} \exp[\pm i(k' - k)x + \epsilon x] \right\} dx$$

$$= \tilde{N}_k^{(\pm)*} \tilde{N}_{k'}^{(\pm)} \left\{ \frac{-1}{\pm i(k' - k) - \epsilon} + \frac{1}{\pm i(k' - k) + \epsilon} \right\}$$

$$= 2\pi \tilde{N}_k^{(\pm)*} \tilde{N}_{k'}^{(\pm)} L_\epsilon(k' - k), \qquad (3.134)$$

where $L_\epsilon(z)$ is the *Lorentzian* defined as

$$L_\epsilon(z) = \frac{1}{\pi}\left(\frac{\epsilon}{z^2+\epsilon^2}\right). \qquad (3.135)$$

We see that for small ϵ the scalar product computed in Equation 3.134 is sharply peaked at $k \approx k'$, i.e., the functions in Equation 3.133 are approximately orthogonal for $k \neq k'$. This orthogonality is better and better fulfilled as $\epsilon \to 0$. In this limit the Lorentzian $L_\epsilon(z)$ approaches a function $\delta(z)$ whose value at $z = 0$ is infinitely large, but whose width around $z = 0$ is infinitely small. The function

$$\delta(z) = \lim_{\epsilon \to 0} L_\epsilon(z) \qquad (3.136)$$

is known as *Dirac's delta function*. Dirac's delta function $\delta(z)$ may be considered a generalization of Kronecker's symbol δ_{nm} defined in Equation 3.109. The analogy becomes evident if we compare the orthogonality relations for quantum systems with a discrete spectrum with square-normalizable wave functions $\varphi_n(x)$,

$$\int_{-\infty}^{\infty} \varphi_n(x)^* \varphi_m(x)\,dx = \delta_{nm} \qquad (3.137)$$

(see, for example, Equations 3.108 and 3.127), with the orthogonality relations for free motion characterized by a continuous spectrum with wave functions that are not square-normalizable:

$$\int_{-\infty}^{\infty} \varphi_k^{(\pm)}(x)^* \varphi_{k'}^{(\pm)}(x)\,dx = \lim_{\epsilon \to 0} \int_{-\infty}^{\infty} \tilde\varphi_k^{(\pm)}(x)^* \tilde\varphi_{k'}^{(\pm)}(x)\,dx$$
$$= 2\pi \tilde N_k^{(\pm)*} \tilde N_{k'}^{(\pm)} \lim_{\epsilon \to 0} L_\epsilon(k'-k) = 2\pi \tilde N_k^{(\pm)*} \tilde N_k^{(\pm)} \delta(k'-k). \qquad (3.138)$$

In both cases the orthogonality relation is expressed with the help of a delta function, the Kronecker delta symbol in the case with a discrete spectrum (see Equation 3.137) and Dirac's delta function in the case of a continuous spectrum (see Equation 3.138). We note that on the right-hand side of Equation 3.138 we set $\tilde N_{k'}^{(\pm)} = \tilde N_k^{(\pm)}$, making use of the property of Dirac's δ function that it is infinitely sharply peaked around argument zero, i.e., around $k' = k$. We obtain complete formal analogy of the orthonormality conditions defined in Equations 3.137 and 3.138 if we set $\tilde N_k^{(\pm)} = 1/\sqrt{2\pi}$. In this case, then, we obtain the properly normalized continuum wave functions of free motion:

$$\varphi_k^{(\pm)}(x) = \frac{1}{\sqrt{2\pi}} \exp(\pm ikx). \qquad (3.139)$$

In this form they satisfy the orthogonality relation

$$\int_{-\infty}^{\infty} \varphi_k^{(\pm)*}(x)\, \varphi_{k'}^{(\mp)}(x)\, dx \;=\; 0 \tag{3.140}$$

and the orthonormality relation

$$\int_{-\infty}^{\infty} \varphi_k^{(\pm)*}(x)\, \varphi_{k'}^{(\pm)}(x)\, dx \;=\; \delta(k'-k), \tag{3.141}$$

where Equation 3.140 may be derived in analogy to Equation 3.138. An immediate consequence of Equations 3.140 and 3.141 is the functional relation

$$\frac{1}{2\pi}\int_{-\infty}^{\infty} \exp(iqx)\, dx \;=\; \delta(q), \tag{3.142}$$

which holds generally for all real q. Equation 3.142 is one of many equivalent formulations of *Fourier's Theorem*.

Before moving on to the case $V_0 \neq 0$ of the finite quantum square-well potential, we explore some more properties of Dirac's delta function. According to its definition, expressed by Equation 3.136, we have

$$\delta(z) = \begin{cases} 0, & \text{for } z \neq 0, \\ \infty, & \text{for } z = 0. \end{cases} \tag{3.143}$$

This is so since the numerator of $L_\epsilon(z)$ is proportional to ϵ, but the denominator of $L_\epsilon(z)$ is finite for finite z, which means that $L_\epsilon(z) \to 0$ for $\epsilon \to 0$ and any $z \neq 0$. At $z = 0$ we have $L_\epsilon(z) = 1/(\pi\epsilon) \to \infty$ for $\epsilon \to 0$.

An immediate consequence of Equation 3.143 is the property

$$f(z)\,\delta(z) \;=\; f(0)\,\delta(z). \tag{3.144}$$

Motivated by heuristic reasoning rather than the more formal mathematics of Dirac's delta function, we have already used the property defined in Equation 3.144 in Equation 3.138 when we set $\tilde{N}_{k'}^{(\pm)}\delta(k'-k) = \tilde{N}_k^{(\pm)}\delta(k'-k)$.

Next, we consider the integral

$$\int_a^b \delta(z)\, dz \tag{3.145}$$

over Dirac's delta function for $a < b$, $a, b \neq 0$. Since, according to Equation 3.144, Dirac's delta function is nonzero only at $z = 0$, the integral in Equation 3.145 is nonzero only if the integration interval $[a, b]$ contains $z = 0$. Since $a, b \neq 0$, this condition may be expressed concisely as $ab < 0$. In this case, substituting $w = z/\epsilon$,

$$\int_a^b \delta(z)\,dz = \int_a^b \lim_{\epsilon \to 0} L_\epsilon(z)\,dz = \lim_{\epsilon \to 0} \int_a^b \frac{1}{\pi}\left(\frac{\epsilon}{z^2 + \epsilon^2}\right)dz$$
$$= \int_{-\infty}^{\infty} \frac{1}{\pi}\left(\frac{1}{w^2+1}\right)dw = \frac{1}{\pi}\arctan(w)\Big|_{-\infty}^{\infty} = 1. \quad (3.146)$$

We obtain the astonishing result that, although Dirac's delta function, according to Equation 3.143 is nonzero only at a single point, it is apparently "infinite enough" to yield a finite integral. In summary, therefore, for $a < b$, we have

$$\int_a^b \delta(z)\,dz = \begin{cases} 1, & \text{if } ab < 0, \\ 0, & \text{if } ab > 0. \end{cases} \quad (3.147)$$

Since $L_\epsilon(z)$ is symmetric with respect to $z = 0$, i.e., $L_\epsilon(-z) = L_\epsilon(z)$, we may even give meaning to integrals of the form $\int_0^b \delta(z)dz$, $b > 0$ and $\int_a^0 \delta(z)dz$, $a < 0$. Since in these two cases we integrate over "half a Lorentzian," we have

$$\int_a^0 \delta(z)\,dz = \int_0^b \delta(z)\,dz = \frac{1}{2}, \quad a < 0, \quad b > 0. \quad (3.148)$$

Combining Equation 3.144 with Equations 3.147 and 3.148, we obtain for $a < b$:

$$\int_a^b f(z)\delta(z)\,dz = \begin{cases} f(0), & \text{if } ab < 0, \\ 0, & \text{if } ab > 0, \\ \frac{1}{2}f(0), & \text{if } ab = 0. \end{cases} \quad (3.149)$$

An immediate consequence of Equation 3.149 is the property:

$$\int_{-\infty}^{\infty} f(z)\,\delta(z - z_0)\,dz = \int_{-\infty}^{\infty} f(w + z_0)\,\delta(w)\,dw = f(z_0). \quad (3.150)$$

Equation 3.150 may be used to *project* the function $f(z)$ onto one of its function values, $f(z_0)$, i.e., given z_0, Dirac's delta function may be used to extract $f(z_0)$ from $f(z)$. This property is most useful in cases where $f(z)$ is not known explicitly or $f(z_0)$ is needed for many arguments z_0 simultaneously.

Following our excursion into the theory of Dirac's delta function, we return to the finite square-well potential and consider the case $V_0 < 0$ as shown in Figure 3.3. Because the potential of the finite square well is piece-wise constant, it is natural to divide the x-axis into three regions,

$$\begin{aligned} \text{Region } I &: \quad x \leq 0, \\ \text{Region } II &: \quad 0 < x < a, \\ \text{Region } III &: \quad x \geq a. \end{aligned} \qquad (3.151)$$

To determine the energy spectrum of $V(x)$ we determine the wave functions in Regions I, II, and III separately, and then match them at $x = 0$ and $x = a$ in analogy to Equations 3.18 and 3.19.

There are three qualitatively different possibilities for the energy E: (1) $V_0 < E < 0$, (2) $E = 0$, and (3) $E > 0$.

Case (1): $V_0 < E < 0$.
Since $V(x) = 0$ in Region I, the wave function $\varphi_I(x)$ in Region I satisfies the stationary Schrödinger equation

$$\varphi_I''(x) - \kappa_I^2 \varphi_I(x) = 0, \qquad (3.152)$$

where

$$\kappa_I = \sqrt{\frac{2m|E|}{\hbar^2}}. \qquad (3.153)$$

The general solution of Equation 3.152 is

$$\varphi_I(x) = A e^{-\kappa_I x} + B e^{\kappa_I x}, \qquad (3.154)$$

where A and B are constants. Since $\exp(-\kappa_I x)$ diverges for $x \to -\infty$, the boundary condition $\varphi_I(-\infty) = 0$ is fulfilled only if we set $A = 0$. Therefore,

$$\varphi_I(x) = B e^{\kappa_I x}. \qquad (3.155)$$

The wave function $\varphi_{II}(x)$ in Region II satisfies:

$$\varphi_{II}''(x) + \kappa_{II}^2 \varphi_{II}(x) = 0, \qquad (3.156)$$

where

$$\kappa_{II} = \sqrt{\frac{2m}{\hbar^2}(E - V_0)}. \qquad (3.157)$$

The general solution is

$$\varphi_{II}(x) = C \sin(\kappa_{II} x) + D \cos(\kappa_{II} x), \qquad (3.158)$$

where C and D are constants. The wave function $\varphi_{III}(x)$ in Region III satisfies the same wave equation as $\varphi_I(x)$ in Region I. Therefore, in Region III:

$$\varphi_{III}(x) = Fe^{-\kappa_I x} + Ge^{\kappa_I x}, \tag{3.159}$$

where F and G are constants. This time, however, the boundary condition $\varphi_{III}(+\infty) = 0$ requires $G = 0$ and therefore:

$$\varphi_{III}(x) = Fe^{-\kappa_I x}. \tag{3.160}$$

We now match $\varphi_I(x)$ and $\varphi_{II}(x)$ at $x = 0$. Continuity of the wave function (the analogue of Equation 3.18) requires:

$$\varphi_I(0) = \varphi_{II}(0), \tag{3.161}$$

which entails:

$$B = D. \tag{3.162}$$

Continuity of the first derivative (the analogue of Equation 3.19) requires:

$$\varphi'_I(0) = \varphi'_{II}(0), \tag{3.163}$$

which entails:

$$B\kappa_I = C\kappa_{II}. \tag{3.164}$$

Continuity of the wave function and its first derivative at $x = a$ require:

$$Fe^{-\kappa_I a} = C\sin(\kappa_{II}a) + D\cos(\kappa_{II}a), \tag{3.165}$$
$$-F\kappa_I e^{-\kappa_I a} = C\kappa_{II}\cos(\kappa_{II}a) - D\kappa_{II}\sin(\kappa_{II}a). \tag{3.166}$$

From Equations 3.162 and 3.164 we obtain

$$\frac{C}{D} = \frac{C}{B} = \frac{\kappa_I}{\kappa_{II}}. \tag{3.167}$$

Dividing Equations 3.165 and 3.166 by D and using Equation 3.167, we obtain:

$$\frac{F}{D}e^{-\kappa_I a} = \frac{\kappa_I}{\kappa_{II}}\sin(\kappa_{II}a) + \cos(\kappa_{II}a), \tag{3.168}$$
$$-\frac{F}{D}\kappa_I e^{-\kappa_I a} = \kappa_I \cos(\kappa_{II}a) - \kappa_{II}\sin(\kappa_{II}a). \tag{3.169}$$

Multiplying Equation 3.168 by κ_{II} and then dividing Equation 3.169 by Equation 3.168 yields:

$$-\frac{\kappa_I}{\kappa_{II}} = \frac{\kappa_I \cos(\kappa_{II}a) - \kappa_{II}\sin(\kappa_{II}a)}{\kappa_I \sin(\kappa_{II}a) + \kappa_{II}\cos(\kappa_{II}a)}. \quad (3.170)$$

Dividing the numerator and denominator of the right-hand side of Equation 3.170 by $\kappa_{II}\cos(\kappa_{II}a)$ and rearranging terms, results in the spectral equation

$$\tan(\kappa_{II}a) = \frac{2\kappa_I \kappa_{II}}{\kappa_{II}^2 - \kappa_I^2} \quad (3.171)$$

for Case (1): ($V_0 < E < 0$) of the finite square-well potential. Reintroducing the energy E and the potential strength V_0 via Equations 3.153 and 3.157 results in

$$\tan\left[a\sqrt{\frac{2m}{\hbar^2}(E - V_0)}\right] = \frac{2\sqrt{E(V_0 - E)]}}{2E - V_0}. \quad (3.172)$$

Contrary to the infinite square well or the harmonic oscillator potential, the Spectral Equation 3.172 is not a simple algebraic equation. This is so, since the energy E in the Spectral Equation 3.172 appears under square roots and in the argument of a trigonometric function. Although an explicit solution formula for the Spectral Equation 3.172 of the finite square-well exists, it contains higher transcendental functions and is not needed for our purposes here. Still, it is interesting to know that even a spectral equation as complicated as Equation 3.172 can be solved analytically.

Further discussion of the Spectral Equation 3.172 is facilitated if we express it with the help of the dimensionless variables

$$v_0 = \sqrt{\frac{2ma^2|V_0|}{\hbar^2}}, \quad \xi = 1 - \left(\frac{E}{V_0}\right), \quad 0 < \xi < 1. \quad (3.173)$$

Then, the Spectral Equation 3.172 acquires its dimensionless form

$$\tan\left(v_0\sqrt{\xi}\right) = \frac{2\sqrt{\xi(1-\xi)}}{2\xi - 1}. \quad (3.174)$$

We notice that $\xi = 0$, which, according to Equation 3.173, corresponds to $E = V_0$, is a formal solution of the Spectral Equation 3.174. According to Equation 3.157, however, $E = V_0$ corresponds to $\kappa_{II} = 0$, which leads to a wave function that is identically zero for all x. Since such a wave function does not represent a particle, the solution $\xi = 0$ of the Spectral Equation 3.174 is unphysical. This is the reason why we excluded $E = V_0$ in the first place and consider $E > V_0$ only. Thus, in line with $V_0 < E < 0$, it is the *positive* solutions $0 < \xi_n < 1, n = 1, 2, \ldots,$ of the Spectral Equation 3.174 that determine the energy levels E_n of the finite square-well potential according to

$$E_n = -\frac{\hbar^2 v_0^2}{2ma^2}(1-\xi_n), \quad n = 1, 2, \ldots. \tag{3.175}$$

The point $\xi = 1$ is another special value that deserves attention. Although it will turn out that the Spectral Equation 3.174 is formally valid in this case, we do not include the case $\xi = 1$ in our discussion right now since the wave functions $\varphi_I(x)$ and $\varphi_{III}(x)$ take a form different from Equations 3.155 and 3.159 and, therefore, the derivations that lead to the Spectral Equation 3.174 would be based on an incorrect asymptotic form of the wave functions. We will look at the case $\xi = 1$ in detail when we discuss the case $E = 0$ below.

Graphically, the solutions of Equation 3.174 are the intersections of the function $\tan(v_0\sqrt{\xi})$ with the function $2\sqrt{\xi(1-\xi)}/(2\xi - 1)$. Both functions are shown in Figure 3.4 for $v_0 = 14$. In this case, as shown in Figure 3.4, there are exactly five intersections,

$$\xi_1 = 0.0385, \quad \xi_2 = 0.153, \quad \xi_3 = 0.3411,$$
$$\xi_4 = 0.5955, \quad \xi_5 = 0.893, \tag{3.176}$$

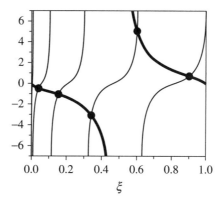

Figure 3.4 Graphical solution of the dimensionless spectral equation of the finite square-well potential with $v_0 = 14$. The five solutions are the intersections (full dots) of the various branches of the function $\tan(v_0\sqrt{\xi})$ (thin lines) with the function $2\sqrt{\xi(1-\xi)}/(2\xi - 1)$ (thick lines). Note that the two curves touch at $\xi = 0$. But, as explained in the text, this does not correspond to an energy level and is therefore not marked with a dot.

which correspond to five energy levels E_n, $n = 1, \ldots, 5$ according to Equation 3.175.

As indicated by our example with $v_0 = 14$, the finite square-well potential has only a finite number of discrete energy levels in the energy interval $V_0 < E < 0$. The deeper the well, i.e., the larger $|V_0|$, the more energy levels there are. Moreover, no matter how shallow the potential well, i.e., no matter how small $|V_0|$, there is always at least one energy level. We see this in the following way. For small $|V_0|$, i.e., small v_0, the argument of the tan function in the Spectral Equation 3.174 is small. Therefore, we may approximate $\tan(v_0\sqrt{\xi})$ by $v_0\sqrt{\xi}$, a function that starts at v_0 for $\xi = 1$ and then, as ξ decreases from $\xi = 1$ to $\xi = 0$, decreases monotonically toward zero. The right-hand side of the Spectral Equation 3.174 starts at zero at $\xi = 1$, increases monotonically for decreasing ξ, and has a singularity at $\xi = 1/2$. Thus, the two functions must intersect somewhere in the interval $1/2 < \xi < 1$. This intersection guarantees at least one energy level for any $V_0 < 0$.

Case (2): $E = 0$.

In this case, in Region I, the stationary Schrödinger equation is

$$\varphi_I''(x) = 0. \tag{3.177}$$

The general solution of this equation is

$$\varphi_I(x) = Ax + B, \tag{3.178}$$

where A and B are constants. Based on our discussion of continuum wave functions, we may tolerate $\varphi_I(x) \sim B$, but we do not tolerate $\varphi_I(x) \sim x$ since $\varphi_I(x) \sim x$ is neither square-normalizable nor "delta-normalizable" such as the continuum wave functions defined in Equation 3.139. Therefore, requiring $A = 0$, we have

$$\varphi_I(x) = B. \tag{3.179}$$

In Region II the stationary Schrödinger equation is

$$\varphi_{II}''(x) + \kappa_{II}^2 \varphi_{II}(x) = 0, \tag{3.180}$$

where

$$\kappa_{II} = \sqrt{\frac{2m|V_0|}{\hbar^2}}. \tag{3.181}$$

The general solution is

$$\varphi_{II}(x) = C\sin(\kappa_{II} x) + D\cos(\kappa_{II} x), \tag{3.182}$$

where C and D are constants. In Region III, $\varphi_{III}(x)$ satisfies the same stationary Schrödinger equation as $\varphi_I(x)$ and the same argument about excluding wave functions of the form $\varphi_{III}(x) \sim x$ applies. Therefore,

$$\varphi_{III}(x) = F, \qquad (3.183)$$

where F is a constant. We now match $\varphi_I(x)$ and $\varphi_{II}(x)$ at $x = 0$. Continuity and continuity of the first derivative require

$$B = D, \qquad (3.184)$$
$$0 = C\kappa_{II}. \qquad (3.185)$$

Since $\kappa_{II} \neq 0$, Equation 3.185 requires $C = 0$. Matching $\varphi_{II}(x)$ and $\varphi_{III}(x)$ at $x = a$, we obtain:

$$F = D\cos(\kappa_{II} a), \qquad (3.186)$$
$$0 = -D\kappa_{II}\sin(\kappa_{II} a), \qquad (3.187)$$

where we used the previously determined condition $C = 0$. Since $\kappa_{II} \neq 0$, Equation 3.187 implies either $D = 0$ or $\sin(\kappa_{II} a) = 0$. According to Equation 3.184, setting $D = 0$ would imply $B = 0$, which, in turn, according to Equation 3.186, would imply $F = 0$. In this case the wave function would be identically zero in all three regions, corresponding to the absence of a quantum particle. Therefore, in order to represent a quantum particle, we need $D \neq 0$. In this case, Equation 3.187 is fulfilled only if

$$\sin(\kappa_{II} a) = \sin(v_0) = 0. \qquad (3.188)$$

This is not a spectral equation, since, via Equation 3.173, Equation 3.188 is a condition on the potential strength V_0, a fixed system parameter, not on the energy E, which can be chosen freely once V_0 is fixed. For given, generic V_0 it is highly unlikely that Equation 3.188 is fulfilled. Therefore, we conclude that only in exceptional cases, where Equation 3.188 is accidentally fulfilled, is there an energy level at $E = 0$. We notice that Equations 3.174 and 3.188 are formally equivalent for $E = 0$, i.e., $\xi = 1$. But, as discussed in Case (1), this agreement is accidental.

A classical particle trapped inside the finite square-well potential oscillates between $x = 0$ and $x = a$. It is never found in the *classically forbidden Regions I* and *III*. However, inspecting Figure 3.3, we see that the wave function $\varphi_1(x)$ associated with the first energy level E_1 leaks into the classically forbidden Regions *I* and *III*. Since $|\varphi_1(x)|^2$ is the probability density of finding a particle with wave function $\varphi_1(x)$ at x, there is a nonzero probability of finding the particle in the classically forbidden Regions I and III. This phenomenon is called *tunneling*. It is a purely quantum mechanical effect, not found in classical, Newtonian mechanics.

Case (3): $E > 0$.
For $E > 0$ and $V_0 < 0$ the finite square-well potential has a purely continuous spectrum whose associated wave functions are delta-normalizable but not square-normalizable. In analogy to the free-particle case ($V_0 = 0$) there are two classes of wave functions, $\varphi^{(+)}(k,x)$, representing particles incident from the left on the potential well, and $\varphi^{(-)}(k,x)$, representing particles incident from the right. Here,

$$k = \sqrt{\frac{2mE}{\hbar^2}}. \tag{3.189}$$

From optics we know that waves incident on a transparent object such as a glass plate, here represented by the finite square-well potential, generate both reflected and transmitted waves. Since $\varphi^{(+)}(k,x)$ represents a wave incident from the left, we impose the boundary condition of a purely outgoing wave in Region III, i.e.,

$$\varphi^{(+)}_{III}(k,x) = \frac{1}{\sqrt{2\pi}} t^{(+)} e^{ikx}, \quad x \geq a, \tag{3.190}$$

where $t^{(+)}$, a constant, is the *transmission amplitude*. In Region I the wave function $\varphi^{(+)}(k,x)$ is a superposition of an incoming wave, $\exp(ikx)$, and a reflected wave, $\exp(-ikx)$, according to:

$$\varphi^{(+)}_{I}(k,x) = \frac{1}{\sqrt{2\pi}} \left(e^{ikx} + r^{(+)} e^{-ikx} \right), \quad x \leq 0, \tag{3.191}$$

where $r^{(+)}$, a constant, is the *reflection amplitude*. In Region II, the wave function is given by

$$\varphi^{(+)}_{II}(k,x) = \frac{1}{\sqrt{2\pi}} \left[C^{(+)} \sin(\kappa_{II} x) + D^{(+)} \cos(\kappa_{II} x) \right], \tag{3.192}$$

where $C^{(+)}$ and $D^{(+)}$ are constants, and

$$\kappa_{II} = \sqrt{\frac{2m}{\hbar^2}(E - V_0)}. \tag{3.193}$$

Since the wave function $\varphi^{(-)}(k,x)$ represents a wave incident from the right, we have reflection in Region III and transmission in Region I according to:

$$\begin{aligned}\varphi_I^{(-)}(k,x) &= \frac{1}{\sqrt{2\pi}} t^{(-)} e^{-ikx}, \\ \varphi_{II}^{(-)}(k,x) &= \frac{1}{\sqrt{2\pi}} \left[C^{(-)} \sin(\kappa_{II} x) + D^{(-)} \cos(\kappa_{II} x) \right], \\ \varphi_{III}^{(-)}(k,x) &= \frac{1}{\sqrt{2\pi}} \left(e^{-ikx} + r^{(-)} e^{ikx} \right). \end{aligned} \quad (3.194)$$

The factor $1/\sqrt{2\pi}$ is included in the wave functions $\varphi^{(\pm)}(k,x)$ to ensure proper normalization according to

$$\int_{-\infty}^{\infty} \varphi^{(\pm)}(k,x)^* \varphi^{(\pm)}(k',x)\, dx = \delta(k'-k). \quad (3.195)$$

The reflection and transmission amplitudes $r^{(\pm)}$ and $t^{(\pm)}$ determine the reflection and transmission probabilities $R^{(\pm)}$ and $T^{(\pm)}$, respectively, according to

$$R^{(\pm)} = |r^{(\pm)}|^2, \quad T^{(\pm)} = |t^{(\pm)}|^2. \quad (3.196)$$

Since particles *scatter* at the finite square-well potential, but are not created or destroyed, we have

$$R^{(\pm)} + T^{(\pm)} = 1. \quad (3.197)$$

Because of the symmetry of the problem, we have

$$R^{(-)} = R^{(+)}, \quad T^{(-)} = T^{(+)}, \quad (3.198)$$

but $r^{(+)}$ and $r^{(-)}$ ($t^{(+)}$ and $t^{(-)}$), complex numbers in general, may still differ by a complex phase factor.

We will now determine $r^{(+)}$ and $t^{(+)}$, leaving the computation of $r^{(-)}$ and $t^{(-)}$ as an exercise. Matching the wave functions and their derivatives at $x=0$ and at $x=a$ yields:

$$\begin{aligned} 1 + r^{(+)} &= D^{(+)}, \\ 1 - r^{(+)} &= \left(\frac{\kappa_{II}}{ik}\right) C^{(+)}, \\ C^{(+)} \sin(\kappa_{II} a) + D^{(+)} \cos(\kappa_{II} a) &= t^{(+)} e^{ika}, \\ C^{(+)} \cos(\kappa_{II} a) - D^{(+)} \sin(\kappa_{II} a) &= \left(\frac{ik}{\kappa_{II}}\right) t^{(+)} e^{ika}. \end{aligned} \quad (3.199)$$

Solving this system of linear equations for $r^{(+)}$ and $t^{(+)}$ yields:

$$r^{(+)} = \frac{\left(\frac{k^2-\kappa_{II}^2}{2ik\kappa_{II}}\right)\sin(\kappa_{II}a)}{\cos(\kappa_{II}a)+\left(\frac{\kappa_{II}^2+k^2}{2ik\kappa_{II}}\right)\sin(\kappa_{II}a)},$$

$$t^{(+)} = \frac{e^{-ika}}{\cos(\kappa_{II}a)+\left(\frac{\kappa_{II}^2+k^2}{2ik\kappa_{II}}\right)\sin(\kappa_{II}a)}. \qquad (3.200)$$

Indeed, the reflection and transmission amplitudes given in Equation 3.200 satisfy

$$|r^{(+)}|^2 + |t^{(+)}|^2 = R^{(+)} + T^{(+)} = 1. \qquad (3.201)$$

Next, we study the square-hump potential, i.e., the potential defined in Equation 3.128 with $V_0 > 0$. It is shown in Figure 3.5. Its spectrum is purely continuous; all energies with $E \geq 0$ are allowed. Technically, this potential does not offer anything new. Given an energy $E \geq 0$ the associated wave functions $\varphi^{(\pm)}(k,x)$ are determined in the usual way by matching the wave functions $\varphi_I^{(\pm)}(k,x)$ and $\varphi_{II}^{(\pm)}(k,x)$ in regions I and II, respectively, at $x = 0$, and by matching the wave functions $\varphi_{II}^{(\pm)}(k,x)$ and $\varphi_{III}^{(\pm)}(k,x)$ in Regions II and III, respectively, at $x = a$. For $E \geq V_0$ even the form of the wave functions is qualitatively the same as the continuum wave functions in the case of a well ($V_0 < 0$), and the same formulas defined in Equation 3.200 for the reflection and transmission amplitudes hold. However, for $0 < E < V_0$, there is a quantum effect that merits attention. Classically, a particle incident from the left on the square-hump potential with energy $E < V_0$ is reflected elastically off the edge of the potential and returns to the left where it came from. Quantum mechanically, however, as shown in Figure 3.5, there is a finite probability for the particle to tunnel through the barrier, emerge on the right-hand side of the barrier, and continue traveling to the right. This

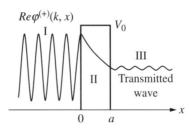

Figure 3.5 The square-hump potential. The real part of the wave function $\varphi(k,x)$ with $E = 0.99\, V_0$ is also shown.

purely quantum effect is called *barrier penetration*. Historically it played an important role when Gamov used this effect to explain the α-decay of some radioactive atomic nuclei.

Even for $E < V_0$, in the tunneling regime, the formulas in Equation 3.200 stay valid. In this case

$$\kappa_{II} = \sqrt{\frac{2m}{\hbar^2}(E - V_0)} = i\kappa_{II}^{(r)}, \qquad (3.202)$$

where

$$\kappa_{II}^{(r)} = \sqrt{\frac{2m}{\hbar^2}(V_0 - E)} \qquad (3.203)$$

is a real number, and κ_{II}, therefore, is an imaginary number. Using the expression in Equation 3.202 for κ_{II} together with

$$\cos(\kappa_{II}a) = \cos\left(i\kappa_{II}^{(r)}a\right) = \cosh\left(\kappa_{II}^{(r)}a\right),$$
$$\sin(\kappa_{II}a) = \sin\left(i\kappa_{II}^{(r)}a\right) = i\sinh\left(\kappa_{II}^{(r)}a\right), \qquad (3.204)$$

in the formulas of Equation 3.200 for the reflection and transmission amplitudes, we obtain

$$r^{(+)} = \frac{\left(\frac{k^2 + \left(\kappa_{II}^{(r)}\right)^2}{2ik\kappa_{II}^{(r)}}\right)\sinh(\kappa_{II}^{(r)}a)}{\cosh(\kappa_{II}^{(r)}a) + \left(\frac{k^2 - \left(\kappa_{II}^{(r)}\right)^2}{2ik\kappa_{II}^{(r)}}\right)\sinh(\kappa_{II}^{(r)}a)},$$

$$t^{(+)} = \frac{e^{-ika}}{\cosh(\kappa_{II}^{(r)}a) + \left(\frac{k^2 - \left(\kappa_{II}^{(r)}\right)^2}{2ik\kappa_{II}^{(r)}}\right)\sinh(\kappa_{II}^{(r)}a)}. \qquad (3.205)$$

For thick barriers $\kappa_{II}^{(r)}a$ is large. Therefore,

$$\cosh(\kappa_{II}^{(r)}a) \approx \sinh(\kappa_{II}^{(r)}a) \approx \frac{1}{2}\exp(\kappa_{II}^{(r)}a). \qquad (3.206)$$

Using this result in Equation 3.205, we obtain approximately

$$t^{(+)} \approx \left[\frac{4ik\kappa_{II}^{(r)}e^{-ika}}{(k + i\kappa_{II}^{(r)})^2}\right]\exp\left[-\kappa_{II}^{(r)}a\right]. \qquad (3.207)$$

We see that, apart from a prefactor, the barrier penetration amplitude is exponentially small in $\kappa_{II}^{(r)}a$. This means that barrier penetration becomes less and less likely as a increases, or as E decreases. Still, as shown by Equation 3.207, the probability $T^{(+)} = |t^{(+)}|^2$ stays finite no matter how thick the barrier, or how close E is to zero.

In this section we introduced Dirac's delta function. An interesting question is whether the delta function may serve as a potential. This is indeed possible. Even more, the delta function is one of the most "popular" model potentials in physics since it is easily solved analytically but nevertheless exhibits most of the salient features of more realistic model potentials. We begin by studying the delta-function potential

$$V(x) = V_0 \, \delta(x) \qquad (3.208)$$

for $V_0 < 0$. Since the delta function is nonzero only at $x = 0$, we divide the x-axis into two regions, Region I for $x < 0$ and Region II for $x > 0$. Since the delta-function potential defined in Equation 3.208 is attractive, we look for bound states with $E < 0$. For $E < 0$ the wave functions $\varphi_I(x)$ and $\varphi_{II}(x)$, with the proper boundary conditions $\varphi_I(-\infty) = 0$ and $\varphi_{II}(+\infty) = 0$, are given by

$$\begin{aligned}\varphi_I(x) &= A \, \exp(\kappa x), \\ \varphi_{II}(x) &= A \, \exp(-\kappa x),\end{aligned} \qquad (3.209)$$

where

$$\kappa = \sqrt{\frac{2m|E|}{\hbar^2}} \qquad (3.210)$$

and A is a normalization constant. Because we chose the same constant in front of the exponential factors in the wave functions defined in Equation 3.209, the wave functions $\varphi_I(x)$ and $\varphi_{II}(x)$ already match at $x = 0$. Since Dirac's delta function is not a piecewise continuous function, we need a different matching condition for the derivatives of the wave function. We derive this matching condition in the following way. Starting with the stationary Schrödinger equation for the potential defined in Equation 3.208,

$$-\frac{\hbar^2}{2m}\varphi''(x) + V_0 \, \delta(x)\varphi(x) = E\varphi(x), \qquad (3.211)$$

we integrate this equation on both sides from $x = -\epsilon$ to $x = +\epsilon$, where $\epsilon > 0$. We obtain:

$$-\frac{\hbar^2}{2m}[\varphi'(\epsilon) - \varphi'(-\epsilon)] + V_0 \, \varphi(0) = E \int_{-\epsilon}^{\epsilon} \varphi(x) \, dx. \qquad (3.212)$$

Taking the limit of $\epsilon \to 0$, the right-hand side of Equation 3.212 vanishes, since $\varphi(x)$ is a continuous function. Therefore, Equation 3.212 becomes:

$$-\frac{\hbar^2}{2m}[\varphi'(0^+) - \varphi'(0^-)] + V_0 \, \varphi(0) = 0, \qquad (3.213)$$

where

$$\varphi'(0^\pm) = \lim_{\epsilon \to 0} \varphi'(\pm\epsilon). \qquad (3.214)$$

In our case this means that

$$\begin{aligned}\varphi'(0^+) &= \varphi'_{II}(0), \\ \varphi'(0^-) &= \varphi'_I(0).\end{aligned} \qquad (3.215)$$

Therefore, using the wave functions defined in Equation 3.209 in Equation 3.213, we obtain:

$$\kappa = \frac{m|V_0|}{\hbar^2}, \qquad (3.216)$$

i.e.,

$$E = -\frac{mV_0^2}{2\hbar^2}. \qquad (3.217)$$

We obtain the important result that the attractive delta function, independent of the strength of the potential (as long as $V_0 < 0$), always has precisely one bound state.

For $E > 0$ the spectrum of the attractive delta function is purely continuous. The associated wave functions may again be computed by matching wave functions in Regions I and II at $x = 0$. For $V_0 = 0$ we obtain the case of free motion already discussed previously. For $V_0 > 0$ the spectrum of the delta-function potential defined in Equation 3.208 is purely continuous; there is no bound state in this case.

Using the techniques we developed in connection with the finite square-well potential, we may compute the reflection and transmission coefficients, $r^{(+)}$ and $t^{(+)}$, respectively, for a quantum particle incident from the left on the delta-function potential defined in Equation 3.208. We obtain:

$$\begin{aligned}r^{(+)} &= \frac{\left(\frac{mV_0}{i\hbar^2 k}\right)}{1 - \left(\frac{mV_0}{i\hbar^2 k}\right)}, \\ t^{(+)} &= \frac{1}{1 - \left(\frac{mV_0}{i\hbar^2 k}\right)}.\end{aligned} \qquad (3.218)$$

The finite square well, the square hump, and the delta-function potential have one important property in common: A quantum particle incident from the left (or the right) on these potentials is split into a reflected and a transmitted wave. Therefore, these potentials may be employed as models for beam splitters of matter waves. Combining any

88 CHAPTER 3 • THE MACHINERY OF QUANTUM MECHANICS

two of these potentials sequentially produces an interferometer for matter waves.

Exercises:

3.4.1 Requiring that the coefficient of the highest power of $H_n(z)$ be 2^n, show that the Differential Equation 3.120 determines the first three Hermite polynomials defined in Equation 3.121 uniquely.

3.4.2 Verify the orthogonality relation 3.122 for $n, m = 0, 1, 2$. Use the Gaussian integrals:

$$\int_{-\infty}^{\infty} e^{-z^2}\, dz = \sqrt{\pi},$$
$$\int_{-\infty}^{\infty} z^2 e^{-z^2}\, dz = \frac{1}{2}\sqrt{\pi},$$
$$\int_{-\infty}^{\infty} z^4 e^{-z^2}\, dz = \frac{3}{4}\sqrt{\pi}. \tag{3.219}$$

3.4.3 Verify that the harmonic oscillator wave functions defined in Equation 3.126 are properly normalized.

3.4.4 Show that

$$\frac{2}{a}\int_0^a \sin(n\pi x/a)\sin(m\pi x/a)\, dx = \delta_{nm},$$

where δ_{nm} is the Kronecker symbol defined in Equation 3.109.

3.4.5 As discussed in the text, for $V_0 < E < 0$ the ground-state wave function $\varphi_1(x)$ of the finite square-well potential leaks into the classically forbidden Regions I and III (see Figure 3.3).

(a) Show that the probability P_I of finding the particle in the classically forbidden Region I equals the probability P_{III} of finding the particle in the classically forbidden Region III.

(b) For $v_0 = 14$ compute the probability $P = P_I = P_{III}$. Use the numerical value of ξ_1 given in Equation 3.176.

3.4.6 For the delta-function potential defined in Equation 3.208

(a) verify $r^{(+)}$ and $t^{(+)}$ in Equation 3.218 and
(b) calculate $r^{(-)}$ and $t^{(-)}$.

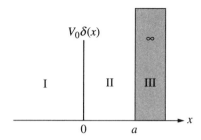

Figure 3.6 Resonance scattering: δ-function potential in front of a potential wall.

3.4.7 A quantum particle of mass m and energy $E = \hbar^2 k^2/(2m)$ is incident from the left on the potential

$$V(x) = \begin{cases} V_0\,\delta(x), & \text{for } x < a, \\ \infty, & \text{for } x \geq a, \end{cases}$$

where $V_0 > 0$ and $a > 0$. The potential is shown in Figure 3.6. This potential naturally divides the x-axis into three regions, Region I: $x < 0$, Region II: $0 < x < a$, and Region III: $x > a$. Since the potential is infinite in Region III, the wave function $\varphi_{III}(x)$ in Region III is identically zero. The wave functions in the other two regions are given by

$$\varphi_I(x) = \frac{1}{\sqrt{2\pi}}\left[e^{ikx} + r e^{-ikx}\right],$$

$$\varphi_{II}(x) = \frac{1}{\sqrt{2\pi}} C \sin[k(a-x)].$$

(a) Confirm that $\varphi_I(x)$ and $\varphi_{II}(x)$ are solutions of the one-dimensional, stationary Schrödinger equation in Regions I and II, respectively.

(b) What is the physical meaning of r? Why is $|r| = 1$?

(c) Compute r and C.

(d) Check your result for r. According to (b) you should obtain $|r| = 1$.

(e) For what energies E_n is the probability of finding the particle in Region II particularly large? To make your task easier, assume that $[mV_0/(\hbar^2 k)] \gg 1$.

(f) Give a physical explanation for the energy values you found in (e).

3.4.8 Show that for real $a \neq 0$:

$$\delta(ax) = \frac{1}{|a|}\delta(x). \tag{3.220}$$

3.5 Dirac Notation

Some of the discussions in Sections 3.3 and 3.4 may have reminded us of material typically covered in a course on linear algebra. In linear algebra, too, we encounter scalar products, Hermitian conjugates, and eigenvalue equations. Let us expand on this analogy a bit further.

The fundamental objects in linear algebra are *vectors*. Typically, at the beginning of a course on linear algebra, vectors are introduced as abstract objects $\boldsymbol{v}$, without any reference to a coordinate system. Vectors may be added,

$$\boldsymbol{w} = \boldsymbol{u} + \boldsymbol{v}, \tag{3.221}$$

they may be multiplied by a scalar,

$$\boldsymbol{w} = \lambda \boldsymbol{v}, \tag{3.222}$$

and there is a *scalar product*, $(\boldsymbol{u}, \boldsymbol{v})$, also called *inner product*, with the following properties:

$$\begin{aligned}
\text{symmetry:} \quad & (\boldsymbol{u}, \boldsymbol{v}) = (\boldsymbol{v}, \boldsymbol{u})^*, \\
\text{linearity:} \quad & (\boldsymbol{u}, \lambda \boldsymbol{v} + \mu \boldsymbol{w}) = \lambda(\boldsymbol{u}, \boldsymbol{v}) + \mu(\boldsymbol{u}, \boldsymbol{w}), \\
\text{definiteness:} \quad & (\boldsymbol{u}, \boldsymbol{u}) \geq 0, \quad (\boldsymbol{u}, \boldsymbol{u}) = 0 \iff \boldsymbol{u} = \boldsymbol{0},
\end{aligned} \tag{3.223}$$

where $\boldsymbol{0}$ is the zero vector and λ, μ are arbitrary complex numbers. Even without introducing coordinates, we may prove important properties of vectors and the scalar product in general. For instance,

$$\begin{aligned}
\text{anti-linearity:} \quad (\lambda \boldsymbol{v} + \mu \boldsymbol{w}, \boldsymbol{u}) &= (\boldsymbol{u}, \lambda \boldsymbol{v} + \mu \boldsymbol{w})^* \\
&= \lambda^*(\boldsymbol{u}, \boldsymbol{v})^* + \mu^*(\boldsymbol{u}, \boldsymbol{w})^* = \lambda^*(\boldsymbol{v}, \boldsymbol{u}) + \mu^*(\boldsymbol{w}, \boldsymbol{u}).
\end{aligned} \tag{3.224}$$

Dirac was the first to exploit the formal similarities between linear algebra and quantum mechanics to their fullest extent. In analogy to abstract, coordinate-free vectors $\boldsymbol{v}$, quantum systems are assigned an abstract *state*. Instead of using boldface notation, like we do when referring to vectors, Dirac used the notation

$$|\psi(t)\rangle \tag{3.225}$$

to refer to the quantum state of a given quantum system. Given two quantum states $|\varphi(t)\rangle$ and $|\psi(t)\rangle$, Dirac defines their scalar product as

$$\langle \varphi(t)|\psi(t)\rangle. \tag{3.226}$$

Notice that this notation is very similar to the $(\boldsymbol{u},\boldsymbol{v})$ notation of the scalar product of ordinary vectors. In the notation used in Equation 3.226, the scalar product of two quantum states may be interpreted as the product of a "bra" state $\langle \varphi(t)|$ and a "ket" state $|\psi(t)\rangle$, which, together, form the "bra-ket" $\langle \varphi(t)|\psi(t)\rangle$, i.e., the bracket that denotes the scalar product. The same rules of symmetry, linearity, anti-linearity, and definiteness apply to quantum states and their scalar products:

$$\begin{aligned}
\text{symmetry:} \quad & \langle \varphi(t)|\psi(t)\rangle \;=\; \langle \psi(t)|\varphi(t)\rangle^*, \\
\text{linearity:} \quad & \langle \varphi(t)|\big[\lambda|\alpha(t)\rangle + \mu|\beta(t)\rangle\big] \;=\; \lambda\langle \varphi(t)|\alpha(t)\rangle + \mu\langle \varphi(t)|\beta(t)\rangle, \\
\text{definiteness:} \quad & \langle \psi(t)|\psi(t)\rangle \geq 0, \quad \langle \psi(t)|\psi(t)\rangle = 0 \Leftrightarrow |\psi(t)\rangle = |\rangle,
\end{aligned} \tag{3.227}$$

where $|\rangle$ denotes the empty state. The empty state $|\rangle$ is the analogue of the zero vector $\boldsymbol{0}$. The empty state $|\rangle$ has to be carefully distinguished from the state $|0\rangle$, which usually denotes a ground state or the vacuum state. For instance, we have $\langle 0|0\rangle = 1$, whereas $\langle |\rangle = 0$. The empty state may arise in several different ways, for instance, $0|\psi(t)\rangle = |\rangle$ or $|\rangle = |\alpha(t)\rangle - |\alpha(t)\rangle$. While it is important, conceptually, to carefully distinguish between 0 and $\boldsymbol{0}$, and 0 and $|\rangle$, this distinction is usually not emphasized in the physics literature and the symbol 0 is frequently used for both $\boldsymbol{0}$ and $|\rangle$ when the context is clear.

Anti-linearity for the scalar product of states may be demonstrated in the following way. Let $|\psi(t)\rangle = \lambda|\alpha(t)\rangle + \mu|\beta(t)\rangle$. Then,

$$\begin{aligned}
\langle \psi(t)|\varphi(t)\rangle \;&=\; \langle \varphi(t)|\psi(t)\rangle^* \;=\; \big[\langle \varphi(t)|\big(\lambda|\alpha(t)\rangle + \mu|\beta(t)\rangle\big)\big]^* \\
&=\; \lambda^*\langle \varphi(t)|\alpha(t)\rangle^* + \mu^*\langle \varphi(t)|\beta(t)\rangle^* \;=\; \lambda^*\langle \alpha(t)|\varphi(t)\rangle + \mu^*\langle \beta(t)|\varphi(t)\rangle.
\end{aligned} \tag{3.228}$$

Coordinates are introduced in linear algebra by choosing a set of "privileged" vectors, for instance $\boldsymbol{e}_1, \boldsymbol{e}_2, \ldots, \boldsymbol{e}_n$. We call these vectors *basis vectors* if they are *linearly independent*, i.e.,

$$\sum_{j=1}^{n} \lambda_j \boldsymbol{e}_j \;=\; \boldsymbol{0} \quad \Longleftrightarrow \quad \lambda_j \;=\; 0, \quad j = 1, 2, \ldots, n. \tag{3.229}$$

Then, there is a unique set of numbers v_j, $j = 1, \ldots, n$, called *coordinates*, for which

$$\boldsymbol{v} = \sum_{j=1}^{n} v_j \, \boldsymbol{e}_j. \tag{3.230}$$

We arrange the coordinates v_j, $j = 1, \ldots, n$ in a *column vector*,

$$\boldsymbol{v} = \begin{pmatrix} v_1 \\ v_2 \\ \cdot \\ \cdot \\ \cdot \\ v_n \end{pmatrix}. \tag{3.231}$$

This vector is a *representation* of $\boldsymbol{v}$ in terms of the basis $\{\boldsymbol{e}_1, \boldsymbol{e}_2, \ldots, \boldsymbol{e}_n\}$.

Many calculations in vector algebra are facilitated if, drawing on the concept of orthonormality introduced in Section 3.4, we choose a set of orthonormal basis vectors. In vector algebra an orthonormal basis $\{\boldsymbol{e}_1, \ldots, \boldsymbol{e}_n\}$ is defined by

$$(\boldsymbol{e}_i, \boldsymbol{e}_j) = \delta_{ij}, \tag{3.232}$$

where δ_{ij} is the Kronecker symbol defined in Equation 3.109. In this case the coordinates v_j, $j = 1, \ldots, n$, of $\boldsymbol{v}$ are given by

$$v_j = (\boldsymbol{e}_j, \boldsymbol{v}). \tag{3.233}$$

In Dirac's theory similar concepts apply. We choose a set of privileged, orthonormal states, $|\varphi_1\rangle, |\varphi_2\rangle, \ldots, |\varphi_n\rangle$ with

$$\langle \varphi_i | \varphi_j \rangle = \delta_{ij}, \tag{3.234}$$

which, for convenience, we choose to be time-independent. Then, defining the scalar products

$$A_j(t) = \langle \varphi_j | \psi(t) \rangle, \quad j = 1, 2, \ldots, n \tag{3.235}$$

of the state $|\psi(t)\rangle$ with respect to the basis states $\{|\varphi_1\rangle, |\varphi_2\rangle, \ldots, |\varphi_n\rangle\}$, the state $|\psi(t)\rangle$ may be expressed uniquely in the form

$$|\psi(t)\rangle = \sum_{j=1}^{n} A_j(t) |\varphi_j\rangle. \tag{3.236}$$

This looks very much like the linear combination defined in Equation 3.230 of an ordinary vector $\boldsymbol{v}$ in terms of the basis vectors $\boldsymbol{e}_j$. The only difference is that instead of calling $A_1(t), A_2(t), \ldots, A_n(t)$ "coordinates," we call $A_1(t), A_2(t), \ldots, A_n(t)$ the *amplitudes* of $|\psi(t)\rangle$ with respect to the basis states $\{|\varphi_1\rangle, |\varphi_2\rangle, \ldots, |\varphi_n\rangle\}$. The amplitudes $A_j(t)$, $j = 1, \ldots, n$, just like in the case of ordinary vectors, are then a *representation* of $|\psi(t)\rangle$ with respect to the basis states $\{|\varphi_1\rangle, \ldots, |\varphi_n\rangle\}$.

In linear algebra the space spanned by the basis vectors $\{\boldsymbol{e}_1,\ldots,\boldsymbol{e}_n\}$ is called a *vector space*; in quantum mechanics, the space spanned by the basis states $\{|\varphi_1\rangle,|\varphi_2\rangle,\ldots,|\varphi_n\rangle\}$ is called a *Hilbert space*.

In the case of ordinary vectors $\boldsymbol{v}$, in addition to their column vector representation, defined in Equation 3.231, we are familiar with row vectors $(v_1, v_2, \ldots, v_n)$. Row vectors are the Hermitian conjugate of column vectors according to

$$\begin{pmatrix} v_1 \\ v_2 \\ \cdot \\ \cdot \\ \cdot \\ v_n \end{pmatrix}^\dagger = (v_1^*, v_2^*, \ldots, v_n^*). \tag{3.237}$$

With the help of the Hermitian conjugate of vectors, we may write the scalar product as the matrix product of an $1 \times n$ row matrix with an $n \times 1$ column matrix in the form

$$(\boldsymbol{u}, \boldsymbol{v}) = \boldsymbol{u}^\dagger \boldsymbol{v} = (u_1^*, u_2^*, \ldots, u_n^*) \begin{pmatrix} v_1 \\ v_2 \\ \cdot \\ \cdot \\ \cdot \\ v_n \end{pmatrix} = \sum_{j=1}^n u_j^* v_j. \tag{3.238}$$

We also introduce the *dot notation* for the scalar product, which is more convenient than the parenthesis notation. We define

$$\boldsymbol{u} \cdot \boldsymbol{v} \equiv (\boldsymbol{u}, \boldsymbol{v}) = \sum_{j=1}^n u_j^* v_j. \tag{3.239}$$

This *dot product* looks much like the familiar dot product we are used to when working with real vectors. For complex vectors, however, as indicated in Equation 3.239, we should not forget to take the complex conjugates of the components of the first vector before forming the sum of the products of corresponding vector components.

The definition of the Hermitian conjugate for ordinary vectors reminds us strongly of the previous use of the same term in Section 3.3 in connection with quantum operators. The use of the same name including the use of the same "dagger" symbol to denote the Hermitian conjugate in both cases is no accident. The basic concepts are the same in both linear algebra and in quantum mechanics. In analogy with the case of ordinary vectors, the Hermitian conjugate of quantum states is:

$$|\psi(t)\rangle^\dagger = \langle\psi(t)|, \quad \langle\psi(t)|^\dagger = |\psi(t)\rangle. \quad (3.240)$$

This way bra states are revealed as the Hermitian conjugates of ket states, and vice versa.

There is one profound difference between ordinary vector spaces and Hilbert spaces. The basis states that span the Hilbert space may not be countable. The most prominent example are the states $|r\rangle$, i.e., the states of a quantum particle with a sharp position r. Since r is a vector formed from three real numbers,

$$r = \begin{pmatrix} x \\ y \\ z \end{pmatrix}, \quad (3.241)$$

and since all values of x, y, and z are allowed, the states $|r\rangle$ form a continuum. Nevertheless, we may use them to represent $|\psi(t)\rangle$ according to:

$$\psi(r,t) = \langle r|\psi(t)\rangle. \quad (3.242)$$

This provides a fundamental insight. What we have so far called the *wave function* of a particle is revealed as the position representation of the abstract quantum state $|\psi(t)\rangle$. Therefore, in addition to its various names encountered so far (e.g., "wave function" and "amplitude"), $\psi(r,t)$ may now also be called the quantum state $|\psi(t)\rangle$ in position representation.

What works for position, works for momentum. Let us denote by $|p\rangle$ the basis states of a quantum particle with definite momentum p. Then, here again, the basis states $|p\rangle$ are uncountable; they form a continuum. But, still, we may use them to compute

$$\psi(p,t) = \langle p|\psi(t)\rangle, \quad (3.243)$$

where, now, $\psi(p,t)$ is the state $|\psi(t)\rangle$ in momentum representation. This provides another fundamental insight. What we have so far called *the* wave function $\psi(r,t)$ of a quantum particle turns out to be only one way of *representing*, i.e., looking at, the abstract quantum state $|\psi(t)\rangle$ of a quantum particle; $\psi(p,t)$ provides another. As we can imagine, in addition to $\psi(r,t)$ and $\psi(p,t)$, there are infinitely many other ways of representing $|\psi(t)\rangle$. We may, for example, represent $|\psi(t)\rangle$ using the states of the infinite square-well potential or the harmonic-oscillator states defined in Section 3.4. We may even use the mixed bound-continuum states of the finite square-well potential (see Section 3.4).

Having discussed the analogy between vectors in ordinary vector algebra and states in quantum mechanics, we will now look at the similarities between operators in vector algebra and quantum mechanics. In vector algebra, an operator assigns an image vector $\boldsymbol{w}$ to each vector $\boldsymbol{v}$ according to:

$$\hat{M}\boldsymbol{v} = \boldsymbol{w}. \tag{3.244}$$

In quantum mechanics a quantum operator $\hat{\theta}$ assigns a quantum state $|\psi(t)\rangle$ to each state $|\varphi(t)\rangle$ according to

$$\hat{\theta}|\varphi(t)\rangle = |\psi(t)\rangle. \tag{3.245}$$

The most important type of operators in both vector algebra and quantum mechanics are *linear operators*. Let $\hat{M}$ be a linear operator in vector algebra. Then:

$$\hat{M}(\lambda \boldsymbol{u} + \mu \boldsymbol{v}) = \lambda \hat{M}\boldsymbol{u} + \mu \hat{M}\boldsymbol{v}, \tag{3.246}$$

where λ and μ are arbitrary complex numbers. The analogous statement for a linear quantum operator $\hat{\theta}$ and states $|\alpha(t)\rangle$ and $|\beta(t)\rangle$ is:

$$\hat{\theta}\left[\lambda|\alpha(t)\rangle + \mu|\beta(t)\rangle\right] = \lambda\hat{\theta}|\alpha(t)\rangle + \mu\hat{\theta}|\beta(t)\rangle. \tag{3.247}$$

In many cases these definitions are enough to derive important results involving vectors and operators. However, as mentioned, calculations are frequently simplified by choosing an orthonormal basis, $\{\boldsymbol{e}_1, \ldots, \boldsymbol{e}_n\}$ in an ordinary vector space or $\{|\varphi_1\rangle, \ldots, |\varphi_n\rangle\}$ in a Hilbert space, and then introducing a *matrix representation* of the operators $\hat{M}$ and $\hat{\theta}$.

We define the matrix M associated with the operator $\hat{M}$ whose matrix elements are defined as

$$M_{ij} = (\boldsymbol{e}_i, \hat{M}\boldsymbol{e}_j). \tag{3.248}$$

Starting with Equation 3.244 and using the coordinate representations

$$\boldsymbol{v} = \sum_{j=1}^{n} v_j \boldsymbol{e}_j, \quad v_j = (\boldsymbol{e}_j, \boldsymbol{v}),$$

$$\boldsymbol{w} = \sum_{j=1}^{n} w_j \boldsymbol{e}_j, \quad w_j = (\boldsymbol{e}_j, \boldsymbol{w}), \tag{3.249}$$

of the vectors $\boldsymbol{v}$ and $\boldsymbol{w}$, we take the scalar product of Equation 3.244 with $\boldsymbol{e}_i$ to obtain the coordinate representation of Equation 3.244 according to

$$(e_i, \hat{M}v) = (e_i, \hat{M}\sum_{j=1}^{n} v_j e_j) = \sum_{j=1}^{n}(e_i, \hat{M}e_j)v_j$$

$$= \sum_{j=1}^{n} M_{ij}v_j = w_i. \qquad (3.250)$$

In quantum mechanics, starting from Equation 3.245, we obtain:

$$\hat{\theta}|\varphi(t)\rangle = \sum_{m=1}^{n} B_m(t)\,\hat{\theta}\,|\varphi_m\rangle = \sum_{j=1}^{n} A_j(t)\,|\varphi_j\rangle, \qquad (3.251)$$

where $B_j(t)$ are the amplitudes of $|\varphi(t)\rangle$ with respect to the basis $\{|\varphi_1\rangle,\ldots,|\varphi_n\rangle\}$ and $A_j(t)$ are the amplitudes of $|\psi(t)\rangle$ defined in Equation 3.235. Now, taking the scalar product of Equation 3.251 with $\langle\varphi_i|$, we obtain:

$$\sum_{m=1}^{n} B_m(t)\,\langle\varphi_i|(\hat{\theta}|\varphi_m\rangle) = \sum_{j=1}^{n} A_j(t)\,\langle\varphi_i|\varphi_j\rangle$$

$$= \sum_{j=1}^{n} A_j(t)\delta_{ij} = A_i(t). \qquad (3.252)$$

Defining the matrix elements

$$\langle\varphi_i|(\hat{\theta}|\varphi_m\rangle) = \theta_{im}, \qquad (3.253)$$

we may then write:

$$\sum_{m=1}^{n} \theta_{im}\,B_m(t) = A_i(t). \qquad (3.254)$$

Instead of writing $\langle\varphi_i|(\hat{\theta}|\varphi_m\rangle)$, it is customary to omit the parentheses and write:

$$\theta_{im} = \langle\varphi_i|\hat{\theta}|\varphi_m\rangle. \qquad (3.255)$$

Using Dirac notation, we may now formulate the time-dependent Schrödinger equation in the following way:

$$i\hbar\frac{\partial|\psi(t)\rangle}{\partial t} = \hat{\mathcal{H}}(\hat{p},\hat{r},t)\,|\psi(t)\rangle. \qquad (3.256)$$

The eigenvalue equation of the stationary Schrödinger equation with time-independent Hamilton operator $\mathcal{H}(p,r)$ may be formulated as:

$$\hat{\mathcal{H}}(\hat{p},\hat{r})\,|\varphi_n\rangle \;=\; E_n\,|\varphi_n\rangle. \tag{3.257}$$

The expectation value of an operator $\hat{\theta}$ in Dirac notation is given by

$$\bar{\theta}(t) \;=\; \langle\psi(t)|\hat{\theta}|\psi(t)\rangle. \tag{3.258}$$

Expressed in the basis $\{|\varphi_1\rangle,\ldots,|\varphi_n\rangle\}$ according to Equation 3.236 and using the matrix elements defined in Equation 3.255, this may be evaluated to

$$\begin{aligned}\bar{\theta}(t) &= \sum_{j=1}^{n}\sum_{m=1}^{n} A_j^*(t)\,\langle\varphi_j|\hat{\theta}|\varphi_m\rangle\,A_m(t) \\ &= \sum_{j=1}^{n}\sum_{m=1}^{n} A_j^*(t)\,A_m(t)\,\theta_{jm}.\end{aligned} \tag{3.259}$$

We now prove an important theorem. Let $\hat{\theta}$ be an observable, i.e.,

$$\hat{\theta}^\dagger \;=\; \hat{\theta}, \tag{3.260}$$

and let $|\vartheta_n\rangle$ be its normalized eigenstates with eigenvalues ϑ_n, i.e.,

$$\hat{\theta}\,|\vartheta_n\rangle \;=\; \vartheta_n\,|\vartheta_n\rangle, \quad \langle\vartheta_n|\vartheta_n\rangle \;=\; 1. \tag{3.261}$$

We also assume that $\vartheta_n \neq \vartheta_m$ for $n \neq m$. Then:

$$\hat{\theta}^\dagger = \hat{\theta}: \quad \vartheta_n \text{ is real and } \langle\vartheta_n|\vartheta_m\rangle = 0 \quad \text{for } n \neq m. \tag{3.262}$$

First, we show that ϑ_n is real:

$$\begin{aligned}\vartheta_n^* &= \langle\vartheta_n|\hat{\theta}|\vartheta_n\rangle^* = \langle\hat{\theta}^\dagger\vartheta_n|\vartheta_n\rangle^* \\ &= \langle\hat{\theta}\vartheta_n|\vartheta_n\rangle^* = \langle\vartheta_n|\hat{\theta}\vartheta_n\rangle = \langle\vartheta_n|\hat{\theta}|\vartheta_n\rangle = \vartheta_n.\end{aligned} \tag{3.263}$$

Then, for $n \neq m$:

$$\langle\vartheta_n|\vartheta_m\rangle \;=\; \frac{1}{\vartheta_m}\langle\vartheta_n|\hat{\theta}\vartheta_m\rangle \;=\; \frac{1}{\vartheta_m}\langle\hat{\theta}\vartheta_n|\vartheta_m\rangle \;=\; \frac{\vartheta_n}{\vartheta_m}\langle\vartheta_n|\vartheta_m\rangle, \tag{3.264}$$

or,

$$\langle\vartheta_n|\vartheta_m\rangle\left[1-\left(\frac{\vartheta_n}{\vartheta_m}\right)\right] \;=\; 0. \tag{3.265}$$

The product is zero if one of its factors is zero. Since $\vartheta_n \neq \vartheta_m$, the second factor is not zero. Therefore, $\langle \vartheta_n | \vartheta_m \rangle = 0$. One may object that ϑ_m may be zero, and, therefore, $1/\vartheta_m$ does not exist. However, if this is the case, and since $\vartheta_n \neq \vartheta_m$, then $\vartheta_n \neq 0$. Therefore, instead of starting our proof with $\langle \vartheta_n | \vartheta_m \rangle$, we start our proof with $\langle \vartheta_m | \vartheta_n \rangle$ and no problems will arise. In summary we have shown that, indeed, as claimed in Equation 3.262, the eigenvalues of a Hermitian operator are real and its eigenstates are mutually orthogonal if they belong to different eigenvalues. A particularly important example are the eigenvalues E_n and the eigenstates $|\varphi_n\rangle$ of the stationary Schrödinger Equation 3.257. Since $\hat{\mathcal{H}}(\hat{\boldsymbol{p}}, \hat{\boldsymbol{r}})^\dagger = \hat{\mathcal{H}}(\hat{\boldsymbol{p}}, \hat{\boldsymbol{r}})$, we now know in full generality that the energy eigenstates of a Hermitian Hamiltonian operator are *always* orthogonal if they correspond to different energy eigenvalues E_n. For example, since the energies E_n of the harmonic oscillator (see Equation 3.124) are all different, and since the Hamiltonian operator of the harmonic oscillator is a Hermitian operator, we know *without calculation* that the oscillator eigenstates $|\varphi_n\rangle$ are all mutually orthogonal.

Suppose any vector $\boldsymbol{v}$ in a certain vector space can be *linearly combined* with the help of n basis states $\boldsymbol{e}_j$, $j = 1, 2, \ldots, n$ according to Equation 3.230 and n is the smallest number of basis states necessary to accomplish this task. Then, we say that the vector space is n dimensional and is *spanned* by the *complete basis* $\boldsymbol{e}_j$, $j = 1, 2, \ldots, n$.

The same procedure works in Hilbert space. A basis $\{|\varphi_1\rangle, \ldots, |\varphi_n\rangle\}$ with $\langle \varphi_i | \varphi_j \rangle = \delta_{ij}$ spans an n-dimensional Hilbert space, and any state $|\psi(t)\rangle$ in this Hilbert space may be linearly combined according to Equation 3.236 with amplitudes defined in Equation 3.235.

In an n-dimensional vector space spanned by the complete, orthonormal basis $\{\boldsymbol{e}_1, \ldots, \boldsymbol{e}_n\}$, we define the operator

$$\hat{\boldsymbol{I}} = \sum_{j=1}^{n} \boldsymbol{e}_j \boldsymbol{e}_j^\dagger. \qquad (3.266)$$

Given any vector $\boldsymbol{v}$ in this n-dimensional vector space, the operator defined in Equation 3.266 acts in the following way:

$$\hat{\boldsymbol{I}} \cdot \boldsymbol{v} = \left(\sum_{j=1}^{n} \boldsymbol{e}_j \boldsymbol{e}_j^\dagger \right) \cdot \boldsymbol{v}$$
$$= \sum_{j=1}^{n} \boldsymbol{e}_j \left(\boldsymbol{e}_j^\dagger \cdot \boldsymbol{v} \right) = \sum_{j=1}^{n} \boldsymbol{e}_j v_j = \boldsymbol{v}. \qquad (3.267)$$

Thus, $\hat{\boldsymbol{I}}$ is revealed as the unit operator in the n-dimensional vector space. The same construction works in Hilbert space. We define

$$\hat{1} = \sum_{j=1}^{n} |\varphi_j\rangle\langle\varphi_j| \qquad (3.268)$$

and show that

$$\hat{1}|\psi(t)\rangle = \sum_{j=1}^{n} |\varphi_j\rangle\langle\varphi_j|\psi(t)\rangle = \sum_{j=1}^{n} |\varphi_j\rangle A_j(t) = |\psi(t)\rangle. \qquad (3.269)$$

Therefore, $\hat{1}$ turns out to be the unit operator in Hilbert space.

Suppose we would like to compute the scalar product $\langle\alpha(t)|\beta(t)\rangle$, but we only know the scalar products $A_j(t) = \langle\varphi_j|\alpha(t)\rangle$ and $B_j(t) = \langle\varphi_j|\beta(t)\rangle$ of the two states $|\alpha(t)\rangle$ and $|\beta(t)\rangle$ with the basis states $|\varphi_j\rangle$. Then, the technique of *"inserting a one"* leads to success:

$$\langle\alpha(t)|\beta(t)\rangle = \langle\alpha(t)|\hat{1}|\beta(t)\rangle = \langle\alpha(t)|\sum_{j=1}^{n} |\varphi_j\rangle\langle\varphi_j|\beta(t)\rangle$$

$$= \sum_{j=1}^{n} \langle\varphi_j|\alpha(t)\rangle^*\langle\varphi_j|\beta(t)\rangle = \sum_{j=1}^{n} A_j(t)^* B_j(t). \qquad (3.270)$$

Let us now look more closely at the eigenstates $|x\rangle$ of the position operator $\hat{x}$. First of all, since the states $|x\rangle$ form a continuum, we need to replace the sum over states by an integral to obtain

$$\hat{1} = \int_{-\infty}^{\infty} dx\, |x\rangle\langle x|. \qquad (3.271)$$

Then, in Section 3.3, we showed that $\hat{x}$ is a Hermitian operator. Therefore, according to Equation 3.262:

$$\langle x|x'\rangle = 0, \quad \text{for } x \neq x'. \qquad (3.272)$$

But what about the case $x' = x$? We reason in the following way. Suppose a quantum particle is in the state $|x'\rangle$. Then, $\varphi(x) = \langle x|x'\rangle$ is its wave function in position representation. But since the particle is sharply localized at x', the wave function $\varphi(x)$ is sharply peaked at x', i.e.,

$$\varphi(x) = \langle x|x'\rangle = C\,\delta(x' - x), \qquad (3.273)$$

where C is a normalization constant and $\delta(x' - x)$ is Dirac's δ function. To determine C, we make use of Equation 3.271 and compute:

$$C\,\delta(x' - x) = \langle x|x'\rangle = \int_{-\infty}^{\infty} \langle x|x''\rangle\langle x''|x'\rangle\, dx''$$

$$= \int_{-\infty}^{\infty} C\,\delta(x'' - x)\, C\,\delta(x' - x'')\, dx'' = C^2\,\delta(x' - x). \qquad (3.274)$$

This equation holds only if $C^2 = C$, which implies that either $C = 0$ or $C = 1$. The case $C = 0$ does not represent a sharply peaked wave function; it has to be discarded. This leaves $C = 1$, and therefore,

$$\langle x|x'\rangle = \delta(x' - x). \tag{3.275}$$

The momentum eigenstates $|p\rangle$ are the eigenstates of the momentum operator $\hat{p}$, i.e.,

$$\hat{p}|p\rangle = p|p\rangle. \tag{3.276}$$

We normalize the momentum eigenstates such that

$$\langle p|p'\rangle = \delta(p' - p) \tag{3.277}$$

and

$$\hat{1} = \int_{-\infty}^{\infty} dp\, |p\rangle\langle p|. \tag{3.278}$$

In position representation we have

$$\langle x|p\rangle = \varphi_p(x) \tag{3.279}$$

and

$$-i\hbar \frac{d}{dx}\varphi_p(x) = p\,\varphi_p(x). \tag{3.280}$$

We construct the functions $\varphi_p(x)$ with the help of the functions $\varphi_k^{(\pm)}(x)$ defined in Equation 3.139. Since the functions $\varphi_k^{(\pm)}(x)$ are eigenfunctions of $\hat{p}$ with eigenvalues $\pm\hbar k$, where $k > 0$, we have $p = \pm\hbar k$ and, valid for all p, positive and negative,

$$\langle x|p\rangle = N_p \exp\left(\frac{i}{\hbar}px\right), \tag{3.281}$$

where N_p is a real normalization constant. We determine N_p in the following way. Inserting a one, using Equations 3.220, the Fourier Theorem stated in Equation 3.142, and Equation 3.281, we obtain:

$$\langle p|p'\rangle = \int_{-\infty}^{\infty} \langle p|x\rangle\langle x|p'\rangle\, dx = N_p N_{p'} \int_{-\infty}^{\infty} \exp\left[\frac{i}{\hbar}(p' - p)x\right] dx$$
$$= 2\pi N_p^2\, \delta\left[\frac{1}{\hbar}(p' - p)\right] = 2\pi N_p^2\, \hbar\, \delta(p' - p). \tag{3.282}$$

Choosing N_p equal to $1/\sqrt{2\pi\hbar}$ yields agreement with Equation 3.277. Therefore, the properly normalized momentum eigenstates in position representation are:

$$\langle x|p\rangle = \frac{1}{\sqrt{2\pi\hbar}} \exp\left(\frac{i}{\hbar}px\right). \qquad (3.283)$$

Exercises:

3.5.1 As an illustration of properties of vectors and the scalar products that can be derived without introducing coordinates, show that

(a) $(\mathbf{u}, \mathbf{0}) = 0$ for all vectors $\mathbf{u}$,

(b) if $(\mathbf{u}, \mathbf{v}) = 0$, $(\mathbf{u}, \mathbf{u}) \neq 0$, $(\mathbf{v}, \mathbf{v}) \neq 0$ then

$$\lambda \mathbf{u} + \mu \mathbf{v} = \mathbf{0} \quad \Leftrightarrow \quad \lambda = \mu = 0.$$

3.5.2 Let $\hat{B} = \hat{A}^\dagger \hat{A}$, $\hat{B}|\beta\rangle = \beta|\beta\rangle$, and $\langle\beta|\beta\rangle = 1$. Show that

(a) β is real,

(b) $\beta \geq 0$.

3.5.3 The expectation values of the position operator $\hat{x}$ and the momentum operator $\hat{p}$ in the state $|\psi(t)\rangle$ are given by:

$$\bar{x}(t) = \langle\psi(t)|\hat{x}|\psi(t)\rangle, \quad \bar{p}(t) = \langle\psi(t)|\hat{p}|\psi(t)\rangle.$$

Using the Schrödinger Equation 3.256, show that the expectation values $\bar{x}(t)$ and $\bar{p}(t)$ fulfill the classical relationship:

$$\dot{\bar{x}}(t) = \frac{d\bar{x}(t)}{dt} = \frac{\bar{p}(t)}{m}.$$

3.5.4 Let $|\varphi\rangle$ and $|\psi\rangle$ be two arbitrary states. Without introducing a basis, prove Schwarz's inequality

$$\langle\varphi|\varphi\rangle \langle\psi|\psi\rangle \geq |\langle\varphi|\psi\rangle|^2. \qquad (3.284)$$

3.6 Heisenberg Picture

In Schrödinger's formulation of quantum mechanics the quantum states, $|\psi(t)\rangle$, are time dependent, whereas the operators, $\hat{\theta}$, are time independent. But this is not the only way to construct a consistent formulation of quantum mechanics. According to Heisenberg we may arrive at an equivalent formulation of quantum mechanics by considering the quantum state of a system as time independent, but the quantum operators as time dependent. An operator is time independent if it satisfies

$$\frac{\partial \hat{\theta}}{\partial t} = 0; \qquad (3.285)$$

an operator is time dependent if it does not satisfy the condition defined in Equation 3.285. Time-independent operators satisfying the condition in Equation 3.285 are called *Schrödinger operators*; time-dependent operators, for which the condition in Equation 3.285 does not hold, are called *Heisenberg operators*. In order to avoid unnecessary complications and to keep the presentation straightforward, we assume that the Hamiltonian $\hat{\mathcal{H}}$ is not explicitly time dependent, i.e., $\hat{\mathcal{H}}$ satisfies Equation 3.285.

In order to make the transition between Schrödinger's and Heisenberg's formulations of quantum mechanics, we need to take a closer look at functions of operators. Functions of operators are not a new concept for us. We already encountered them in Section 3.3. Examples are the angular momentum operator $\hat{\boldsymbol{L}} = \hat{\boldsymbol{r}} \times \hat{\boldsymbol{p}}$ and the potential energy operator $V(\hat{\boldsymbol{r}})$. In these cases it was always clear to us from the context how to interpret these operator functions. But what about more complicated functions of operators, such as, $\exp(\hat{\theta})$, $\cos(\hat{\theta})$, or $\sin(\hat{\theta})$? We give meaning to these operator functions via their Taylor series expansions:

$$\exp(\hat{\theta}) = \sum_{n=0}^{\infty} \frac{1}{n!} \hat{\theta}^n,$$

$$\cos(\hat{\theta}) = \sum_{n=0}^{\infty} \frac{(-1)^n}{(2n)!} \hat{\theta}^{2n},$$

$$\sin(\hat{\theta}) = \sum_{n=0}^{\infty} \frac{(-1)^n}{(2n+1)!} \hat{\theta}^{2n+1}, \qquad (3.286)$$

where

$$\hat{\theta}^n = \underbrace{\hat{\theta}\hat{\theta}\ldots\hat{\theta}}_{n \text{ times}}, \qquad (3.287)$$

means that we apply the operator $\hat{\theta}$ consecutively n times.

We are now ready to discuss Heisenberg's formulation of quantum mechanics. Just like in Schrödinger's approach, we start from

$$i\hbar \frac{\partial |\psi(t)\rangle}{\partial t} = \hat{\mathcal{H}} |\psi(t)\rangle, \qquad (3.288)$$

where we assume that $\hat{\mathcal{H}}$ is time independent, i.e., $\hat{\mathcal{H}}$ is a Schrödinger operator. If $\hat{\mathcal{H}}$ were a number, and not an operator, we would immediately write down the solution of the Schrödinger Equation 3.288 as

$$|\psi(t)\rangle = \exp\left(-\frac{i}{\hbar}\hat{\mathcal{H}}t\right)|\psi(0)\rangle. \tag{3.289}$$

This, actually, works for operators, too! We show this by inserting the state defined in Equation 3.289 into the Schrödinger Equation 3.288 and using Equation 3.286 to give meaning to the operator $\exp(-i\hat{\mathcal{H}}t/\hbar)$:

$$\begin{aligned}
i\hbar\frac{\partial}{\partial t}|\psi(t)\rangle &= i\hbar\frac{d}{dt}\left[\exp\left(-\frac{i}{\hbar}\hat{\mathcal{H}}t\right)\right]|\psi(0)\rangle \\
&= i\hbar\frac{d}{dt}\left[\sum_{n=0}^{\infty}\frac{1}{n!}\left(-\frac{i}{\hbar}\hat{\mathcal{H}}\right)^n t^n\right]|\psi(0)\rangle \\
&= i\hbar\left[\sum_{n=1}^{\infty}\frac{1}{(n-1)!}\left(-\frac{i}{\hbar}\hat{\mathcal{H}}\right)\left(-\frac{i}{\hbar}\hat{\mathcal{H}}\right)^{n-1} t^{n-1}\right]|\psi(0)\rangle \\
&= \hat{\mathcal{H}}\left[\sum_{n=0}^{\infty}\frac{1}{n!}\left(-\frac{i}{\hbar}\hat{\mathcal{H}}\right)^n t^n\right]|\psi(0)\rangle \\
&= \hat{\mathcal{H}}\exp\left(-\frac{i}{\hbar}\hat{\mathcal{H}}t\right)|\psi(0)\rangle = \hat{\mathcal{H}}|\psi(t)\rangle. \tag{3.290}
\end{aligned}$$

This shows that $|\psi(t)\rangle$ as constructed according to Equation 3.289 is a solution of Schrödinger's Equation 3.288. According to Equation 3.289 the operator $\exp(-i\hat{\mathcal{H}}t/\hbar)$ evolves the initial state $|\psi(0)\rangle$ at time $t = 0$ into the state $|\psi(t)\rangle$ at time t. Therefore, we call

$$\hat{U}(t) = \exp\left(-\frac{i}{\hbar}\hat{\mathcal{H}}t\right) \tag{3.291}$$

the *time-evolution operator* of the quantum system governed by the Hamiltonian $\hat{\mathcal{H}}$ and write according to Equation 3.289:

$$|\psi(t)\rangle = \hat{U}(t)|\psi(0)\rangle. \tag{3.292}$$

Since $\hat{\mathcal{H}}$ is Hermitian, i.e., $\hat{\mathcal{H}}^{\dagger} = \hat{\mathcal{H}}$, the time-evolution operator $\hat{U}(t)$ is unitary, i.e.,

$$\hat{U}(t)^{\dagger}\hat{U}(t) = \exp\left(\frac{i}{\hbar}\hat{\mathcal{H}}t\right)\exp\left(-\frac{i}{\hbar}\hat{\mathcal{H}}t\right) = \hat{1}. \tag{3.293}$$

In the Schrödinger picture the expectation value of a time-independent Schrödinger operator $\hat{o}$ in the state $|\psi(t)\rangle$ is given by Equation 3.258.

With the help of the time-evolution operator defined in Equation 3.291 and the explicit form of $|\psi(t)\rangle$ in Equation 3.292, this may be written as

$$\begin{aligned}\bar{\theta}(t) &= \langle\psi(t)|\hat{\theta}|\psi(t)\rangle = \langle\psi(0)|\hat{U}(t)^\dagger\hat{\theta}\hat{U}(t)|\psi(0)\rangle \\ &= \langle\psi(0)|\hat{\theta}(t)|\psi(0)\rangle,\end{aligned} \quad (3.294)$$

where we defined the *Heisenberg operator*

$$\hat{\theta}(t) = \hat{U}^\dagger(t)\,\hat{\theta}\,\hat{U}(t). \quad (3.295)$$

Equation 3.294 shows that instead of using time-dependent states and time-independent operators, we may equally well work with time-independent states and time-dependent operators, where the time dependence is transferred from the states to the operators according to Equation 3.295. Taking the time derivative of the Heisenberg operator defined in Equation 3.295 and using the explicit representation of the time-evolution operator $\hat{U}(t)$ (see Equation 3.291), we obtain:

$$\begin{aligned}\frac{d}{dt}\hat{\theta}(t) &= \frac{d}{dt}\exp\left(\frac{i}{\hbar}\hat{\mathcal{H}}t\right)\hat{\theta}\exp\left(-\frac{i}{\hbar}\hat{\mathcal{H}}t\right) \\ &= \frac{i}{\hbar}\hat{\mathcal{H}}\hat{\theta}(t) + \hat{\theta}(t)\left(-\frac{i}{\hbar}\hat{\mathcal{H}}\right) \\ &= \frac{i}{\hbar}[\hat{\mathcal{H}},\hat{\theta}(t)].\end{aligned} \quad (3.296)$$

This is the Heisenberg equation of motion for the Heisenberg operator $\hat{\theta}(t)$.

Exercises:

3.6.1 Show that

$$\exp(\hat{A})\hat{B}\exp(-\hat{A}) = \hat{B} + \frac{1}{1!}[\hat{A},\hat{B}] + \frac{1}{2!}[\hat{A},[\hat{A},\hat{B}]] + \ldots. \quad (3.297)$$

Then, show that

$$\exp(ia\hat{p}/\hbar)\hat{x}\exp(-ia\hat{p}/\hbar) = \hat{x} + a, \quad (3.298)$$

where a is real.

3.6.2 Show that

$$e^{i\alpha\hat{p}}e^{i\beta\hat{x}} = e^{i(\alpha\hat{p}+\beta\hat{x}+\hbar\alpha\beta/2)}. \quad (3.299)$$

Hint: Define the operator function

$$\hat{g}(\lambda) = e^{i\alpha\hat{p}\lambda} e^{i\beta\hat{x}\lambda},$$

where λ is a real parameter. Differentiating $\hat{g}(\lambda)$ with respect to λ, derive a differential equation for $\hat{g}(\lambda)$. In doing so, Equations 3.297 and 3.298 will come in handy. Solve the differential equation for $\hat{g}(\lambda)$. For $\lambda = 1$ we will then obtain Equation 3.299.

3.6.3 Show that

$$e^{ia\hat{p}/\hbar} f(x) = f(x+a).$$

This means that $\exp(ia\hat{p}/\hbar)$ is the generator of translations in space.

3.7 Two-Level Systems

We already encountered electrons in Chapters 1 and 2. They are elementary particles that have a rest mass, $m_e = 9.109\ldots \times 10^{-31}$ kg and carry $-e = -1.602\ldots \times 10^{-19}$ C of electric charge. But they also have another fundamental property: Each electron, independent of its state of motion, carries a spin angular momentum, a vector denoted by $\boldsymbol{s}$. Since $\boldsymbol{s}$ is a vector, we are interested in its components, s_x, s_y, and s_z with respect to a Cartesian coordinate system. Choosing such a system, and measuring the projection s_z of $\boldsymbol{s}$ on the z-axis, we either find $s_z = +\hbar/2$, or $s_z = -\hbar/2$. We never find any other values! Even if we rotate our coordinate system in space, the projection s_z of $\boldsymbol{s}$ on the new coordinate system's z-axis is $\pm\hbar/2$. No other values occur.

It is customary to quote as the spin S of a quantum particle the maximum value of the projection of the particle's spin on the z-axis, in units of $\hbar$. In the case of the electron this is $S = 1/2$. Therefore, electrons are spin-1/2 particles.

Since the projection of a classical angular momentum vector on the z-axis of a Cartesian coordinate system changes continuously if we reorient the z-axis of the coordinate system, the invariant result $\pm\hbar/2$ for electrons is a mind-boggling quantum effect. Since with respect to a given axis there are just two possibilities, $s_z = \pm\hbar/2$, for the projection of the spin on this axis, the entire Hilbert space of the electron's spin is just two-dimensional. In this space we define two basis states, $|\uparrow\rangle$ and $|\downarrow\rangle$, which we call "spin-up" and "spin-down," respectively. If an electron is in the state $|\uparrow\rangle$, its spin projection on the z-axis is guaranteed to be $+\hbar/2$. If an electron is in the state $|\downarrow\rangle$, its spin projection on the

z-axis is guaranteed to be $-\hbar/2$. The two basis states are orthonormal, i.e.,

$$\langle \uparrow | \uparrow \rangle = 1, \quad \langle \downarrow | \downarrow \rangle = 1, \quad \langle \uparrow | \downarrow \rangle = \langle \downarrow | \uparrow \rangle = 0. \qquad (3.300)$$

Since the spin s of an electron is an observable, we represent it by the operator $\hat{s}$ with components $\hat{s}_x$, $\hat{s}_y$, and $\hat{s}_z$. Since the z component of the electron's spin is sharp in the states $|\uparrow\rangle$ and $|\downarrow\rangle$, these two states are eigenstates of the operator $\hat{s}_z$ with eigenvalues $\pm\hbar/2$ according to

$$\hat{s}_z |\uparrow\rangle = +\frac{\hbar}{2}|\uparrow\rangle, \quad \hat{s}_z |\downarrow\rangle = -\frac{\hbar}{2}|\downarrow\rangle. \qquad (3.301)$$

Since $\hat{s}$ is an angular momentum operator, its components satisfy commutator identities analogous to those of the orbital angular momentum (see Equations 3.67 and 3.68):

$$[\hat{s}_x, \hat{s}_y] = i\hbar \hat{s}_z, \quad [\hat{s}_y, \hat{s}_z] = i\hbar \hat{s}_x, \quad [\hat{s}_z, \hat{s}_x] = i\hbar \hat{s}_y. \qquad (3.302)$$

In addition to the operators $\hat{s}_x$, $\hat{s}_y$, and $\hat{s}_z$ we define the operators

$$\hat{\sigma}_x = 2\hat{s}_x/\hbar, \quad \hat{\sigma}_y = 2\hat{s}_y/\hbar, \quad \hat{\sigma}_z = 2\hat{s}_z/\hbar. \qquad (3.303)$$

Then,

$$\hat{\sigma}_z |\uparrow\rangle = |\uparrow\rangle, \quad \hat{\sigma}_z |\downarrow\rangle = -|\downarrow\rangle \qquad (3.304)$$

and

$$[\hat{\sigma}_x, \hat{\sigma}_y] = 2i\hat{\sigma}_z, \quad [\hat{\sigma}_y, \hat{\sigma}_z] = 2i\hat{\sigma}_x, \quad [\hat{\sigma}_z, \hat{\sigma}_x] = 2i\hat{\sigma}_y. \qquad (3.305)$$

Working with the operators $\hat{\sigma}_{x,y,z}$ instead of the original operators $\hat{s}_{x,y,z}$ is frequently more convenient, since they are dimensionless. Since s is an observable, the components $\hat{s}_x$, $\hat{s}_y$, and $\hat{s}_z$ of its associated operator $\hat{s}$ are Hermitian operators. Since the operators $\hat{\sigma}_{x,y,z}$ differ from the operators $\hat{s}_{x,y,z}$ by a scale factor only, the operators $\hat{\sigma}_{x,y,z}$ are Hermitian as well.

We are now going to calculate the matrix elements of the operators $\hat{\sigma}_x$, $\hat{\sigma}_y$, and $\hat{\sigma}_z$ in the $\{|\uparrow\rangle, |\downarrow\rangle\}$ basis. Because of Equation 3.304 we have:

$$\langle \uparrow |\hat{\sigma}_z| \uparrow \rangle = 1, \quad \langle \uparrow |\hat{\sigma}_z| \downarrow \rangle = 0, \quad \langle \downarrow |\hat{\sigma}_z| \uparrow \rangle = 0, \quad \langle \downarrow |\hat{\sigma}_z| \downarrow \rangle = -1. \qquad (3.306)$$

Therefore, in 2×2 matrix notation:

$$\sigma_z = \begin{pmatrix} \langle \uparrow |\hat{\sigma}_z| \uparrow \rangle & \langle \uparrow |\hat{\sigma}_z| \downarrow \rangle \\ \langle \downarrow |\hat{\sigma}_z| \uparrow \rangle & \langle \downarrow |\hat{\sigma}_z| \downarrow \rangle \end{pmatrix} = \begin{pmatrix} 1 & 0 \\ 0 & -1 \end{pmatrix}. \qquad (3.307)$$

Using Equation 3.304, Equation 3.305, and the fact that $\hat{\sigma}_z$ is Hermitian, we compute:

$$\langle \uparrow |\hat{\sigma}_x| \uparrow \rangle = \langle \uparrow |\hat{\sigma}_x\hat{\sigma}_z| \uparrow \rangle = \langle \uparrow |\hat{\sigma}_z\hat{\sigma}_x - 2i\hat{\sigma}_y| \uparrow \rangle$$
$$= (\hat{\sigma}_z| \uparrow \rangle)^\dagger \hat{\sigma}_x| \uparrow \rangle - 2i\langle \uparrow |\hat{\sigma}_y| \uparrow \rangle = \langle \uparrow |\hat{\sigma}_x| \uparrow \rangle - 2i\langle \uparrow |\hat{\sigma}_y| \uparrow \rangle$$
$$\Rightarrow \quad \langle \uparrow |\hat{\sigma}_y| \uparrow \rangle = 0. \tag{3.308}$$

In the same way, by computing $\langle \downarrow |\hat{\sigma}_x| \downarrow \rangle$, we establish that $\langle \downarrow |\hat{\sigma}_y| \downarrow \rangle = 0$. Replacing $\hat{\sigma}_x$ by $\hat{\sigma}_y$ in the first matrix element of Equation 3.308, we derive $\langle \uparrow |\hat{\sigma}_x| \uparrow \rangle = 0$ and $\langle \downarrow |\hat{\sigma}_x| \downarrow \rangle = 0$. Therefore, taking into account the Hermiticity of $\hat{\sigma}_x$ and $\hat{\sigma}_y$, which implies $\sigma_x^\dagger = \sigma_x$ and $\sigma_y^\dagger = \sigma_y$, we established:

$$\sigma_x = \begin{pmatrix} 0 & \alpha^* \\ \alpha & 0 \end{pmatrix}, \quad \sigma_y = \begin{pmatrix} 0 & \beta^* \\ \beta & 0 \end{pmatrix}, \tag{3.309}$$

where α and β are complex numbers. The commutator $[\hat{\sigma}_z, \hat{\sigma}_x] = 2i\hat{\sigma}_y$ (see Equation 3.305), which implies the commutator $[\sigma_z, \sigma_x] = 2i\sigma_y$ for the associated matrices, provides a connection between α and β:

$$[\sigma_z, \sigma_x] = \begin{pmatrix} 1 & 0 \\ 0 & -1 \end{pmatrix}\begin{pmatrix} 0 & \alpha^* \\ \alpha & 0 \end{pmatrix} - \begin{pmatrix} 0 & \alpha^* \\ \alpha & 0 \end{pmatrix}\begin{pmatrix} 1 & 0 \\ 0 & -1 \end{pmatrix}$$
$$= \begin{pmatrix} 0 & 2\alpha^* \\ -2\alpha & 0 \end{pmatrix} = 2i\sigma_y = \begin{pmatrix} 0 & 2i\beta^* \\ 2i\beta & 0 \end{pmatrix}. \tag{3.310}$$

Comparing matrix elements we obtain:

$$\beta = i\alpha. \tag{3.311}$$

Considering the commutator $[\hat{\sigma}_x, \hat{\sigma}_y] = 2i\hat{\sigma}_z$ (see Equation 3.305), which implies the commutator $[\sigma_x, \sigma_y] = 2i\sigma_z$ for the associated matrices, we obtain:

$$[\sigma_x, \sigma_y] = \begin{pmatrix} 0 & \alpha^* \\ \alpha & 0 \end{pmatrix}\begin{pmatrix} 0 & -i\alpha^* \\ i\alpha & 0 \end{pmatrix} - \begin{pmatrix} 0 & -i\alpha^* \\ i\alpha & 0 \end{pmatrix}\begin{pmatrix} 0 & \alpha^* \\ \alpha & 0 \end{pmatrix}$$
$$= \begin{pmatrix} 2i|\alpha|^2 & 0 \\ 0 & -2i|\alpha|^2 \end{pmatrix} = 2i\sigma_z = \begin{pmatrix} 2i & 0 \\ 0 & -2i \end{pmatrix}. \tag{3.312}$$

Comparing matrix elements we obtain:

$$|\alpha|^2 = 1. \tag{3.313}$$

Thus, up to the phase of α, the matrix elements of $\hat{\sigma}_x$, $\hat{\sigma}_y$, and $\hat{\sigma}_z$ are now determined. Since the phase of α cannot be determined from the commutator relations, we choose the *phase convention* $\alpha = 1$. Then, in summary, we obtain:

$$\sigma_x = \begin{pmatrix} 0 & 1 \\ 1 & 0 \end{pmatrix}, \quad \sigma_y = \begin{pmatrix} 0 & -i \\ i & 0 \end{pmatrix}, \quad \sigma_z = \begin{pmatrix} 1 & 0 \\ 0 & -1 \end{pmatrix}. \quad (3.314)$$

The matrices σ_x, σ_y, and σ_z in Equation 3.314 are known as the *Pauli matrices*. Together with the 2×2 unit matrix

$$\mathbf{1} = \begin{pmatrix} 1 & 0 \\ 0 & 1 \end{pmatrix}, \quad (3.315)$$

they form a complete set of operators in the two-dimensional Hilbert space of spin-1/2 electrons in the sense that any 2×2 matrix Ω in this space can be linearly combined from σ_x, σ_y, σ_z, and $\mathbf{1}$ according to

$$\Omega = \mu \sigma_x + \lambda \sigma_y + \nu \sigma_z + \gamma \mathbf{1}, \quad (3.316)$$

where μ, λ, ν, and γ are complex constants. The same holds for the associated operator $\hat{\Omega}$, which can be combined in analogy to Equation 3.316 in the form

$$\hat{\Omega} = \mu \hat{\sigma}_x + \lambda \hat{\sigma}_y + \nu \hat{\sigma}_z + \gamma \hat{\mathbf{1}}. \quad (3.317)$$

Using the explicit forms of the Pauli matrices defined in Equation 3.314, it is straightforward to show that the Pauli matrices satisfy

$$\begin{aligned} \sigma_x^2 &= \sigma_y^2 = \sigma_z^2 = \mathbf{1}, \\ \sigma_x \sigma_y &= i\sigma_z, \quad \sigma_y \sigma_z = i\sigma_x, \quad \sigma_z \sigma_x = i\sigma_y, \\ \sigma_y \sigma_x &= -i\sigma_z, \quad \sigma_z \sigma_y = -i\sigma_x, \quad \sigma_x \sigma_z = -i\sigma_y. \end{aligned} \quad (3.318)$$

The associated operators $\hat{\sigma}_x$, $\hat{\sigma}_y$, and $\hat{\sigma}_z$ satisfy the same relations.

As an application of the properties defined in Equation 3.318 of the Pauli operators let us evaluate $\exp(i\beta\hat{\sigma}_x)$, where β is a real number. Using $\hat{\sigma}_x^2 = \hat{\mathbf{1}}$, we obtain

$$\begin{aligned} \exp(i\beta\hat{\sigma}_x) &= \sum_{n=0}^{\infty} \frac{i^n}{n!} (\beta\hat{\sigma}_x)^n \\ &= \sum_{n=0}^{\infty} \frac{i^{2n}}{(2n)!} (\beta\hat{\sigma}_x)^{2n} + \sum_{n=0}^{\infty} \frac{i^{2n+1}}{(2n+1)!} (\beta\hat{\sigma}_x)^{2n+1} \\ &= \sum_{n=0}^{\infty} \frac{(-1)^n}{(2n)!} \beta^{2n} \hat{\mathbf{1}} + \sum_{n=0}^{\infty} \frac{i(-1)^n}{(2n+1)!} \beta^{2n+1} \hat{\sigma}_x \\ &= \hat{\mathbf{1}} \cos(\beta) + i\hat{\sigma}_x \sin(\beta). \end{aligned} \quad (3.319)$$

A useful generalization of Equation 3.319 is

$$\exp(i\beta \boldsymbol{u} \cdot \hat{\boldsymbol{\sigma}}) = \hat{\mathbf{1}} \cos(\beta) + i(\boldsymbol{u} \cdot \hat{\boldsymbol{\sigma}}) \sin(\beta), \qquad (3.320)$$

where $\boldsymbol{u}$ is a real unit vector.

Apart from spin, the electron has another intrinsic property: It has a magnetic moment given by:

$$\boldsymbol{\mu} = -g_s \mu_B (\boldsymbol{s}/\hbar), \qquad (3.321)$$

where

$$g_s = 2.002319\ldots \qquad (3.322)$$

is the spin g-factor, which can be calculated very accurately using the methods of quantum electrodynamics,

$$\mu_B = e\hbar/(2m_e) = 9.274\ldots \times 10^{-24}\,\text{J/T} \qquad (3.323)$$

is the Bohr magneton, a convenient unit of magnetic moment, and $\boldsymbol{s}$ is the spin of the electron. Notice that because of the minus sign in Equation 3.321, the electron's magnetic moment and its spin are oriented opposite to each other. The reason is the negative charge of the electron. The electron's magnetic moment allows the electron to interact with an external magnetic field. The interaction energy is

$$V = -\boldsymbol{\mu} \cdot \boldsymbol{B} = g_s \mu_B \frac{1}{\hbar} (\boldsymbol{s} \cdot \boldsymbol{B}), \qquad (3.324)$$

where $\boldsymbol{B}$ is the applied magnetic field. By canonical quantization we obtain the operator $\hat{V}$ of potential energy of the electron according to:

$$\hat{V} = g_s \mu_B \frac{1}{\hbar} (\hat{\boldsymbol{s}} \cdot \boldsymbol{B}). \qquad (3.325)$$

If $\boldsymbol{B}$ is in the z-direction, we obtain:

$$\hat{V} = \frac{1}{2} E_0 \hat{\sigma}_z, \qquad (3.326)$$

where

$$E_0 = g_s \mu_B B. \qquad (3.327)$$

For an electron at rest in a magnetic field oriented in the $+z$-direction ($B > 0$), the energy of the state $|\downarrow\rangle$ is $\langle \downarrow | E_0 \hat{\sigma}_z/2 | \downarrow \rangle = -E_0/2$; the energy of the state $|\uparrow\rangle$ is $\langle \uparrow | E_0 \hat{\sigma}_z/2 | \uparrow \rangle = E_0/2$. We say that the state $|\downarrow\rangle$ is the *ground state* of the electron in a magnetic field oriented in

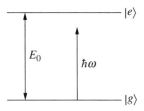

Figure 3.7 Sketch of a general two-level system. The situation shown corresponds to a red-detuned applied field, i.e., $\hbar\omega < E_0$.

the $+z$-direction, and $|\uparrow\rangle$ is the *excited state*. Since there are only two states, one ground state and one excited state, the electron's spin in an external magnetic field is an ideal *two-level system*.

Figure 3.7 shows a general two-level system with a ground state $|g\rangle$ and an excited state $|e\rangle$. Our electron in a magnetic field fits neatly into this scheme if we identify

$$|g\rangle \equiv |\downarrow\rangle, \quad |e\rangle \equiv |\uparrow\rangle. \tag{3.328}$$

In fact, this identification allows us to use our electron-spin machinery generally for all two-level systems. With the identification defined in Equation 3.328 we have, for example,

$$\begin{aligned}
\langle g|\sigma_x|g\rangle &= \langle e|\sigma_x|e\rangle = 0, \quad \langle g|\sigma_x|e\rangle = \langle e|\sigma_x|g\rangle = 1,\\
\langle g|\sigma_y|g\rangle &= \langle e|\sigma_y|e\rangle = 0, \quad \langle g|\sigma_y|e\rangle = i, \quad \langle e|\sigma_y|g\rangle = -i,\\
\langle g|\sigma_z|g\rangle &= -1, \quad \langle g|\sigma_z|e\rangle = \langle e|\sigma_z|g\rangle = 0, \quad \langle e|\sigma_z|e\rangle = 1.
\end{aligned} \tag{3.329}$$

The energy splitting between the ground state and the excited state is

$$E_0 = \hbar\omega_0, \tag{3.330}$$

where we introduced the frequency ω_0, called the *resonance frequency*. Quantum mechanical transitions between the ground state $|g\rangle$ and the excited state $|e\rangle$ are induced by an externally applied field of frequency ω. Usually the applied field is slightly detuned from resonance, such that the *detuning* $\Delta\omega = \omega - \omega_0$ is not zero. If $\Delta\omega < 0$ we call the externally applied field *red detuned*. If $\Delta\omega > 0$ we call the externally applied field *blue detuned*.

Apart from an electron in a magnetic field there are many other "pure" two-level systems. In addition there are many quantum systems that may be approximately described as two-level systems. For example, all atoms have infinitely many energy levels. Thus, atoms are never "pure" two-level systems. However, in a situation where weak laser light interacts with a pair of atomic levels, and the laser's frequency ω is very close to resonance, i.e., $\omega \approx \omega_0$, the other atomic levels are only weakly coupled and may be neglected to a very good approximation. Thus, two atomic levels coupled by a near-resonant laser may be considered an excellent two-level system.

The interaction of a weak field of frequency ω with a two-level system may be modeled by the Hamiltonian

$$\hat{\mathcal{H}}(t) = \frac{1}{2}E_0\hat{\sigma}_z + V_0\hat{\sigma}_x\cos(\omega t), \tag{3.331}$$

where E_0 is the splitting of the two energy levels and V_0 is the strength of the interaction. The time-dependent Schrödinger equation is:

$$i\hbar\frac{\partial|\psi(t)\rangle}{\partial t} = \left[\frac{1}{2}E_0\hat{\sigma}_z + V_0\hat{\sigma}_x\cos(\omega t)\right]|\psi(t)\rangle. \tag{3.332}$$

We expand $|\psi(t)\rangle$ in the two states, $|g\rangle$ and $|e\rangle$, of the two-level system according to:

$$|\psi(t)\rangle = A(t)|g\rangle + B(t)|e\rangle. \tag{3.333}$$

Inserting this equation into Equation 3.332 yields:

$$i\hbar\left[\dot{A}(t)|g\rangle + \dot{B}(t)|e\rangle\right]$$
$$= \left[\frac{1}{2}E_0\hat{\sigma}_z + V_0\hat{\sigma}_x\cos(\omega t)\right]\left[A(t)|g\rangle + B(t)|e\rangle\right], \tag{3.334}$$

where the dot indicates differentiation with respect to the time t. Taking the scalar product of Equation 3.334 with the ground state $\langle g|$ and using the matrix elements of Equation 3.329, we obtain:

$$i\hbar\dot{A}(t) = -\frac{1}{2}E_0A(t) + V_0\cos(\omega t)B(t). \tag{3.335}$$

Taking the scalar product of Equation 3.334 with the excited state $\langle e|$ and using the matrix elements of Equation 3.329 yields:

$$i\hbar\dot{B}(t) = \frac{1}{2}E_0B(t) + V_0\cos(\omega t)A(t). \tag{3.336}$$

Let us introduce new amplitudes $\alpha(t)$ and $\beta(t)$ according to

$$A(t) = \exp[iE_0 t/(2\hbar)]\,\alpha(t),$$
$$B(t) = \exp[-iE_0 t/(2\hbar)]\,\beta(t). \tag{3.337}$$

Used in Equations 3.335 and 3.336, we obtain:

$$i\hbar\dot{\alpha}(t) = V_0 \exp(-i\omega_0 t) \cos(\omega t) \beta(t),$$
$$i\hbar\dot{\beta}(t) = V_0 \exp(i\omega_0 t) \cos(\omega t) \alpha(t), \quad (3.338)$$

where we used ω_0 defined in Equation 3.330. The set of Equations 3.338 is difficult to solve in general. However, writing it in the form

$$i\hbar\dot{\alpha}(t) = \frac{V_0}{2} \{\exp[i(\omega - \omega_0)t] + \exp[-i(\omega + \omega_0)t]\} \beta(t),$$
$$i\hbar\dot{\beta}(t) = \frac{V_0}{2} \{\exp[i(\omega + \omega_0)t] + \exp[-i(\omega - \omega_0)t]\} \alpha(t), \quad (3.339)$$

suggests an approximate solution technique. We notice that close to resonance the functions $\exp[\pm(\omega-\omega_0)t]$ are slowly varying functions of t, while the functions $\exp[\pm(\omega+\omega_0)t]$ are rapidly varying in time. Assuming that the effect of the rapidly varying functions is small, we discard these functions and keep only the slowly varying functions $\exp[\pm(\omega - \omega_0)t]$. This procedure is called the *rotating wave approximation*. It results in

$$i\hbar\dot{\alpha}(t) = \frac{V_0}{2} \exp[i(\omega - \omega_0)t] \beta(t),$$
$$i\hbar\dot{\beta}(t) = \frac{V_0}{2} \exp[-i(\omega - \omega_0)t] \alpha(t). \quad (3.340)$$

Precisely on resonance, i.e., for $\omega = \omega_0$, we obtain the set of equations

$$i\hbar\dot{\alpha}(t) = \frac{V_0}{2} \beta(t),$$
$$i\hbar\dot{\beta}(t) = \frac{V_0}{2} \alpha(t). \quad (3.341)$$

Differentiating the first equation with respect to time, and then using the second equation for $\dot{\beta}(t)$, results in the following second-order differential equation for $\alpha(t)$:

$$\ddot{\alpha}(t) + \frac{1}{4}\Omega^2 \alpha(t) = 0, \quad (3.342)$$

where

$$\Omega = \frac{V_0}{\hbar} \quad (3.343)$$

is the *Rabi frequency*. Equation 3.342 is the harmonic oscillator equation whose general solution is

$$\alpha(t) = a\cos(\Omega t/2) + b\sin(\Omega t/2), \quad (3.344)$$

where a and b are constants. The function $\beta(t)$ can be obtained from the first equation of Equation 3.341:

$$\beta(t) = \frac{2i\hbar}{V_0}\dot\alpha(t) = i\left[-a\sin(\Omega t/2) + b\cos(\Omega t/2)\right]. \quad (3.345)$$

With $\alpha(t)$ and $\beta(t)$ computed, we now know the on-resonance time dependence of the state $|\psi(t)\rangle$ of the two-level system. According to Equations 3.333 and 3.337 we have:

$$\begin{aligned}|\psi(t)\rangle &= \exp(i\omega_0 t/2)[a\cos(\Omega t/2) + b\sin(\Omega t/2)]\,|g\rangle \\ &+ i\exp(-i\omega_0 t/2)[-a\sin(\Omega t/2) + b\cos(\Omega t/2)]\,|e\rangle.\end{aligned} \quad (3.346)$$

If at time $t = 0$ we have $|\psi(t)\rangle = |g\rangle$, then $a = 1$, $b = 0$, and

$$\begin{aligned}|\psi(t)\rangle &= \exp(i\omega_0 t/2)\cos(\Omega t/2)|g\rangle \\ &- i\exp(-i\omega_0 t/2)\sin(\Omega t/2)|e\rangle.\end{aligned} \quad (3.347)$$

This is a most interesting result. According to Equation 3.347 the probabilities of the two-level system to be in the ground state $|g\rangle$ and the excited state $|e\rangle$ are

$$\begin{aligned}P_g(t) &= |\exp(i\omega_0 t/2)\cos(\Omega t/2)|^2 = \cos^2(\Omega t/2) \\ &= \frac{1}{2} + \frac{1}{2}\cos(\Omega t), \\ P_e(t) &= |-i\exp(-i\omega_0 t/2)\sin(\Omega t/2)|^2 = \sin^2(\Omega t/2) \\ &= \frac{1}{2} - \frac{1}{2}\cos(\Omega t),\end{aligned} \quad (3.348)$$

respectively. From Equation 3.348 we see that as a function of time t, the quantum particle "flops" periodically, with frequency Ω, between the ground state $|g\rangle$ and the excited state $|e\rangle$. This is why the Rabi frequency Ω is also known as the *Rabi flopping frequency* and the periodic oscillations of the occupation probability between $|g\rangle$ and $|e\rangle$ are known as *Rabi flopping*.

Carefully tuning the product of the Rabi frequency Ω and the time t during which the interaction is switched on, we may produce superposition states of $|g\rangle$ and $|e\rangle$ with any predetermined weights $\cos^2(\Omega t/2)$ and $\sin^2(\Omega t/2)$, respectively. This way we can *control* the two-level system! In particular, choosing

$$\Omega t = \frac{V_0 t}{\hbar} = \pi, \quad (3.349)$$

we may change the state of the two-level system from $|g\rangle$ to $|e\rangle$. In the case of an electron this would correspond to a *spin flip* from $|\downarrow\rangle$ to $|\uparrow\rangle$.

Because of the form of the condition defined in Equation 3.349 with "π" on the right-hand side, and also because the change from $|g\rangle \equiv |\downarrow\rangle$ to $|e\rangle \equiv |\uparrow\rangle$ corresponds to a "180° flip" in the direction of the spin, a pulse that changes $|g\rangle$ to $|e\rangle$, or $|\downarrow\rangle$ to $|\uparrow\rangle$ is called a π-*pulse*. Equally important is the production of a state that is 50% $|g\rangle$ and 50% $|e\rangle$. Such a state is produced by a $\pi/2$-*pulse*. Because of Equation 3.349, a π-pulse may be engineered in two different ways: For a given interaction time one adjusts the interaction strength V_0 such that the criterion of Equation 3.349 is fulfilled; or, for given interaction strength, one adjusts the interaction time t. Complete control over two-level systems is important because, as we will see later, two-level systems play a fundamental role in quantum computers.

For a continuous drive, $\cos(\omega t)$, we were able to solve the Schrödinger equation for the two-level system only on resonance and in the rotating-wave approximation. For impulsive drive, we may solve the two-level system exactly, which gives us complete control over all aspects of the two-level system. Consider the Hamiltonian

$$\hat{\mathcal{H}}(t) = \frac{1}{2}E_0\hat{\sigma}_z + W_0(\boldsymbol{u}\cdot\hat{\boldsymbol{\sigma}})\,\delta(t), \qquad (3.350)$$

where $\boldsymbol{u}$ is a real unit vector, $\hat{\boldsymbol{\sigma}}$ is the vector

$$\hat{\boldsymbol{\sigma}} = \begin{pmatrix} \hat{\sigma}_x \\ \hat{\sigma}_y \\ \hat{\sigma}_z \end{pmatrix} \qquad (3.351)$$

of Pauli spin operators, and $\delta(t)$ is Dirac's delta function. The Hamiltonian defined in Equation 3.350 is constructed in analogy to the Hamiltonian defined in Equation 3.331 with two important differences. (1) We now allow any mixture of Pauli operators to be present in the Hamiltonian defined in Equation 3.350, and (2) the continuous drive $\cos(\omega t)$ in the Hamiltonian defined in Equation 3.331 is replaced with an impulsive drive, $\delta(t)$, in the Hamiltonian defined in Equation 3.350. Due to the nature of Dirac's delta function, the impulsive drive in the Hamiltonian in Equation 3.350 is switched on only for the briefest of moments at $t = 0$.

Suppose we prepare our two-level system in a starting state $|\psi(t_i)\rangle$ at the initial time $t = t_i < 0$ and evolve this starting state into the state $|\psi(t_f)\rangle$ at time $t = t_f > 0$ under the influence of the Hamiltonian defined in Equation 3.350. Due to the nature of the interaction part $W_0(\boldsymbol{u}\cdot\hat{\boldsymbol{\sigma}})\delta(t)$ of the Hamiltonian, which is switched on only at time $t = 0$, we break the time evolution into three stages. Stage I: Evolution from t_i to $t = 0^-$, i.e., a time just before the pulse at $t = 0$ occurs.

Stage II: Evolution from $t = 0^-$ to $t = 0^+$, i.e., from just before the pulse is switched on to just after the pulse is switched off again. Stage III: Evolution from $t = 0^+$ to $t = t_f$.

Stage I: In $t_i \leq t < 0$ the δ function is zero. Therefore, the Hamiltonian in this time interval is

$$\hat{\mathcal{H}} = \frac{1}{2} E_0 \hat{\sigma}_z. \tag{3.352}$$

This Hamiltonian is a time-independent Schrödinger operator. Therefore, we may use Equation 3.289 to propagate $|\psi(t_i)\rangle$ to $|\psi(0^-)\rangle$. Using the duration $0 - t_i = -t_i$ of the propagation in Equation 3.289, we obtain:

$$|\psi(0^-)\rangle = \exp\left[i\left(\frac{E_0 t_i}{2\hbar}\right)\hat{\sigma}_z\right]|\psi(t_i)\rangle. \tag{3.353}$$

Stage II: We have arrived at $|\psi(0^-)\rangle$, when at $t = 0$ the δ pulse occurs, which propagates $|\psi(0^-)\rangle$ into $|\psi(0^+)\rangle$. In order to solve this propagation problem properly, we use the following representation of Dirac's δ function:

$$\delta(t) = \lim_{\tau \to 0} \begin{cases} 0, & \text{for } t < -\tau, \\ \frac{1}{2\tau}, & \text{for } -\tau < t < \tau, \\ 0, & \text{for } t > \tau, \end{cases} \tag{3.354}$$

where $\tau > 0$. Then, in the time interval $-\tau < t < \tau$, the Schrödinger equation is given by

$$i\hbar \frac{\partial}{\partial t}|\psi(t)\rangle = \hat{\mathcal{H}}|\psi(t)\rangle, \tag{3.355}$$

where

$$\hat{\mathcal{H}} = \frac{1}{2} E_0 \hat{\sigma}_z + W_0 (\boldsymbol{u} \cdot \hat{\boldsymbol{\sigma}}) \frac{1}{2\tau}. \tag{3.356}$$

Again, the Hamilton operator governing the time evolution is time independent. Using Equation 3.289 with propagation time 2τ, we obtain:

$$|\psi(0^+)\rangle = \lim_{\tau \to 0} \exp\left\{-\frac{i}{\hbar}\left[\frac{1}{2}E_0\hat{\sigma}_z + W_0(\boldsymbol{u} \cdot \hat{\boldsymbol{\sigma}})\frac{1}{2\tau}\right]2\tau\right\}|\psi(0^-)\rangle$$

$$= \exp\left[-i\left(\frac{W_0}{\hbar}\right)(\boldsymbol{u} \cdot \hat{\boldsymbol{\sigma}})\right]|\psi(0^-)\rangle. \tag{3.357}$$

Stage III: In the interval $0 < t \leq t_f$ the δ function is again zero and we obtain:

$$|\psi(t_f)\rangle = \exp\left[-i\left(\frac{E_0 t_f}{2\hbar}\right)\hat{\sigma}_z\right]|\psi(0^+)\rangle. \tag{3.358}$$

Substituting $|\psi(0^+)\rangle$ from Equation 3.357 into Equation 3.358 and $|\psi(0^-)\rangle$ from Equation 3.353 into Equation 3.357, we obtain the exact expression for $|\psi(t_f)\rangle$ in terms of $|\psi(t_i)\rangle$ in the form

$$|\psi(t_f)\rangle = \hat{U}(t_f, t_i)|\psi(t_i)\rangle, \tag{3.359}$$

where the time-evolution operator $\hat{U}(t_f, t_i)$ is

$$\hat{U}(t_f, t_i) = \exp\left[-i\left(\frac{\omega_0 t_f}{2}\right)\hat{\sigma}_z\right] \exp\left[-i\left(\frac{W_0}{\hbar}\right)(\boldsymbol{u}\cdot\hat{\boldsymbol{\sigma}})\right] \exp\left[i\left(\frac{\omega_0 t_i}{2}\right)\hat{\sigma}_z\right], \tag{3.360}$$

where we used ω_0 defined in Equation 3.330. Suppose we start our two-level system at time $t = t_i$ in the ground state $|g\rangle$, i.e., $|\psi(t_i)\rangle = |g\rangle$, and choose $\boldsymbol{u}\cdot\hat{\boldsymbol{\sigma}} = \hat{\sigma}_y$. Then, with Equations 3.320, 3.359, and 3.360 we obtain:

$$|\psi(t_f)\rangle = \cos(\beta)e^{i(\gamma+\alpha)}|g\rangle - \sin(\beta)e^{i(\gamma-\alpha)}|e\rangle, \tag{3.361}$$

where

$$\alpha = \frac{\omega_0 t_f}{2}, \quad \beta = \frac{W_0}{\hbar}, \quad \gamma = -\frac{\omega_0 t_i}{2}. \tag{3.362}$$

Since, by careful tuning of t_i, t_f, and W_0, we can generate any β and any $\alpha, \gamma > 0$, Equation 3.361 implies that we can construct any imaginable two-level state. For instance, choosing t_i, t_f, and W_0 such that $\alpha = \gamma$ and $\beta = -\pi/2$, we obtain

$$|\psi(t_f)\rangle = |e\rangle. \tag{3.363}$$

Choosing $\alpha = \gamma = \pi$ and $\beta = -\pi/4$, we obtain the superposition state

$$|\psi(t_f)\rangle = \frac{1}{\sqrt{2}}\Big(|g\rangle + |e\rangle\Big), \tag{3.364}$$

while for $\alpha = \gamma = \pi$ and $\beta = \pi/4$ we obtain:

$$|\psi(t_f)\rangle = \frac{1}{\sqrt{2}}\Big(|g\rangle - |e\rangle\Big). \tag{3.365}$$

As we will see later in Chapters 8 and 10, the three states defined in Equations 3.363, 3.364, and 3.365 play a fundamental role in quantum computing. As we have just demonstrated, a two-level system driven with a zero-width δ-pulse is capable of producing these states on command. One may object that zero-width δ functions cannot be produced technically in a real quantum computer. However, it can be shown that finite-width pulses work just as well. However, they do not allow for an analytical solution of Schrödinger's equation and numerical methods are required. Thus, our choice of a zero-width δ function is merely an analytical convenience that takes nothing away from the physics of a driven two-level system. In fact, it can be shown that no matter what we choose for the shape of the pulse, for instance a zero-width δ function, a finite-width pulse, or a $\cos(\omega t)$ function switched on over a well-chosen, predetermined time interval, we may generate any possible two-level state. Thus we have demonstrated complete control over two-level systems.

To conclude this section we need to discuss situations where several particles are present. Consider a system of two particles, labeled particle number 1 and particle number 2. If particle number 1 is in the state $|\psi_1(t)\rangle$ and particle number 2 is in the state $|\psi_2(t)\rangle$, then the two-particle system as a whole is in the state

$$|\psi(t)\rangle = |\psi_1(t)\rangle |\psi_2(t)\rangle. \tag{3.366}$$

Thus, in a case in which the two individual particles are in definite states themselves, the total state of the system is the product of the two individual states. In this case we say that $|\psi(t)\rangle$ is *separable*. This procedure may be generalized to N particles resulting in

$$|\psi(t)\rangle = |\psi_1(t)\rangle |\psi_2(t)\rangle \ldots |\psi_N(t)\rangle \tag{3.367}$$

as a separable N-particle state. Suppose an operator $\hat{\vartheta}$ acts only on particle number m. Then we indicate this by a superscript on $\hat{\vartheta}$ and write $\hat{\vartheta}^{(m)}$. Thus,

$$\begin{aligned} \hat{\vartheta}^{(m)}|\psi(t)\rangle = |\psi_1(t)\rangle |\psi_2(t)\rangle \ldots |\psi_{m-1}(t)\rangle \\ \left[\hat{\vartheta}^{(m)}|\psi_m(t)\rangle\right]|\psi_{m+1}(t)\rangle \ldots |\psi_N(t)\rangle. \end{aligned} \tag{3.368}$$

System operators $\hat{\theta}$ that act on all particles of a system are usually sums or products of operators $\hat{\vartheta}^{(m)}$ that act on individual particles. An example is the total spin $\hat{s}$ of an N-particle system,

$$\hat{s} = \sum_{m=1}^{N} \hat{s}^{(m)}, \tag{3.369}$$

where $\hat{s}^{(m)}$ acts on particle number m. For instance, let us evaluate the z component of the spin of an N-particle system, which, at time $t = 0$, is in the state

$$|\psi(t=0)\rangle = |\mu_1\rangle|\mu_2\rangle\ldots|\mu_N\rangle, \qquad (3.370)$$

where $\mu_m \in \{-\hbar/2, +\hbar/2\}$ is the z component of the spin of particle number m. Then,

$$\hat{s}_z|\psi(t=0)\rangle = \sum_{m=1}^{N} \hat{s}_z^{(m)}|\psi(t=0)\rangle =$$
$$= \left[s_z^{(1)}|\mu_1\rangle\right]|\mu_2\rangle\ldots|\mu_N\rangle + |\mu_1\rangle\left[s_z^{(2)}|\mu_2\rangle\right]\ldots|\mu_N\rangle + \ldots$$
$$+ |\mu_1\rangle|\mu_2\rangle\ldots\left[s_z^{(N)}|\mu_N\rangle\right]$$
$$= \left(\sum_{m=1}^{N} \mu_m\right)|\mu_1\rangle|\mu_2\rangle\ldots|\mu_N\rangle. \qquad (3.371)$$

Thus, in an N-particle system in which each particle is in a definite spin state, the z component of the total spin of the system is the sum of the z components of the spins of the individual particles.

Not all states are separable. Consider, for example, the two-particle state

$$|\psi(t)\rangle = \frac{1}{\sqrt{2}}\left(|\varphi_1(t)\rangle|\psi_2(t)\rangle + |\psi_1(t)\rangle|\varphi_2(t)\rangle\right). \qquad (3.372)$$

For $|\varphi(t)\rangle \neq |\psi(t)\rangle$ this state cannot be written as a product state. This means that neither of the particles is entirely in the state $|\varphi(t)\rangle$ or in the state $|\psi(t)\rangle$. The two particles form a compound, which we call an *entangled state*. An entangled state is a purely quantum feature. There is not even a conceptual analogue of it in the (fictitious) world of classical particles.

Exercises:

3.7.1 Are the following states separable or entangled? "Separable" means that we can write the states as product states. "Entangled" means that we cannot write the states as product states.

(a) $|\psi(t)\rangle = |\varphi_1(t)\rangle|\alpha_2(t)\rangle + 3|\varphi_1(t)\rangle|\beta_2(t)\rangle$,

(b) $|\psi(t)\rangle = |\varphi_1(t)\rangle|\beta_2(t)\rangle + |\psi_1(t)\rangle|\alpha_2(t)\rangle$,

(c) $|\psi(t)\rangle = |\varphi_1(t)\rangle|\alpha_2(t)\rangle - 2|\psi_1(t)\rangle|\beta_2(t)\rangle - |\varphi_1(t)\rangle|\beta_2(t)\rangle + 2|\psi_1(t)\rangle|\alpha_2(t)\rangle$,

(d) $|\psi(t)\rangle = |\psi_1(t)\rangle|\alpha_2(t)\rangle + |\varphi_1(t)\rangle|\alpha_2(t)\rangle + 2|\varphi_1(t)\rangle|\beta_2(t)\rangle + 3|\psi_1(t)\rangle|\beta_2(t)\rangle$.

3.7.2 Show that the three Pauli matrices σ_x, σ_y, and σ_z satisfy:

(a) Equation 3.318 and

(b)
$$\{\sigma_x, \sigma_y\} = \{\sigma_y, \sigma_z\} = \{\sigma_z, \sigma_x\} = 0,$$

where
$$\{\hat{A}, \hat{B}\} = \hat{A}\hat{B} + \hat{B}\hat{A}$$

is the *anti-commutator* of the operators $\hat{A}$ and $\hat{B}$.

3.7.3 While, according to Equation 3.304, the operator $\hat{\sigma}_z$ leaves the spin state of an electron invariant (up to a sign), the operators $\hat{\sigma}_x$ and $\hat{\sigma}_y$ may be used to change the spin state of an electron. Show that

$$\hat{\sigma}_x|\uparrow\rangle = |\downarrow\rangle, \quad \hat{\sigma}_x|\downarrow\rangle = |\uparrow\rangle,$$
$$\hat{\sigma}_y|\uparrow\rangle = i|\downarrow\rangle, \quad \hat{\sigma}_y|\downarrow\rangle = -i|\uparrow\rangle.$$

3.7.4 According to Equation 3.316 any 2×2 matrix Ω can be written as a linear combination of the three Pauli matrices σ_x, σ_y, σ_z, and the unit matrix **1**. Given

$$\Omega = \begin{pmatrix} \Omega_{\uparrow\uparrow} & \Omega_{\uparrow\downarrow} \\ \Omega_{\downarrow\uparrow} & \Omega_{\downarrow\downarrow} \end{pmatrix},$$

compute μ, λ, ν, and γ.

3.7.5 An electron is at rest in a magnetic field of $B = 1\,\text{T}$, oriented in the z-direction.

(a) What is the energy splitting E_0 between the states $|\downarrow\rangle$ and $|\uparrow\rangle$?

(b) What is the resonance frequency ω_0?

3.7.6 Prove Equation 3.320.

3.7.7 Prove Equation 3.361.

3.8 Summary

There are two layers of quantum mechanics: The conceptual and the machinery. In Chapters 1 and 2 we encountered many baffling quantum effects and learned how to interpret them with the help of the three fundamental rules of quantum mechanics: Feynman's Rule, Born's Rule, and the Composition Rule. While these three rules are sufficient to explain many quantum effects qualitatively, we need the machinery of quantum mechanics to make quantitative predictions. The most powerful tool in the quantum shed is Schrödinger's equation (see Section 3.2). Given a quantum system, Schrödinger's theory allows us to formulate an equation for the amplitude of the system. Solving Schrödinger's equation, we obtain this amplitude explicitly, and with it possess the maximal information about the system allowed by quantum mechanics. Since all the information about the system is encoded in its wave function, the natural question to ask is how to extract this information and how to predict the values of physical observables such as position and momentum on the basis of the wave function. This question is answered in Section 3.3, where we connect the formal, mathematical machinery of quantum mechanics to the real world. This is accomplished by assigning Hermitian operators to physical observables. The operators may then be used to "operate" on the wave function to extract physical system information. In Section 3.4 we studied several examples of how to do this in practice. We also introduced and studied the important concept of the spectrum of a quantum system. We learned that the spectrum is computed by solving certain eigenvalue equations that directly derive from Schrödinger's equation. The model systems we studied in Section 3.4 were carefully selected to exemplify the most important types of spectra encountered in real-life quantum systems. In Section 3.5 we took a closer look at Dirac's notation already introduced in Chapter 2. Not only did we develop it into a handy tool to perform actual quantum calculations, we also encountered many parallels to ordinary linear algebra, which made Dirac's notation more palatable, since it suggests that Dirac's notation is nothing but a convenient adaptation of linear algebra to the specific needs of quantum mechanics. In Section 3.6 we made extensive use of Dirac's notation when we discussed Heisenberg's formulation of quantum mechanics. We also formulated some new operator-based techniques for the solution of Schrödinger's equation. The most useful one is the time-evolution operator $\hat{U}(t)$, which also plays a central role in the transition from Schrödinger's formulation to Heisenberg's formulation of quantum mechanics. In Section 3.7, in order to prepare the ground for quantum computing, we studied quantum two-level systems. The central point here was to show that we have complete control over two-level systems. This means that if we have a physical means to induce transitions from the ground state to the excited state of a two-level system, for instance a magnetic field or a laser, then we are able to produce any two-level quantum state whatsoever, at will. This is a fundamental enabling technology for the construction of quantum computers.

Chapter Review Exercises:

1. Show that the expectation value $\bar{\theta}$ of an operator $\hat{\theta}$ computed with the wave function $\psi(\mathbf{r}, t)$ satisfies

$$\frac{d}{dt}\bar{\theta} = \frac{i}{\hbar}\bar{C},$$

where $\bar{C}$ is the expectation value of the commutator

$$\hat{C} = [\hat{\mathcal{H}}, \hat{\theta}]$$

computed with the same wave function $\psi(\mathbf{r}, t)$, and $\hat{\mathcal{H}}$ is the Hamiltonian operator.

2. Show that the spin state

$$|\psi\rangle = |\uparrow_1\rangle|\uparrow_2\rangle$$

of a two-electron system is an eigenstate of

$$\hat{s}^2 = \left[\hat{\mathbf{s}}^{(1)} + \hat{\mathbf{s}}^{(2)}\right]^2$$

with eigenvalue $2\hbar^2$.

3. Prove the operator identity

$$[\hat{x}_j, \hat{p}_l] = i\hbar\delta_{jl}, \quad j,l = 1,2,3,$$

where $x_1 \equiv x$, $x_2 \equiv y$, $x_3 \equiv z$, $p_1 \equiv p_x$, $p_2 \equiv p_y$, $p_3 \equiv p_z$.

4. Use only the operator identities $[\hat{p}, \hat{\mathbf{1}}] = 0$ and $[\hat{p}, \hat{x}] = -i\hbar$ to prove that

$$[\hat{p}, \hat{x}^n] = \begin{cases} 0, & \text{for } n = 0, \\ -i\hbar n \hat{x}^{n-1}, & \text{for } n \geq 1. \end{cases}$$

Use this result to show that

$$[\hat{p}, f(\hat{x})] = -i\hbar f'(\hat{x}).$$

Assume that f has a Taylor series expansion.

5. For $V_0 > 0$ and $a > 0$ compute the spectrum of a quantum particle of mass m moving in the potential $V(x) = \infty$ for $x \leq -a$, $V(x) = V_0 \delta(x)$ for $-a < x < a$, and $V(x) = \infty$ for $x \geq a$. Since the potential $V_0 \delta(x)$ is symmetric with respect to $x = 0$, there are two classes of stationary states: symmetric and anti-symmetric. The symmetric states, denoted by $\psi_n^{(+)}(x)$, have energy $E_n^{(+)}$, and satisfy $\psi_n^{(+)}(-x) = \psi_n^{(+)}(x)$. The anti-symmetric states, denoted by $\psi_n^{(-)}(x)$, have energy $E_n^{(-)}$, and satisfy $\psi_n^{(-)}(-x) = -\psi_n^{(-)}(x)$. We call $\{E_n^{(+)}\}$ the even-parity spectrum and $\{E_n^{(-)}\}$ the odd-parity spectrum.

   (a) Compute the odd-parity spectrum $\{E_n^{(-)}\}$ analytically.

   (b) Derive the spectral equation that determines the even-parity spectrum $\{E_n^{(+)}\}$. The spectral equation for $\{E_n^{(+)}\}$ cannot be solved algebraically, but may still be solved approximately. Assuming $\lambda = mV_0 a/\hbar^2 \ll 1$, compute an approximate, analytical solution of the spectral equation for $E_n^{(+)}$ and use it to derive an explicit, approximate expression for $E_n^{(+)}$.

6. A quantum particle of mass m and energy E is incident from the left on a matter-wave interferometer modeled by the potential

$$V(x) = V_0 \delta(x) + W_0 \delta(x - a), \quad V_0, W_0, a > 0.$$

   (a) Compute reflection and transmission amplitudes r and t, respectively. Express your results with the help of the two dimensionless parameters

$$\alpha = \frac{mV_0}{\hbar^2 k}, \quad \beta = \frac{mW_0}{\hbar^2 k},$$

   where $k = \sqrt{2mE/\hbar^2}$ is the wave number.

   (b) Check the unitarity of your results, i.e., make sure that $|r|^2 + |t|^2 = 1$.

   (c) For given a, V_0, and W_0, and in the special case $\alpha = \beta$, compute the transmission resonances of the matter-wave interferometer. Transmission resonances are energy values E_n, $n = 1, 2, \ldots$ of the quantum particle that result in perfect transmission of the particle through the matter-wave interferometer, i.e., $r = 0$ or, equivalently, $|t| = 1$.

7. A spin-up particle is incident from the left on the spin-dependent δ-function scattering potential

$$\hat{V}(x) = V_0 \hat{\sigma}_x \delta(x).$$

(a) Compute the reflection amplitudes $r_\uparrow$ and $r_\downarrow$ and the transmission amplitudes $t_\uparrow$ and $t_\downarrow$ for the spin-up and spin-down components of the scattered wave function.

(b) Check that the absolute squares of the scattering amplitudes $r_\uparrow$, $r_\downarrow$, $t_\uparrow$, and $t_\downarrow$ add up to 1 (unitarity check).

8. In the text we solved the system of equations 3.340 on resonance, i.e., for $\omega = \omega_0$. Solve the system of equations 3.340 for the general case $\omega \neq \omega_0$.

9. Use Equation 3.318 to show that

$$(\boldsymbol{a} \cdot \boldsymbol{\sigma})(\boldsymbol{b} \cdot \boldsymbol{\sigma}) = \boldsymbol{a} \cdot \boldsymbol{b} + i(\boldsymbol{a} \times \boldsymbol{b}) \cdot \boldsymbol{\sigma},$$

where $\boldsymbol{a}$ and $\boldsymbol{b}$ are vectors whose components commute with each other and with the Pauli matrices i.e., $[a_m, b_n] = 0$, $[a_m, \sigma_n] = 0$, $[b_m, \sigma_n] = 0$ for all $m, n \in \{x, y, z\}$. The symbol $\times$ denotes the usual cross product (outer product, vector product) of two vectors.

chapter 4

Measurement

In this chapter:

- ◆ Introduction
- ◆ von Neumann Measurement
- ◆ Uncertainty Principle
- ◆ No-Cloning Theorem
- ◆ Quantum Zeno Effect
- ◆ Summary

4.1 Introduction

Typically, as shown in Figure 4.1, a quantum mechanical experiment unfolds in three stages. (1) A quantum state $|\psi(t_i)\rangle$ is prepared at time $t = t_i$, (2) evolves under the influence of a Hamiltonian operator, and (3) is *measured* at time $t = t_f$, i.e., its structure and composition are analyzed. Measurement of $|\psi(t_f)\rangle$ means that part or all of its information content is communicated to macroscopic observers, i.e., us. Thus,

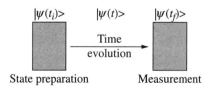

Figure 4.1 Key stages of a quantum mechanical experiment.

measurement provides the interface between the micro-world, governed by the laws of quantum mechanics, and the macro-world, effectively governed by the laws of classical physics. Here the word "effectively" has been chosen deliberately and indicates trouble. For all we know, there is no magic object size that separates the micro-world, strictly governed by quantum mechanics, from our macro-world that appears to be governed by classical physics. Our world, in its entirety, micro-world and macro-world, is governed by quantum mechanics. Cars follow the laws of quantum mechanics just like electrons do. But if this is the case, how can "measurement" ever occur? How can we ever stand outside the quantum world and make objective measurements on quantum objects when we, ourselves, are part of the quantum world, and therefore part of the system? If we cannot stand outside of the system, if we are part of the system, it is hard, if not impossible, to tell where the observer stops and the observed object begins! This conundrum is one of the central puzzles of quantum measurement theory, which has been debated without conclusive results since the mid 1920s when Heisenberg, Schrödinger, and many other scientists formulated the foundations of modern quantum theory. Among others there are two prominent schools of thought. One group of scientists agrees that the whole world is quantum mechanical, and placing ourselves outside of this quantum world is impossible. We are quantum, the measurement equipment is quantum, and the system to be "observed" follows quantum mechanical laws as well. Thus, measurement has to happen *inside* of the quantum world. According to this point of view, an approximate separation between the observed object and the observer is then accomplished by various (self) organization mechanisms that conceptually work according to the principle that "more is different." To be more explicit: A tiny quantum object, such as an electron, may be prepared in a pure quantum state and shielded from external perturbations such as radio-frequency noise and thermal radiation to an extent that, over a given time interval $\Delta t = t_f - t_i$, this particle may be considered isolated to an excellent and *controllable* approximation. While even in the presence of noise the electron would follow the quantum mechanical laws, getting rid of the noise has the effect that the electron's time evolution may be described by a simple Hamiltonian operator that ignores all extraneous influences and focuses only on the electron's spin and space degrees of freedom. Thus, the electron's quantum properties are emphasized. The measuring apparatus, on the other hand, is large, usually consists of more than 10^{20} particles, and noise–mostly produced by the apparatus itself– is very hard to control. Thus, by its sheer size, the measuring apparatus deemphasizes quantum aspects and appears "classical." Therefore, while still fundamentally quantum mechanical in nature, the measurement apparatus emulates a classical piece of machinery to an excellent degree. This way "observers," who are manifestly part of the system, manage to introduce a separation between themselves and the observed. This view of measurement is conceptually elegant, since quantum mechanics is all we need, and the process of measurement is contained within the framework of quantum mechanics. In practice, however, it is very difficult to explain quantum measurement within quantum mechanics itself, and debate continues as to whether this is possible at all.

Starting with John von Neumann, another school of thought holds that the act of measurement in quantum mechanics is strictly outside of quantum mechanics, and cannot be explained by quantum mechanics at all. In more concrete terms: Measurement requires a separate *axiom of measurement* in addition to the formalism of quantum mechanics. In von Neumann's view a quantum particle evolves according to the *deterministic* Schrödinger equation when it is not observed, and is forced *nondeterministically* into one of the possible measurement states of the measurement apparatus when a measurement occurs. In this view, as a result of the measurement, the state $|\psi(t_f)\rangle$ is usually destroyed and replaced by one of the measurement states $|\psi_n\rangle$ that the measurement apparatus offers the quantum particle to be measured to jump into. The destruction of $|\psi(t_f)\rangle$ by the measurement apparatus is referred to as the *collapse of the wave function*. Von Neumann's view of measurement, in particular since it invokes the "collapse of the wave function," may seem conceptually less appealing than the idea of measurement as a dynamical process within quantum mechanics itself. However, von Neumann's theory of measurement works in practice and is technically straightforward. Moreover, even if eventually the act of measurement can be explained dynamically within quantum mechanics itself, von Neumann's theory will never be outdated, since it may be viewed as an effective, phenomenological theory that works for all practical purposes. For this reason we will develop the theory of measurement in Section 4.2 on the basis of von Neumann's theory. In Section 4.3 we discuss the fact that unlike in classical mechanics, where we may measure all the dynamical variables of a classical particle simultaneously to any prespecified accuracy at any instant in time, in quantum mechanics this is not possible. For instance, if we know the position of a quantum particle accurately, we know nothing about its momentum—in principle—not because we made a sloppy measurement. Heisenberg called this property of quantum mechanics the *uncertainty principle*. In Section 4.4 we prove the remarkable fact that it is impossible to make copies of an unknown quantum state. Since, because of the statistical nature of the quantum measurement process, we need many copies of an unknown quantum state to determine its properties, the impossibility of making such copies prohibits us from ever learning anything about this unknown state. The no-cloning theorem is one of the most striking quantum effects that has absolutely no counterpart in classical physics, where it is always possible to make any number of copies of any given, classical system. In Section 4.5 we encounter another striking quantum effect connected with quantum measurements: By repeatedly measuring a quantum system we can slow down, or even halt, its time evolution! This reminds us of the famous paradoxes constructed by Zeno more than two millennia ago that purport to show that motion is impossible. In Zeno's honor the corresponding quantum effect is called the *quantum Zeno effect*.

4.2 von Neumann Measurement

We already encountered measurements in Chapter 2 when we recorded photons on the observation screen Ω, which serves as a position-sensitive measurement device. In this section, following von Neumann's theory of measurement, we will look more closely at what can be measured and how measurements come about.

As discussed in Section 3.3, measurable quantities in quantum mechanics are called observables. Each observable θ is assigned a Hermitian operator $\hat{\theta}$ whose eigenvalues and eigenstates play a fundamental role in the theory of quantum measurement. In analogy to the energy spectrum of the Hamiltonian $\hat{\mathcal{H}}$, introduced in Section 3.4, we call the set of eigenvalues of $\hat{\theta}$ the spectrum of $\hat{\theta}$. Since $\hat{\theta}$ is Hermitian, its spectrum is real (see Equation 3.262). To simplify our discussion, we assume that (1) the spectrum of $\hat{\theta}$ is discrete and countable, i.e.,

$$\hat{\theta}|\vartheta_n\rangle = \vartheta_n|\vartheta_n\rangle, \quad n = 1, 2, \ldots, \qquad (4.1)$$

(2) the eigenstates of $\hat{\theta}$ are normalizable and normalized to 1, i.e.,

$$\langle\vartheta_n|\vartheta_n\rangle = 1, \qquad (4.2)$$

and (3) the spectrum of $\hat{\theta}$ is nondegenerate, i.e.,

$$\vartheta_n \neq \vartheta_m, \quad \text{for} \quad n \neq m. \qquad (4.3)$$

In this case, as guaranteed by Equation 3.262, the eigenstates of $\hat{\theta}$ are orthonormal, i.e.,

$$\langle\vartheta_n|\vartheta_m\rangle = \delta_{nm}. \qquad (4.4)$$

In addition, we assume that the eigenstates of $\hat{\theta}$ span the Hilbert space, i.e.,

$$\hat{1} = \sum_{n=1}^{\infty} |\vartheta_n\rangle\langle\vartheta_n|. \qquad (4.5)$$

4.2 • VON NEUMANN MEASUREMENT

Because of Equation 4.5, we may expand a given, normalized state $|\psi\rangle$, $\langle\psi|\psi\rangle = 1$, into the states $|\vartheta_n\rangle$ according to

$$|\psi\rangle = \hat{\mathbf{1}}|\psi\rangle = \sum_{n=1}^{\infty} |\vartheta_n\rangle\langle\vartheta_n|\psi\rangle = A_n|\vartheta_n\rangle, \qquad (4.6)$$

where

$$A_n = \langle\vartheta_n|\psi\rangle. \qquad (4.7)$$

According to von Neumann, a measurement of the observable θ in the state $|\psi\rangle$ *collapses* the state $|\psi\rangle$ into one of the eigenstates $|\vartheta_n\rangle$ of $\hat{\theta}$, where the state $|\vartheta_n\rangle$ is selected *at random* with probability

$$P_n = |A_n|^2 = |\langle\vartheta_n|\psi\rangle|^2. \qquad (4.8)$$

Thus, according to von Neumann, the collapse of the state $|\psi\rangle$ into $|\vartheta_n\rangle$ is a probabilistic process that does not follow any dynamical, deterministic rules. In summary, according to von Neumann, predicting the outcome of a quantum measurement proceeds along the following three steps:

1. Determine the operator $\hat{\theta}$ that corresponds to the observable quantity θ.

2. Determine the eigenvalues ϑ_n and eigenstates $|\vartheta_n\rangle$ of $\hat{\theta}$.

3. Determine the probability P_n of obtaining ϑ_n as the result of a θ measurement according to Equation 4.8.

Let us illustrate these three steps with the help of an example. Assume that the spin of an electron is in the state

$$|\psi\rangle = |\uparrow\rangle, \qquad (4.9)$$

and we would like to measure the z component of the electron's spin. According to the three-step program above, we first have to find the operator that corresponds to the observable s_z of the z component of spin. This is $\hat{s}_z$. Then, we have to find the eigenvalues and eigenvectors of $\hat{s}_z$. According to Equation 3.301 we have:

$$\hat{s}_z|\uparrow\rangle = +\frac{\hbar}{2}|\uparrow\rangle, \quad \hat{s}_z|\downarrow\rangle = -\frac{\hbar}{2}|\downarrow\rangle. \qquad (4.10)$$

Thus, the eigenvalues of $\hat{s}_z$, i.e., the possible outcomes of an s_z measurement on $|\psi\rangle$, are $\pm\hbar/2$. According to the final step of the three-step program we now compute the probabilities

$$\begin{aligned} P_\uparrow &= |\langle\uparrow|\psi\rangle|^2 = 1, \\ P_\downarrow &= |\langle\downarrow|\psi\rangle|^2 = 0 \end{aligned} \qquad (4.11)$$

for measuring $s_z = \hbar/2$ and $s_z = -\hbar/2$, respectively. According to Equation 4.11, there is no chance of obtaining $s_z = -\hbar/2$ as a result of s_z measurement. Making *repeated measurements* on electrons prepared in $|\psi\rangle = |\uparrow\rangle$, always yields the result $s_z = \hbar/2$. We say that the value of s_z is *sharp* in the state $|\psi\rangle = |\uparrow\rangle$. We also notice that since we never measure the value $s_z = |\downarrow\rangle$, the state $|\psi\rangle = |\uparrow\rangle$ never collapses into the state $|\downarrow\rangle$ after the measurement is completed. This means that after the measurement, we always have $|\psi'\rangle = |\psi\rangle = |\uparrow\rangle$, i.e., the measurement leaves the quantum state of the electron invariant. We call this type of measurement a *nondestructive measurement*.

Having prepared our electron in $|\psi\rangle = |\uparrow\rangle$, we could equally well have decided to measure the x component of the spin, i.e., the observable s_x. In this case the operator corresponding to the observable s_x is $\hat{s}_x$. With Equations 3.303 and 3.314 we have:

$$\hat{s}_x |\uparrow\rangle = \frac{\hbar}{2} |\downarrow\rangle,$$
$$\hat{s}_x |\downarrow\rangle = \frac{\hbar}{2} |\uparrow\rangle. \qquad (4.12)$$

Adding both equations and using the linearity of $\hat{s}_x$ yields:

$$\hat{s}_x \left(|\uparrow\rangle + |\downarrow\rangle \right) = \frac{\hbar}{2} \left(|\uparrow\rangle + |\downarrow\rangle \right). \qquad (4.13)$$

This means that

$$|+\rangle = \frac{1}{\sqrt{2}} \left(|\uparrow\rangle + |\downarrow\rangle \right) \qquad (4.14)$$

is a normalized eigenstate of $\hat{s}_x$ with eigenvalue $\hbar/2$. Similarly, subtracting the second equation from the first in Equation 4.12 produces

$$|-\rangle = \frac{1}{\sqrt{2}} \left(|\uparrow\rangle - |\downarrow\rangle \right) \qquad (4.15)$$

as a normalized eigenstate of $\hat{s}_x$ with eigenvalue $-\hbar/2$. Since the two eigenvalues are different, the states $|\pm\rangle$ are orthogonal. Expanding $|\psi\rangle = |\uparrow\rangle$ into the states $|\pm\rangle$, we obtain

$$|\psi\rangle = \frac{1}{\sqrt{2}} |+\rangle + \frac{1}{\sqrt{2}} |-\rangle. \qquad (4.16)$$

Thus, the two possible results of an s_x measurement of $|\psi\rangle$ are $\pm\hbar/2$. According to Equation 4.8 the probability to obtain the result $+\hbar/2$ is

$$P_+ = |\langle +|\psi\rangle|^2 = \frac{1}{2}, \qquad (4.17)$$

and the probability to obtain $-\hbar/2$ is

$$P_- = |\langle -|\psi\rangle|^2 = \frac{1}{2}. \qquad (4.18)$$

We see that s_x measurement of $|\psi\rangle$ is qualitatively different from s_z measurement of $|\psi\rangle$. Instead of obtaining a single, sharp result, there is the possibility of obtaining two different results, $+\hbar/2$ and $-\hbar/2$. Moreover, according to Equation 4.18 we obtain both results with equal probability. In concrete terms this means the following: Preparing an electron in $|\psi\rangle = |\uparrow\rangle$ and measuring s_x we (1) have no way of predicting which of the two possible results, $\pm\hbar/2$, we will obtain, i.e., we will obtain *one* of these two possibilities *at random*, and (2) repeated measurements on identically prepared electrons in the quantum state $|\psi\rangle = |\uparrow\rangle$ will yield each of the results $+\hbar/2$ or $-\hbar/2$ in 50% of the measurements. Thus, since repeated s_x measurements of the same state $|\psi\rangle = |\uparrow\rangle$ yield different results with nonzero probabilities, the value of s_x in $|\psi\rangle = |\uparrow\rangle$ is *not sharp*.

Suppose that as a result of s_x measurement of $|\psi\rangle = |\uparrow\rangle$ we obtain $\hbar/2$. Then, the state after the measurement is $|\psi'\rangle = |+\rangle$; if we obtain $-\hbar/2$, we have $|\psi'\rangle = |-\rangle$. Thus, s_x measurement of $|\psi\rangle = |\uparrow\rangle$ is an example of $|\psi'\rangle \neq |\psi\rangle$. This means that s_x measurement destroys the state $|\psi\rangle$ and replaces it with a substantially altered state, i.e., $|+\rangle$ or $|-\rangle$. We call this type of measurement a *destructive measurement*.

Suppose an observable θ has a sharp value in a given state $|\psi\rangle$. Since the possible results of θ measurements are the eigenvalues of $\hat\theta$, this sharp value, necessarily, must be one of the eigenvalues of $\hat\theta$. Therefore, let us assume that θ measurement of $|\psi\rangle$ yields the value $\theta = \vartheta_N$, and ϑ_N, therefore, is sharp in $|\psi\rangle$. As a consequence this means that repeated measurements of $|\psi\rangle$ always collapse $|\psi\rangle$ into the same eigenstate $|\vartheta_N\rangle$ of $\hat\theta$, i.e., $|\psi'\rangle = |\vartheta_N\rangle$. If $|\psi\rangle$ had a nonvanishing component $|\vartheta_n\rangle$ with $n \neq N$, then $\langle\vartheta_n|\psi\rangle \neq 0$ and there would be an associated nonzero probability $P_n = |\langle\vartheta_n|\psi\rangle|^2 \neq 0$ for collapse of $|\psi\rangle$ into $|\vartheta_n\rangle \neq |\vartheta_N\rangle$. But since ϑ_N is sharp, i.e., ϑ_n, $n \neq N$, is not possible as a result of θ measurement, we have $P_n = 0$ for all $n \neq N$, which means that $|\psi\rangle$ cannot have any other components but $|\vartheta_N\rangle$. This means that $|\psi\rangle$ is identical with $|\vartheta_N\rangle$, i.e., $|\psi\rangle$ is an eigenstate of $\hat\theta$. Since this logic may be applied to any situation in which an observable θ has a sharp value in a state $|\psi\rangle$, we conclude that

> if the observable θ has a sharp value ϑ_N in $|\psi\rangle$,
> then $|\psi\rangle$ is an eigenstate $|\vartheta_N\rangle$ of $\hat\theta$ with eigenvalue ϑ_N. $\qquad (4.19)$

In this case

$$|\psi'\rangle = |\psi\rangle \qquad (4.20)$$

and the measurement is nondestructive.

A natural question to ask is whether the values of two observables, A and B, may be simultaneously sharp in $|\psi\rangle$. Since the value of A is sharp in $|\psi\rangle$ if $|\psi\rangle$ is an eigenstate of $\hat{A}$, and B is sharp in $|\psi\rangle$ if $|\psi\rangle$ is an eigenstate of $\hat{B}$, in order for both to be sharp, $|\psi\rangle$ must be a simultaneous eigenstate of $\hat{A}$ and $\hat{B}$. Assume that the value of A in $|\psi\rangle$ is a and the value of B in $|\psi\rangle$ is b. Then:

$$\hat{A}|\psi\rangle = a|\psi\rangle,$$
$$\hat{B}|\psi\rangle = b|\psi\rangle, \qquad (4.21)$$

and

$$\hat{A}\hat{B}|\psi\rangle = ab|\psi\rangle = ba|\psi\rangle = \hat{B}\hat{A}|\psi\rangle$$
$$\Rightarrow [\hat{A}, \hat{B}]|\psi\rangle = 0, \qquad (4.22)$$

i.e., the commutator of $\hat{A}$ and $\hat{B}$ vanishes in the state $|\psi\rangle$. In this case we say that the observables A and B are *compatible in the state* $|\psi\rangle$. If $[\hat{A}, \hat{B}] = 0$, always, independently of a particular state $|\psi\rangle$, we call A and B *compatible*. On the other hand, if the commutator of $\hat{A}$ and $\hat{B}$ never vanishes, i.e., $[\hat{A}, \hat{B}] \neq 0$ in general, then the observables A and B cannot be simultaneously sharp in any state. In this case we call A and B *incompatible*. For instance, since $[\hat{x}, \hat{p}] = i\hbar \neq 0$, position and momentum are *incompatible observables*. It follows that x and p cannot have simultaneously sharp values in any quantum state $|\psi\rangle$. This observation is the foundation for Heisenberg's Uncertainty Principle to be discussed in Section 4.3.

For $|\psi\rangle$ to be an eigenstate of an observable θ is obviously a very special situation. For example, there are only two eigenstates of $\hat{s}_z$, $|\uparrow\rangle$ and $|\downarrow\rangle$, but there are infinitely many superposition states $|\psi\rangle = \alpha|\uparrow\rangle + \beta|\downarrow\rangle$, $\alpha, \beta \neq 0$, $|\alpha|^2 + |\beta|^2 = 1$, which are not eigenstates of $\hat{s}_z$. Therefore, most measurements in quantum mechanics will result in $|\psi'\rangle \neq |\psi\rangle$, i.e., most measurements in quantum mechanics are destructive.

The generally destructive nature of measurements in quantum mechanics distinguishes quantum measurements sharply from classical measurements. In classical mechanics we always assume that measurement has a negligible influence on a classical system and that the disturbance of a classical system due to measurement can be made arbitrarily small. This is not the case in quantum mechanics where the effect of measurement is large and usually destroys the quantum state to be measured.

This is illustrated by the s_x measurements of $|\psi\rangle = |\uparrow\rangle$ discussed previously. After the measurement there are only two possibilities for $|\psi\rangle$ to collapse into, $|+\rangle$ and $|-\rangle$, and neither of these two possibilities is a state that would in any sense be "close" to $|\psi\rangle = |\uparrow\rangle$. This has important consequences for the meaning of *reality* in classical and quantum mechanics to be discussed in further detail in Chapter 6. Since measurement in classical mechanics is a perturbation that can be minimized, we may associate an *objective reality* with a classical system. Thus, a classical measurement merely *discovers* a preexisting, objective reality that is independent of the observer(s) and the measurement equipment, while a quantum measurement, drastically changing the quantum state, *creates* reality. Thus, measurement plays a fundamentally different role in classical and quantum mechanics: Passive and nonintrusive in classical mechanics, active and creative in quantum mechanics.

The active role of measurement in quantum mechanics is illustrated with *chains of measurements*. In contrast with repeated measurements of a given state $|\psi\rangle$ discussed previously, a chain of measurements results when we subject the same quantum object to a chain of sequential measurements. For example, if A and B are two observables, we may be interested in the iterative chain of measurements

$$|\psi^{(0)}\rangle \xrightarrow{A} |\psi^{(1)}\rangle \xrightarrow{B} |\psi^{(2)}\rangle \xrightarrow{A} |\psi^{(3)}\rangle \xrightarrow{B} |\psi^{(4)}\rangle \ldots, \quad (4.23)$$

where an A measurement alternates with a B measurement. To be specific, let us choose $|\psi^{(0)}\rangle = |\uparrow\rangle$, $A = s_x$, and $B = s_z$. Then, as discussed previously, the first measurement, i.e., an s_x, measurement, may result in $|\psi^{(1)}\rangle = |+\rangle$ or $|\psi^{(1)}\rangle = |-\rangle$. We do not know which, and either one of these two possibilities may be realized with a probability of 1/2. Since both $|+\rangle$ and $|-\rangle$ are superposition states of $|\uparrow\rangle$ and $|\downarrow\rangle$ with equal weights (up to a sign), the second measurement, a s_z measurement, will result in $|\psi^{(2)}\rangle = |\uparrow\rangle$ or $|\psi^{(2)}\rangle = |\downarrow\rangle$. Again, we do not know which, and either of these two possibilities will be realized with a probability of 1/2. Continuing this sequence of measurements, we generate a *random string* of states, for instance,

$$|\psi^{(1)}\rangle, |\psi^{(2)}\rangle, |\psi^{(3)}\rangle, |\psi^{(4)}\rangle, |\psi^{(5)}\rangle, \ldots = |-\rangle, |\uparrow\rangle, |-\rangle, |\downarrow\rangle, |+\rangle, \ldots . \quad (4.24)$$

Whenever a s_x or s_z measurement occurs, and one of two states will be chosen as the result of the corresponding measurement, there is no way of predicting which one of the two alternatives will actually occur. The realization of the measurement, i.e., the actual choice made, is as random as a coin toss. In fact, it is "more random" than a coin toss, since by using the classical equations of motion the result of a coin toss may, in principle, be predicted. Not so with the quantum choices. Since we lack a dynamical theory of the measurement process, the results of generic measurements are *truly random*, a kind of randomness that is also called *irreducibly random*. Thus, starting from a single state, in our case the state $|\uparrow\rangle$, an infinite string of irreducibly random states can be generated. If we identify $|\uparrow\rangle$ and $|+\rangle$ with the binary digit "1," and $|\downarrow\rangle$ and $|-\rangle$ with the binary digit "0," our s_x, s_z chain of measurements is a perfect random number generator.

Since each measurement outcome involves the choice between two alternatives, i.e., between $|\uparrow\rangle$ and $|\downarrow\rangle$ and $|+\rangle$ and $|-\rangle$, using the language of information theory, one bit of information is revealed in each act of measurement. Continuing the string of measurements *ad infinitum*, we see that the simple starting state $|\uparrow\rangle$ contains an infinite amount of information that is revealed to us by consecutive measurements. We will see in Chapter 6 that this information could not possibly have been encoded in the starting state with the help of hidden "gears and wheels" or "genes" that determine the outcome of each successive measurement. Thus, this information could have been introduced only in the act of measurement, emphasizing the active, information generating role of measurement in quantum theory.

Exercises:

4.2.1 Show that the measurement probabilities P_n as defined in Equation 4.8 satisfy

$$\sum_{n=1}^{\infty} P_n = 1.$$

4.2.2 With respect to the z-axis as quantization axis, an electron is in the spin state $|\uparrow\rangle$. You measure the electron's spin with respect to a new quantization axis defined by the unit vector

$$\boldsymbol{e} = \begin{pmatrix} 3/5 \\ 0 \\ 4/5 \end{pmatrix}.$$

What is the probability that your measurement finds the electron's spin oriented in the $+\boldsymbol{e}$ direction?

4.2.3 Assume that the spin of an electron is in the superposition state

$$|\psi\rangle = \alpha|\uparrow\rangle + \beta|\downarrow\rangle.$$

(a) Determine $P_\uparrow$ and $P_\downarrow$ for an s_z measurement on $|\psi\rangle$.

(b) Determine P_+ and P_- for an s_x measurement on $|\psi\rangle$.

4.2.4 Starting with the state $|\psi^{(0)}\rangle = |\uparrow\rangle$ consider two (finite) chains of measurements:

$$\text{Chain I:} \quad |\psi^{(0)}\rangle \xrightarrow{s_z} |\psi^{(1)}\rangle, \quad |\psi^{(1)}\rangle \xrightarrow{s_x} |\psi^{(2)}\rangle,$$
$$\text{Chain II:} \quad |\psi^{(0)}\rangle \xrightarrow{s_x} |\psi^{(1)}\rangle, \quad |\psi^{(1)}\rangle \xrightarrow{s_z} |\psi^{(2)}\rangle.$$

(a) Compute and list all possible results for $|\psi^{(2)}\rangle$ according to Chain I.

(b) Compute and list all possible results for $|\psi^{(2)}\rangle$ according to Chain II.

(c) Comparing the results for $|\psi^{(2)}\rangle$ you obtained for Chain I and Chain II in (a) and (b), respectively, show that the possible results for $|\psi^{(2)}\rangle$ according to Chain I are always different from the possible results for $|\psi^{(2)}\rangle$ according to Chain II. This shows that successive quantum measurements do not usually commute.

4.3 Uncertainty Principle

Many operators of practical importance in physics satisfy the commutator relation

$$[\hat{A}, \hat{B}] = i\hat{C}, \tag{4.25}$$

where $\hat{A}$, $\hat{B}$, and $\hat{C}$ are Hermitian operators. Examples are

$$[\hat{x}, \hat{p}] = i\hbar \hat{\mathbf{1}} \tag{4.26}$$

($\hat{A} = \hat{x}$, $\hat{B} = \hat{p}$, $\hat{C} = \hbar\hat{\mathbf{1}}$),

$$[\hat{s}_x, \hat{s}_y] = i\hbar \hat{s}_z \tag{4.27}$$

($\hat{A} = \hat{s}_x$, $\hat{B} = \hat{s}_y$, $\hat{C} = \hbar\hat{s}_z$), and

$$[\hat{L}_x, \hat{L}_y] = i\hbar \hat{L}_z \tag{4.28}$$

($\hat{A} = \hat{L}_x$, $\hat{B} = \hat{L}_y$, $\hat{C} = \hbar\hat{L}_z$). Defining, as we did in Sections 3.3 and 3.5,

$$\bar{A} = \langle\psi|\hat{A}|\psi\rangle, \quad \bar{B} = \langle\psi|\hat{B}|\psi\rangle \tag{4.29}$$

and

$$\begin{aligned}(\Delta A)^2 &= \langle\psi|(\hat{A} - \bar{A})^2|\psi\rangle = \langle\psi|\hat{A}^2|\psi\rangle - \bar{A}^2,\\ (\Delta B)^2 &= \langle\psi|(\hat{B} - \bar{B})^2|\psi\rangle = \langle\psi|\hat{B}^2|\psi\rangle - \bar{B}^2,\end{aligned} \tag{4.30}$$

an exact uncertainty relationship exists between ΔA and ΔB of the form

$$(\Delta A)(\Delta B) \geq \frac{1}{2} \langle \psi | \hat{C} | \psi \rangle, \qquad (4.31)$$

where $|\psi\rangle$ is an arbitrary quantum state. We now proceed to prove the uncertainty relation stated in Equation 4.31 in full generality, without introducing a basis. Define the Hermitian operators

$$\hat{\alpha} = \hat{A} - \bar{A}, \quad \hat{\beta} = \hat{B} - \bar{B}. \qquad (4.32)$$

Then:

$$\langle \psi | \hat{\alpha}^2 | \psi \rangle = \langle \psi | (\hat{A} - \bar{A})^2 | \psi \rangle = (\Delta A)^2,$$
$$\langle \psi | \hat{\beta}^2 | \psi \rangle = \langle \psi | (\hat{B} - \bar{B})^2 | \psi \rangle = (\Delta B)^2, \qquad (4.33)$$

and

$$[\hat{\alpha}, \hat{\beta}] = [\hat{A} - \bar{A}, \hat{B} - \bar{B}] = [\hat{A}, \hat{B}] = i\hat{C}. \qquad (4.34)$$

Using Schwarz's inequality stated in Equation 3.284, we obtain:

$$\begin{aligned}(\Delta A)^2 (\Delta B)^2 &= \langle \psi | \hat{\alpha}^2 | \psi \rangle \langle \psi | \hat{\beta}^2 | \psi \rangle = \langle \hat{\alpha}\psi | \hat{\alpha}\psi \rangle \langle \hat{\beta}\psi | \hat{\beta}\psi \rangle \\ &\geq |\langle \hat{\alpha}\psi | \hat{\beta}\psi \rangle|^2 = \langle \hat{\alpha}\psi | \hat{\beta}\psi \rangle \langle \hat{\beta}\psi | \hat{\alpha}\psi \rangle \\ &= \langle \psi | \hat{\alpha}\hat{\beta} | \psi \rangle \langle \psi | \hat{\beta}\hat{\alpha} | \psi \rangle. \end{aligned} \qquad (4.35)$$

We introduce the commutator of $\hat{\alpha}$ and $\hat{\beta}$ by writing:

$$\hat{\alpha}\hat{\beta} = \hat{\omega} + \frac{1}{2}[\hat{\alpha}, \hat{\beta}],$$
$$\hat{\beta}\hat{\alpha} = \hat{\omega} - \frac{1}{2}[\hat{\alpha}, \hat{\beta}], \qquad (4.36)$$

where

$$\hat{\omega} = \frac{1}{2}(\hat{\alpha}\hat{\beta} + \hat{\beta}\hat{\alpha}). \qquad (4.37)$$

With Equation 4.34 we may write Equation 4.36 as:

$$\hat{\alpha}\hat{\beta} = \hat{\omega} + \frac{1}{2}i\hat{C},$$
$$\hat{\beta}\hat{\alpha} = \hat{\omega} - \frac{1}{2}i\hat{C}. \qquad (4.38)$$

4.3 • UNCERTAINTY PRINCIPLE

Using Equation 4.38 in Equation 4.35 we obtain:

$$(\Delta A)^2 (\Delta B)^2 \geq \langle\psi|\hat{\omega} + \frac{1}{2}i\hat{C}|\psi\rangle \langle\psi|\hat{\omega} - \frac{1}{2}i\hat{C}|\psi\rangle$$
$$= \langle\psi|\hat{\omega}|\psi\rangle^2 - \frac{i}{2}\langle\psi|\hat{\omega}|\psi\rangle\langle\psi|\hat{C}|\psi\rangle + \frac{i}{2}\langle\psi|\hat{C}|\psi\rangle\langle\psi|\hat{\omega}|\psi\rangle + \frac{1}{4}\langle\psi|\hat{C}|\psi\rangle^2$$
$$= \langle\psi|\hat{\omega}|\psi\rangle^2 + \frac{1}{4}\langle\psi|\hat{C}|\psi\rangle^2 \geq \frac{1}{4}\langle\psi|\hat{C}|\psi\rangle^2. \tag{4.39}$$

Taking the square root of both sides of Equation 4.39 we obtain the uncertainty relation stated in Equation 4.31.

Heisenberg called the uncertainty relationship in Equation 4.31 the *Uncertainty Principle*. The most important consequence of the Uncertainty Principle is the fact that position and momentum of a quantum particle cannot be both determined simultaneously with arbitrary accuracy. Indeed, because of Equations 4.26 and 4.31:

$$\Delta x \, \Delta p \geq \frac{\hbar}{2}. \tag{4.40}$$

This is a fundamental departure from the principles of classical mechanics where both position and momentum of a classical particle can always be determined simultaneously with arbitrary accuracy. The Uncertainty Principle is often cited as the distinguishing hallmark that sets quantum mechanics apart from classical mechanics.

The commutators in Equations 4.27 and 4.28 yield two more Uncertainty Principles for spin and orbital angular momentum:

$$(\Delta s_x)(\Delta s_y) \geq \frac{\hbar}{2}\langle\psi|\hat{s}_z|\psi\rangle,$$
$$(\Delta L_x)(\Delta L_y) \geq \frac{\hbar}{2}\langle\psi|\hat{L}_z|\psi\rangle. \tag{4.41}$$

Exercises:

4.3.1 Show that the ground state $|0\rangle$ of the harmonic oscillator is a minimum uncertainty state, i.e., in $|0\rangle$:

$$(\Delta x)(\Delta p) = \frac{\hbar}{2}.$$

4.3.2 Use the Uncertainty Principle to estimate the ground-state energy of a quantum particle inside the infinite square-well potential.

4.3.3 Compute an exact expression for the uncertainty product $(\Delta x)(\Delta p)$ for the ground state of the one-dimensional infinite square-well potential of width a. Check your result: According to Equation 4.40 the uncertainty product is expected to be larger than $\hbar/2$.

4.4 No-Cloning Theorem

Suppose we have an electron e^- in an unknown spin state,

$$|\psi\rangle = a|\uparrow\rangle + b|\downarrow\rangle, \qquad (4.42)$$

with unknown amplitudes a and b. Our aim is to determine the amplitudes a and b. A straightforward s_z measurement of $|\psi\rangle$ will not do, since this will collapse $|\psi\rangle$ into either $|\uparrow\rangle$ or $|\downarrow\rangle$ with probabilities $|a|^2$ and $|b|^2$, respectively. But since $|\psi\rangle$ is destroyed after the measurement, we cannot learn anything about a and b, since we would have to repeat the measurement many times to determine the probabilities $|a|^2$ and $|b|^2$. If we are not told how the state $|\psi\rangle$ was produced in the first place, our only hope of learning anything about a and b is to make exact copies of $|\psi\rangle$ and subject each of the copies to a measurement. Given a large number of copies, this would allow us to determine $|a|^2$ and $|b|^2 = 1-|a|^2$ with any prescribed accuracy.

The process of making exact, identical copies of a quantum state is called *cloning*. We clone $|\psi\rangle$ by imprinting it onto the spins of other electrons. To start the process, let us assume that we work with just two electrons, e_1^- and e_2^-, respectively. We assume further that initially electron e_1^- carries the state $|\psi\rangle$ and electron e_2^- is in the known state $|\downarrow\rangle$. Therefore, the initial two-particle state of our two-electron system is

$$|\Psi_{12}\rangle = |\psi_1\rangle|\downarrow_2\rangle, \qquad (4.43)$$

where the subscript "12" refers to the combined system, and the subscripts "1" and "2" refer to the electrons e_1^- and e_2^-, respectively. We would like to make a clone of the state $|\psi\rangle$ by copying it from electron e_1^- to electron e_2^- without destroying the state $|\psi_1\rangle \equiv |\psi\rangle$ of electron e_1^-. We accomplish the transfer with the help of a quantum copying machine C, represented by the copy operator $\hat{C}$. Operating $\hat{C}$ onto $|\Psi_{12}\rangle$, the desired result is:

$$\hat{C}|\Psi_{12}\rangle = \hat{C}|\psi_1\rangle|\downarrow_2\rangle = |\psi_1\rangle|\psi_2\rangle, \qquad (4.44)$$

i.e., two identical copies of the state $|\psi\rangle$ carried by both e_1^- and e_2^-. The operator $\hat{C}$ has to be unitary, i.e.,

$$\hat{C}^\dagger \hat{C} = \hat{1}, \qquad (4.45)$$

because (1) we do not want to lose any quantum information and (2) the states on the left-hand side and the right-hand side of Equation 4.44, respectively, are normalized. The copy operator $\hat{C}$ is assumed to be universal in the sense that it can clone *any* quantum state. To use this property of $\hat{C}$ explicitly, we choose a quantum state $|\phi\rangle$, which is arbitrary except for the condition

$$\langle\phi|\psi\rangle \neq 0, 1. \qquad (4.46)$$

Infinitely many states $|\phi\rangle$ with the property defined in Equation 4.46 exist. If e_1^- is in state $|\phi_1\rangle$ we have

$$\hat{C}|\phi_1\rangle|\downarrow_2\rangle = |\phi_1\rangle|\phi_2\rangle. \qquad (4.47)$$

Taking the scalar products of the left- and right-hand sides of Equations 4.44 and 4.47 and using Equation 4.45, we obtain:

$$\text{LHS} = \langle\downarrow_2|\langle\phi_1|\hat{C}^\dagger\hat{C}|\psi_1\rangle|\downarrow_2\rangle = \langle\downarrow_2|\langle\phi_1|\psi_1\rangle|\downarrow_2\rangle = \langle\phi|\psi\rangle,$$
$$\text{RHS} = \langle\phi_2|\langle\phi_1|\psi_1\rangle|\psi_2\rangle = \langle\phi|\psi\rangle^2. \qquad (4.48)$$

Notice that we omitted the subscripts "1" and "2" in the scalar products $\langle\phi|\psi\rangle$ in Equation 4.48. The subscripts are no longer needed, since the scalar product to be performed refers to a *single* particle. Therefore, irrespectively of whether we perform this scalar product with particle 1 or particle 2 in mind, the numerical value of the scalar product will not change. Since LHS = RHS in Equation 4.48, it follows that

$$\langle\phi|\psi\rangle = \langle\phi|\psi\rangle^2 \implies \langle\phi|\psi\rangle[\langle\phi|\psi\rangle - 1] = 0. \qquad (4.49)$$

This equation is fulfilled only in the two cases where $\langle\phi|\psi\rangle = 0, 1$, which is excluded according to Equation 4.46. Therefore, Equation 4.49 cannot be fulfilled and our universal quantum copy machine C and its associated operator $\hat{C}$ do not exist. As a consequence it follows that cloning of unknown quantum states is impossible. This proves the *no-cloning theorem* of quantum mechanics.

Inspecting the cloning operation defined in Equation 4.44 we notice that we required that the right-hand side is a separable product state of the two individual single-particle states of the two electrons, respectively. This is a necessary requirement of cloning. If the states were entangled, we would not have *cloned* the initial state, since only in a separable two-particle product state the individual particles are in fact in a *state* and separately available for individual quantum measurements. A state cannot be assigned to the individual particles in a nonseparable superposition state.

At the beginning of this section we set out to reveal information about the amplitudes a and b. We needed to generate many identical copies of $|\psi\rangle$ to reach this goal. Previously we showed that it is impossible to make even a single clone of a given, unknown quantum state. Therefore, if even making a single clone is impossible, we can certainly not make many. But since this is a necessary condition for revealing any information about $|\psi\rangle$, we failed. As a consequence we have the following theorem:

> Quantum mechanics does not allow us to learn anything about an unknown quantum state $|\psi\rangle$. (4.50)

The Theorem 4.50 is a strong disappointment as far as investigating unknown quantum states is concerned, but it does not forbid *handling* of unknown quantum states. In fact, we will see later that unknown quantum states can be *swapped* from one particle to another, and even *teleported* in "disembodied form" from one location to another. We will discuss this possibility further in Section 7.6.

Exercises:

4.4.1 When we derived the no-cloning theorem, we arbitrarily chose the initial state of the second electron to be $|\downarrow_2\rangle$. Excluding the possibility that the no-cloning theorem holds only in this special case, prove the no-cloning theorem for an arbitrary initial condition $|\varphi_2\rangle$, $\langle\varphi_2|\varphi_2\rangle = 1$, of the second electron.

4.4.2 In the text, for the sake of definiteness, we used electrons to prove the no-cloning theorem. This assumption, however, is not necessary. Show that the no-cloning theorem holds generally for any type of particles, not just for spin-1/2 electrons.

4.4.3 Perhaps the problem with the apparent impossibility of cloning the state of a particle is that in the proof of the no-cloning theorem presented in the text we worked with only two particles and thus had to insist that the state $|\psi\rangle$ of particle 1 remains intact. What if we worked with three particles? This would allow us to destroy the state $|\psi\rangle$ of particle 1, but re-create it on particles 2 and 3. Show that this idea does not work either, which corroborates the no-cloning theorem.

4.4.4 Suppose $\hat{C}$ clones $|\alpha\rangle$ according to

$$\hat{C}|\alpha_1\rangle|\downarrow_2\rangle = |\alpha_1\rangle|\alpha_2\rangle.$$

Show that $\hat{C}$ cannot clone

$$|\phi\rangle = \frac{1}{\sqrt{2}}[|\alpha\rangle + |\beta\rangle],$$

where $\langle\alpha|\beta\rangle = 0$.

4.5 Quantum Zeno Effect

In our discussion of measurement so far we have disregarded the time evolution of the quantum state between measurements. This is perfectly justified in cases where no external fields are switched on that may alter the quantum state between measurements. However, some of the most intriguing quantum effects result if the quantum state is allowed to evolve between measurements under the influence of an external perturbation. Thus, in this section, we study the interplay between quantum dynamics and measurement. Suppose that at time $t = 0$ we prepare a quantum state $|\psi\rangle$ in the spin-down state $|\downarrow\rangle$, i.e., $|\psi(t=0)\rangle = |\downarrow\rangle$, and let it evolve under the influence of the Hamiltonian

$$\hat{\mathcal{H}} = -V_0\hat{\sigma}_y, \quad V_0 > 0. \tag{4.51}$$

Then, according to Equation 3.289:

$$\begin{aligned}|\psi(t)\rangle &= \exp\left[\frac{i}{\hbar}V_0 t\hat{\sigma}_y\right]|\psi(t=0)\rangle \\ &= \left[\cos\left(\frac{V_0 t}{\hbar}\right) + i\hat{\sigma}_y\sin\left(\frac{V_0 t}{\hbar}\right)\right]|\downarrow\rangle \\ &= \cos\left(\frac{V_0 t}{\hbar}\right)|\downarrow\rangle + \sin\left(\frac{V_0 t}{\hbar}\right)|\uparrow\rangle. \end{aligned} \tag{4.52}$$

If we let $|\psi(t)\rangle$ evolve from $t = 0$ to $t = T$ and determine T such that $V_0 T/\hbar = \pi/2$, i.e.,

$$T = \frac{\pi\hbar}{2V_0}, \tag{4.53}$$

the cosine function in Equation 4.52 is zero, the sine function is one, and

$$|\psi(T)\rangle = |\uparrow\rangle. \tag{4.54}$$

Therefore, after time $t = T$, the spin of $|\psi\rangle$ has flipped from "down" to "up" and an s_z measurement of $|\psi(T)\rangle$ will "collapse" $|\psi(T)\rangle$ into $|\uparrow\rangle$ with certainty.

Let us now assume that midway between $t = 0$ and $t = T$, at $t = T/2$, we perform an additional s_z measurement. At $t = T/2$, according to Equation 4.52, the state $|\psi(0)\rangle$ has evolved into

$$|\psi(T/2)\rangle = \frac{1}{\sqrt{2}}|\downarrow\rangle + \frac{1}{\sqrt{2}}|\uparrow\rangle. \qquad (4.55)$$

Therefore, an s_z measurement may collapse $|\psi(T/2)\rangle$ either into $|\downarrow\rangle$ or into $|\uparrow\rangle$ with probability $1/2$. Suppose $|\psi(T/2)\rangle$ collapses into $|\downarrow\rangle$. Then, $|\psi\rangle$ is back in its starting state and effectively no time evolution has taken place. Between $t = T/2$ and $t = T$, after the s_z measurement has taken place, the state $|\psi\rangle$ continues to evolve, and starting from $|\psi\rangle = |\downarrow\rangle$, after the s_z measurement at $t = T/2$, evolves into $|\psi(T)\rangle = (|\downarrow\rangle + |\uparrow\rangle)/\sqrt{2}$ at $t = T$. Thus, there is a 50% chance that an s_z measurement on $|\psi(T)\rangle$ will result in $|\downarrow\rangle$ at $t = T$. Since there was a 50% chance to obtain $|\downarrow\rangle$ at the midpoint $t = T/2$, and another 50% chance to again obtain $|\downarrow\rangle$ at $t = T$, there is a sizable chance of 25% that a chain of two s_z measurements, one at $t = T/2$ and another at $t = T$, will result in $|\downarrow\rangle$ at $t = T$. In these cases it looks like no time evolution has taken place at all between $t = 0$ and $t = T$, since we started with $|\psi\rangle = |\downarrow\rangle$ at $t = 0$ and ended with $|\psi\rangle = |\downarrow\rangle$ at $t = T$. This is in sharp contrast with the case where we did not interfere with a measurement at $t = T/2$, and $|\psi(0)\rangle = |\downarrow\rangle$, in this case, evolved into $|\psi(T)\rangle = |\uparrow\rangle$ with certainty.

What happens if we perform more s_z measurements in the time interval $0 < t \leq T$? Suppose we perform N measurements of s_z in $0 < t \leq T$ at times

$$t_j = j\,\Delta t, \quad j = 1, 2, 3, \ldots, N, \qquad (4.56)$$

where

$$\Delta t = \frac{T}{N}. \qquad (4.57)$$

We are interested in the case where, as a result of s_z measurement, the state $|\psi\rangle$ collapses into $|\downarrow\rangle$ after each s_z measurement. Between measurements the state $|\psi\rangle$, starting from $|\psi\rangle = |\downarrow\rangle$ at $t = t_j$, evolves into

$$\begin{aligned}|\psi(t_{j+1})\rangle &= \exp\left(\frac{i}{\hbar}V_0\Delta t\hat{\sigma}_y\right)|\downarrow\rangle \\ &= \cos\left(\frac{V_0\Delta t}{\hbar}\right)|\downarrow\rangle + \sin\left(\frac{V_0\Delta t}{\hbar}\right)|\uparrow\rangle \qquad (4.58)\end{aligned}$$

4.5 • QUANTUM ZENO EFFECT

at $t = t_{j+1}$. Therefore, the probability of collapse of $|\psi(t_{j+1})\rangle$ into $|\downarrow\rangle$, when starting in $|\downarrow\rangle$ at time t_j, is

$$P^{(\downarrow)} = \cos^2\left(\frac{V_0 \Delta t}{\hbar}\right). \tag{4.59}$$

Since there are N measurements performed, the total probability $P_N^{(\downarrow)}$ of obtaining the result $|\downarrow\rangle$ for each of the N measurements of s_z is

$$P_N^{(\downarrow)} = \left[P^{(\downarrow)}\right]^N = \cos^{2N}\left(\frac{V_0 \Delta t}{\hbar}\right). \tag{4.60}$$

Using Equations 4.53 and 4.57, this may also be written as

$$P_N^{(\downarrow)} = \cos^{2N}\left(\frac{\pi}{2N}\right). \tag{4.61}$$

As a check we may want to compute $P_1^{(\downarrow)}$, i.e., the probability of finding $|\psi(T)\rangle$ in the state $|\downarrow\rangle$ if only one s_z measurement, the one at $t = T$, is performed. From Equation 4.61 we obtain $P_1^{(\downarrow)} = \cos^2(\pi/2) = 0$. This is consistent, since we already know that in the absence of measurements in $0 < t < T$, we obtain $|\psi(T)\rangle = |\uparrow\rangle$ with certainty, i.e., the probability of obtaining $|\downarrow\rangle$ as a result of s_z measurement is zero.

Table 4.1 shows the numerical values of $P_N^{(\downarrow)}$ for up to five measurements in $0 < t \leq T$. We see that $P_N^{(\downarrow)}$ increases monotonically with N. Therefore, we are interested in the limit of $N \to \infty$, i.e., the case where infinitely many s_z measurements occur in $0 < t \leq T$. This case is referred to as the case of *continuous measurement*.

Table 4.1 Comparison between the exact values of $P_N^{(\downarrow)}$, $N = 1, \ldots, 5$, defined in Equation 4.61, and the approximate values obtained via the analytical approximation formula $F_N^{(\downarrow)}$ defined in Equation 4.65. First column: Index (number of measurements) N; Second column: Exact values for $P_N^{(\downarrow)}$; Third column: Approximate values $F_N^{(\downarrow)}$ for $P_N^{(\downarrow)}$ computed according to the analytical formula 4.65; Fourth column: Absolute error $|P_N^{(\downarrow)} - F_N^{(\downarrow)}|$ between the exact result $P_N^{(\downarrow)}$ and the approximate result $F_N^{(\downarrow)}$; Fifth column: Relative error $|P_N^{(\downarrow)} - F_N^{(\downarrow)}|/P_N^{(\downarrow)}$ in percent.

N	$P_N^{(\downarrow)}$	$F_N^{(\downarrow)}$	error	% error
1	0.000	0.085	0.085	–
2	0.250	0.291	0.041	16.5
3	0.422	0.439	0.017	4.1
4	0.531	0.540	0.009	1.7
5	0.605	0.610	0.005	0.8

In order to determine $\lim_{N\to\infty} P_N^{(\downarrow)}$, the probability of obtaining $|\downarrow\rangle$ at $t = T$ in the case of continuous measurement, we first compute $\lim_{N\to\infty} \ln[P_N^{(\downarrow)}]$, from which we obtain $\lim_{N\to\infty} P_N^{(\downarrow)}$ by a simple exponentiation. We have

$$\begin{aligned}
\lim_{N\to\infty} \ln\left[P_N^{(\downarrow)}\right] &= \lim_{N\to\infty} \ln\left[\cos^{2N}\left(\frac{\pi}{2N}\right)\right] \\
&= \lim_{N\to\infty} 2N \ln\left[\cos\left(\frac{\pi}{2N}\right)\right] \\
&= 2 \lim_{N\to\infty} \frac{\ln\left[\cos\left(\frac{\pi}{2N}\right)\right]}{\left(\frac{1}{N}\right)}.
\end{aligned} \quad (4.62)$$

To determine this limit we notice that both the numerator and denominator of the fraction in Equation 4.62 go to zero for $N \to \infty$. Therefore, applying the rule of de l'Hospital, we obtain

$$\begin{aligned}
\lim_{N\to\infty} \ln\left[P_N^{(\downarrow)}\right] &= 2 \lim_{N\to\infty} \frac{\frac{1}{\cos(\pi/2N)}[-\sin(\pi/2N)](-\pi/2N^2)}{\left(-\frac{1}{N^2}\right)} \\
&= -\pi \lim_{N\to\infty} \tan\left(\frac{\pi}{2N}\right) = 0.
\end{aligned} \quad (4.63)$$

Therefore,

$$\lim_{N\to\infty} \left[P_N^{(\downarrow)}\right] = \exp\left\{\lim_{N\to\infty} \ln\left[P_N^{(\downarrow)}\right]\right\} = 1. \quad (4.64)$$

This means that in the limit of infinitely many measurements the quantum state $|\psi\rangle$ is locked into the starting state $|\downarrow\rangle$; the time evolution toward $|\uparrow\rangle$ never commences.

This observation reminds us of one of Zeno's paradoxes according to which a runner will never be able to commence running. If the runner has to cover a distance of, say, 100 m, the runner first has to reach the half-distance mark, i.e., 50 m. Before that the runner has to reach the 25 m mark, and so on. Therefore, in a finite time, the runner has to reach infinitely many markers, which, in Zeno's time, was considered a logical impossibility. Since no matter how far the runner wants to go, there are always infinitely many "half-distance markers" in between the starting point and the target distance, the runner will never commence moving.

The parallel of Zeno's runner with the above described effect of frozen quantum dynamics due to continuous measurement is obvious. The quantum motion never commences, since in order to cover any time interval, there are always infinitely many measurements in between. Because of this parallel the phenomenon of arrested motion due to continuous measurement is called the *quantum Zeno effect*. Since "measurement" is the same as "observation" the quantum Zeno effect is also called the *watched pot effect* since, according to an old proverb, "a watched pot never boils."

Exercises:

4.5.1 A spin-1/2 particle is governed by the Hamiltonian $\hat{\mathcal{H}} = V_0 \hat{\sigma}_x$. The initial state of the particle is $|\psi(t=0)\rangle = |\downarrow\rangle$.

(a) Compute T such that $|\langle\uparrow|\psi(T)\rangle|^2 = 1$.

(b) At times $t_j = jT/N$, $j = 1, \ldots, N$, s_z measurements are performed. What is the probability of obtaining $|\downarrow\rangle$ as a result of each of the N s_z measurements?

4.5.2 Use the representation

$$\exp(x) = \lim_{n \to \infty} \left(1 + \frac{x}{n}\right)^n$$

of the exponential function and a truncated Taylor series expansion of $\cos(x)$ to derive the approximation formula

$$F_N^{(\downarrow)} = \exp\left(-\frac{\pi^2}{4N}\right) \quad (4.65)$$

for $P_N^{(\downarrow)}$ defined in Equation 4.61. As is evident from Table 4.1, $F_N^{(\downarrow)}$ is an excellent aproximation to $P_N^{(\downarrow)}$ already for relatively small N, and the quality of the approximation increases with increasing N both in absolute and relative terms. Since $\lim_{N \to \infty} \pi^2/(4N) = 0$, and $P_N^{(\downarrow)} \to F_N^{(\downarrow)}$ for $N \to \infty$, Equation 4.65 provides an alternative, more direct, proof of Equation 4.64.

4.5.3 For the quantum Zeno problem discussed in the text, the probability of measuring $|\downarrow\rangle$ in each of N consecutive s_z measurements is $P_N^{(\downarrow)}$ given explicitly in Equation 4.61. How large do you have to choose N to ensure $P_N^{(\downarrow)} > 0.9$? The approximation formula defined in Equation 4.65 may come in handy.

4.6 Summary

In classical physics the theory of measurement plays a minor role and is not usually emphasized in textbooks on classical mechanics. The reason is that in classical physics it is assumed that measurement is a process that can be minimized to such an extent that it negligibly influences the dynamical state of the system. In contrast to the subordinate role of measurement in classical physics, measurement plays a central role in quantum mechanics whose influence on the system to be measured is usually large. Moreover, the act of measurement itself is poorly understood and to this day measurement is the most controversial part of quantum mechanics. Sidestepping the controversy, we adopted von Neumann's view of measurement as a nondeterministic collapse of the wave function, a process strictly outside the realm of Schrödinger's equation that requires its own axiom of measurement. We argued that even if someday the act of measurement is explained microscopically as a purely deterministic effect amenable to a dynamical description via the deterministic Schödinger equation, this would not diminish the value of the von Neumann picture of measurement. The relation between the "new," microscopic theory of measurement and the "old" von Neumann theory would be akin to the relationship between statistical mechanics and thermodynamics, where statistical mechanics plays the role of the microscopic theory that explains the effective, phenomenological theory of thermodynamics without invalidating it. In Section 4.3 we turned to one of the most profound differences between classical and quantum mechanics: While in classical mechanics all system variables can *always* be measured simultaneously with arbitrary accuracy, this is not the case in quantum mechanics. Heisenberg's Uncertainty Principle imposes a fundamental limit on the accuracy with which the values of two incompatible observables can be measured simultaneously. Using our quantum machinery developed in Chapter 3, we were able to derive the exact formulation of Heisenberg's Uncertainty Principle without making any assumptions. In Section 4.4 we encountered another counterintuitive quantum effect: It is impossible to make copies of an unknown quantum state. This fact is known as the quantum no-cloning theorem and, again, we proved it in full generality. In Section 4.5 we studied yet another quantum effect intimately connected with quantum measurement: The quantum Zeno effect. Here, a sequence of measurements is used to slow down the time evolution of a quantum system. The quantum Zeno effect is one of the most fascinating quantum effects. It has already found a practical application in connection with interaction-free measurements. We will encounter the quantum Zeno effect again in Section 5.4.

Chapter Review Exercises:

1. Suppose there exists an operator $\hat{C}$ that clones the spin-state of particle 1 according to

$$\hat{C}|\alpha_1\rangle|\varphi_2\rangle = |\alpha_1\rangle|\alpha_2\rangle,$$

where $|\varphi_2\rangle$ is the initial state of particle 2. Show that $\hat{C}$ cannot clone any other state

$$|\psi\rangle = A|\alpha\rangle + B|\beta\rangle,$$

where $\langle\alpha|\beta\rangle = 0$, and $|A|, |B| \neq 0, 1$ are constants with $|A|^2 + |B|^2 = 1$.

2. For a normalized state $|\varphi\rangle$, $\langle\varphi|\varphi\rangle = 1$, define the uncertainty product

$$P_\varphi = (\Delta x)_\varphi (\Delta p)_\varphi = \sqrt{\langle\varphi|(\hat{x}-\bar{x})^2|\varphi\rangle} \sqrt{\langle\varphi|(\hat{p}-\bar{p})^2|\varphi\rangle}.$$

Then, show by constructing an explicit example, that given two normalized states, $|\alpha\rangle$ and $|\beta\rangle$, $\langle\alpha|\alpha\rangle = \langle\beta|\beta\rangle = 1$, the uncertainty product P_γ of a normalized superposition state

$$|\gamma\rangle = A|\alpha\rangle + B|\beta\rangle, \quad |A|^2 + |B|^2 = 1,$$

may be smaller than the uncertainty products P_α and P_β of each of the original states $|\alpha\rangle$ and $|\beta\rangle$, i.e.,

$$P_\gamma < \min(P_\alpha, P_\beta).$$

This means that in this limited sense we can "beat" the Uncertainty Principle.

3. In order to measure the mean value $\bar{p} = \langle\psi|\hat{p}|\psi\rangle$ of the momentum of a quantum particle in the state $|\psi\rangle$, an experimenter has to prepare many identical copies of $|\psi\rangle$. Why does this not contradict the no-cloning theorem?

4. A spin-1/2 particle is governed by the Hamiltonian $\hat{\mathcal{H}} = -V_0 \hat{\sigma}_y$. N s_z measurements are performed at times $t_j = jT/N$, $j = 1, \ldots, N$, where $T = \hbar\pi/(2V_0)$.

    (a) Compute the probability $P_N^{(\uparrow)}$ to obtain $|\uparrow\rangle$ as a result of each of the N s_z measurements.
    (b) Show that $\lim_{N\to\infty} P_N^{(\uparrow)} = 0$.

5. An electron source S is capable of producing two different types of electron beams. In source mode A the electron source produces a beam in which 50% of the electrons are in spin-up and 50% of the electrons are in spin-down. In source mode B the electron source produces a beam in which all electrons are in the superposition state $|\psi\rangle = (|\uparrow\rangle + |\downarrow\rangle)/\sqrt{2}$.

(a) Explain why you cannot tell the difference between the two types of beams using s_z measurements.

(b) Explain how you can use s_x measurements to tell which type of electron beam the source is producing.

chapter 5

Interaction-Free Measurements

In this chapter:

- Introduction
- Seeing in the Dark: Conceptual Scheme
- Elitzur-Vaidman Scheme
- Optimal Interaction-Free Measurements
- Summary

5.1 Introduction

The impression we got in Chapter 4 is that in order to make a measurement the measuring apparatus has to interact with the object to be measured and that it is this interaction that influences the future behavior of the measured object and results in Heisenberg's Uncertainty Principle. We will learn in this chapter that, surprisingly, direct interaction between measuring apparatus and object is not always necessary for performing a measurement. This type of measurement is called *interaction-free measurement*.

The first example of an interaction-free measurement was presented by Renninger in 1958 in the form of a thought experiment. Renninger considered the setup shown in Figure 5.1. An α-particle-emitting, radioactive source S is located at the center of two detectors D_1 and D_2. Detector D_1 is a hemispherical shell of radius R_1, covering the right hemisphere around the source S; detector D_2 is a spherical shell of radius $R_2 > R_1$, completely enclosing the source S and the detector D_1.

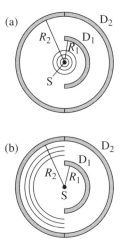

Figure 5.1 Setup of Renninger's thought experiment illustrating interaction-free measurement.

Both detectors are made of scintillating material and are assumed to be 100% effective. Whenever an α particle strikes one of the detectors, the α particle is absorbed with certainty and the corresponding detector emits a flash of light. Whenever a radioactive decay occurs, the source S emits an α particle with a spherical wave function $\psi_{\text{sph}}(r,t)$. α particles have a fairly constant energy when they are emitted, which means that their velocity v is fairly well determined. Let us assume that the α particles are emitted with a sharp velocity v. Then, if a decay occurs at time $t = 0$, the corresponding α particle will reach detector D_1 at time $t_1 = R_1/v$. Since D_1 covers only half of the solid angle around the source S, there is only a 50% chance that the α particle triggers detector D_1. A flash originating from detector D_1 indicates that a measurement of the α particle has occurred. The α particle interacted with the detector D_1, produced a flash of light, and as a result we know the position of the α particle at time t_1. At the very least, as a result of this measurement, we know that the α particle was detected in the right hemisphere. However, if *no* flash is produced at time t_1, something interesting happens. In this case we know that the α particle is *not* in the right hemisphere and is headed for detector D_2. So, even in the *absence* of a flash, we gain knowledge about the state of the particle. Renninger called this a *negative-result experiment*. The knowledge we gained should reflect itself in the wave function of the α particle. And indeed, the absence of a flash in D_1 causes the spherical wave function $\psi_{\text{sph}}(r,t)$, valid for $0 < t < t_1$, to collapse into the wave function $\psi_{\text{lhs}}(r,t)$, a hemispherical wave function, which is zero in the right hemisphere and normalized in the left hemisphere.

It propagates outward toward the detector D_2, where it will eventually produce a flash at time $t_2 = R_2/v$ with certainty. In Chapter 4 we learned that a measurement, i.e., the interaction of the measuring apparatus with the measured object, produces a collapse of the wave function. Here we learn that even the *absence* of a recorded measurement result can produce a collapse of the wave function. This is what Renninger called a *negative-result experiment* and what we call an *interaction-free measurement*. This idea, which seemed like an esoteric, philosophical point in 1958, has recently been verified experimentally and resulted in experimental schemes that allow us to make meaningful interaction-free measurements on microscopic objects. Even interaction-free imaging has recently been proposed and experimentally demonstrated. Interaction-free imaging is an important, new method of microscopy that may be used whenever the object to be imaged, a living cell or micro-organism for instance, is particularly sensitive to light.

In this chapter we are developing the topic of interaction-free measurement in a systematic way. In Section 5.2 we will learn how to see an object without photons striking it, i.e., we will learn the conceptual foundations of how to "see in the dark." Elitzur and Vaidman turned this concept into a viable detection scheme. We will discuss it in Section 5.3. The Elitzur-Vaidman scheme is not very efficient. However, combining the Elitzur-Vaidman idea with the quantum Zeno effect (see Section 4.5), turns the Elitzur-Vaidman scheme into a successful detection and imaging technique (see Section 5.4).

5.2 Seeing in the Dark: Conceptual Scheme

What is seeing? Within the photon picture of light the conventional process of seeing involves the following five steps:

Step 1: A photon is generated by a light source and travels toward the object.

Step 2: The photon hits the object, scatters off the object, and starts traveling toward the observer's eye.

Step 3: The photon hits the observer's eye and gets focused onto the retina.

Step 4: In the retina the photon is converted into an electrical signal that travels toward the observer's brain.

Step 5: The observer's brain registers the presence of the object.

Therefore, according to these five steps, it is necessary that at least one photon hits a given object in order to see the object, i.e., register its presence. But, there is another way of seeing *without* a photon ever hitting the object. This possibility was first pointed out by Elitzur and Vaidman in 1993. It is known as "seeing in the dark," as the "Elitzur-Vaidman scheme," or, "interaction-free measurement."

In preparation for a detailed discussion of the Elitzur-Vaidman scheme in Section 5.3, and to illustrate the essence of seeing in the dark, let us turn our familiar double-slit experiment (see Chapter 2) into a device for making interaction-free measurements. The setup is shown in Figure 5.2. The measurement proceeds according to the following five steps:

Step 1: Use a strong light source to establish the locations of the interference fringes on the screen Ω (see Figure 5.2(a)).

Step 2: Mark the location x_0 of one of the minima of the fringes (arrow in Figure 5.2(a)). In an ideal experiment, no light ever reaches this spot.

Step 3: Turn down the light to the single-photon level. Observe that no photon is ever registered at x_0.

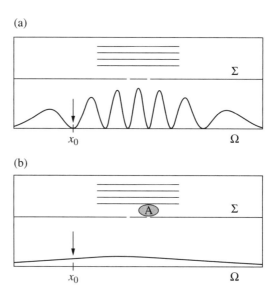

Figure 5.2 Conceptual scheme of seeing in the dark: (a) A strong light source establishes the locations of interference fringes on the observation screen Ω of a two-slit experiment. (b) An absorber A, placed in front of one of the two slits, changes the intensity distribution on Ω, which allows the detection of the presence of A without a photon striking it: An interaction-free measurement has taken place.

Step 4: Now turn up the light again, but place a strongly absorbing object A (absorber) between the light source and the screen Σ such that the light is blocked from going through the right hole (see Figure 5.2(b)). Observe that the fringes are gone and that a strong light intensity is registered at x_0 (arrow in Figure 5.2(b)), where in the absence of A no light at all was detected. Notice that the presence of A is detected in two different ways:

(i) Directly, because we see its shadow on Σ.

(ii) Indirectly, because the intensity pattern on Ω has changed, indicating the presence of A.

Step 5: We use observation (ii) of Step 4 to see A *without* a photon ever striking A. This is how it works: Turn down the light to the single-photon level. Because of Step 3, we know that in the absence of A a photon will never hit x_0. If we register a single photon at x_0, we have proof for the presence of A, even though this photon has never hit A. We have seen in the dark! We have made an interaction-free measurement!

Why can we draw this astonishing conclusion from the presence of a photon at x_0? The answer is the following:

(1) Since the photon has actually arrived at x_0, it cannot possibly have interacted with A. If it had, it would have been absorbed, and thus been destroyed, before reaching x_0.

(2) The presence of a photon at x_0 also means that something is present between S and Σ, since without its presence, no photon would ever have been recorded at x_0.

Points (1) and (2) show that (a) our measurement is interaction-free and (b) that we actually "saw" something, i.e., we detected the presence of an object between S and Σ.

Obviously the two-slit setup is very crude and our resolution leaves much to be desired. However, the result of this thought experiment is powerful enough to draw an important conclusion: In order to "see" an object, i.e., to record its presence, it is not necessary that a photon strikes the object. This implies that there are alternative ways of seeing that do not follow all of the "five steps of seeing" introduced at the beginning of this section.

We are now ready to study a more refined setup for interaction-free measurement proposed by Elitzur and Vaidman.

Exercises:

5.2.1 In what sense is Renninger's "negative-result experiment" discussed in the Introduction an example for "seeing in the dark"?

5.2.2 Suppose that instead of absorbing 100% of incident photons, the object covering one of the slits of the double-slit experiment absorbs only 50% of the photons striking it. Can you still make interaction-free observations?

5.3 Elitzur-Vaidman Scheme

As a refinement of the two-slit interaction-free measurement scheme discussed in Section 5.2, we discuss now a detection scheme proposed by Elitzur and Vaidman. It is also an excellent illustration and application of the following three rules of quantum mechanics that we encountered in Chapter 2 and repeat here for convenience:

(1) If we cannot tell which way a photon takes, we add the amplitudes of all alternatives (Feynman's Rule).

(2) If we *can* tell which way a photon takes, we add probabilities (central conclusion from our "which-way" experiments).

(3) Whenever a "conventional" measurement is performed that results in the destruction of the photon, the photon has to "take a stand," i.e., reveal its presence by revealing its location (manifest, for instance, in the marks on the observation screen Ω).

The thought experiment of Elitzur and Vaidman may be formulated in many different ways. We state it in the following way.

For the purpose of making position-sensitive measurements of photon impacts of extraordinary efficiency and accuracy, a company produces tiny grains of photosensitive material, which they call *photonic nanograins*. The grains are so sensitive that, struck by a single photon, they immediately, and permanently, blacken the grain. It is very difficult to produce these grains. Therefore some of the grains work (they are *live*); some of the grains do not work (they are *duds*). Because of the nature of the production process it is impossible to predict which of the grains are live and which of the grains are duds. The customers buy only live grains. Therefore, the factory has to find out which of the manufactured grains are live and which are duds. Thus, the factory's task is not only to make the grains, but also to *certify* the live ones. However, the only way to certify a grain is to expose it to a photon to see whether it works. Unfortunately, if a photon hits a live grain, it permanently blackens the grain, which is then destroyed and can no longer be sold as a photon detector. Therefore, limited to the use of photons for its certification process, the factory has to separate the live grains from the duds without triggering the live grains—a seemingly impossible task. The factory's manufacturing success seems to be its own downfall: The factory's grains are so exquisitely sensitive that a single photon triggers a live grain with 100% efficiency. In looking for a solution to the factory's conundrum, we assume that when hit by a photon, live grains are 100% absorptive, whereas duds are 100% transparent. In the case of a dud, it is as if the grain isn't even there, although it may be located squarely in a photon's path.

Elitzur and Vaidman suggest using an interaction-free measurement scheme to solve the grain certification problem based on a Mach-Zehnder interferometer. According to Figure 5.3 (a) a Mach-Zehnder interferometer consists of a light source S, two beam splitters (B_1 and B_2), two mirrors (M_1 and M_2), and two detectors (D_1 and D_2). A beam of light, emitted by the source S, is split by B_1 into two beams that travel along the *arms a* and *b* of the interferometer. The light in arm b is then directed into arm c by M_1 and then into arm d by M_2, where it recombines coherently with the light in arm a at the beam splitter B_2. Depending on the difference between the optical path lengths of arm a and the total optical path length of arms b, c, and d, we can arrange for all of the light to arrive at detector D_2 and none of the light to arrive at detector D_1. In this case we call D_2 the *bright detector* and D_1 the *dark detector*.

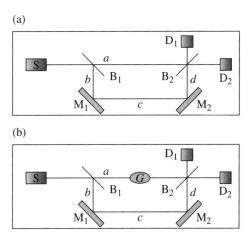

Figure 5.3 Elitzur-Vaidman setup for certifying photonic nano grains interaction-free without blackening them. (a) Mach-Zehnder interferometer, the centerpiece of the Elitzur-Vaidman certification scheme. (b) A grain G is located in arm a of the Mach-Zehnder interferometer. In this configuration it is possible to detect its presence without a photon striking it.

Simply changing the optical path length in one or all of the arms a, b, c, and d, we can arrange for D_1 to be the bright detector and D_2 to be the dark detector. A continuous change of the optical path length between these two settings will realize any in-between situation in which part of the light arrives at D_1 and part of the light arrives at D_2. For interest to us, however, are only the two situations in which one detector is bright and the other is dark. Choosing a strong, coherent light source for S, such as a laser, we may operate the Mach-Zehnder interferometer in *classical mode*. This is the most convenient mode for performing the necessary fine adjustments of the optical path lengths that assure that one of the detectors is dark and the other detector is bright. It is, however, much more interesting to operate the Mach-Zehnder interferometer in *single-photon mode*. In this case the source S emits single photons, one-by-one, with time intervals between emissions chosen long enough such that there is only at most one photon in the interferometer at any given time. The Elitzur-Vaidman photonic grain certification scheme is based on the single-photon Mach-Zehnder interferometer and works in the following way.

Step 1: Operating the Mach-Zehnder interferometer shown in Figure 5.3(a) in classical mode, tune the optical path lengths of the arms of the Mach-Zehnder interferometer such that D_1 is dark and D_2 is bright.

Step 2: Switch to quantum single-photon mode. Since D_2 is the bright detector, we register clicks in D_2, but none in D_1, since D_1 is the dark detector.

Step 3: As shown in Figure 5.3(b), place a grain into arm a of the Mach-Zehnder interferometer. There are two possibilities: (1) The grain is live or (2) the grain is a dud.

Step 4: Let us focus on the dud case first: According to assumption, a dud grain acts as if it is not there at all. No matter how many photons we inject into the interferometer, D_1 never clicks; only D_2 clicks.

Step 5: Now assume that the grain we placed into arm a of the Mach-Zehnder interferometer is live. Trigger a single photon in S and send it toward B_1. Having arrived at B_1, the photon has two choices: It may take arm a or arm b. If we assume that B_1 is a 50-50 beam splitter, i.e., it has 50% reflectivity and 50% transmissivity, the photon may choose arm a or arm b with 50% probability, respectively.

Step 6: Assume the photon takes arm a, which contains the live grain. Since, according to assumption, the grain is 100% efficient, the photon is absorbed, the grain blackened, and, as a consequence, cannot be sold anymore.

Step 7: Assume that the photon takes arm b, which is part of the path that does *not* contain the live grain. Since in this case the photon is *not* absorbed, we *know* that the photon took arm b. Because of this knowledge, there is no interference to be considered at B_2. It is as if the photon was created in arm b and sees the beam splitter B_2 for the first time. This photon now has the choice of either reflecting off of B_2 or transmitting through B_2. If we assume that B_2 is a 50-50 beam splitter, the photon will transmit or reflect with 50% probability, respectively, i.e., it will enter either D_1 or D_2 with 50% probability.

Step 8: Assume that the photon chooses to reflect off of B_2 and triggers detector D_2. In this case we are unable to certify that G is live, since a dud grain also sends photons into detector D_2.

Step 9: Assume that the photon chooses to transmit through B_2 and triggers detector D_1. This event never happens if G is a dud. We have certified that G is live without a photon ever striking it! We have made an interaction-free measurement with practical implications. The grain in arm a of the interferometer is still live and can be sold.

Now that we have a viable scheme for certifying grains without destroying them, we need to characterize its efficiency. We define the certification yield C_n as the conditional probability of certifying a grain using n photons under the condition that the grain is live. To compute C_n of the Elitzur-Vaidman scheme, we place a live grain in arm a of the Mach-Zehnder interferometer (see Figure 5.3(b)) and launch a photon. The following are the three possible outcomes:

$$
\begin{aligned}
&50\%\text{ probability:} &&\text{The photon takes arm } a: \\
&&&\text{the grain is destroyed;} \\
&25\%\text{ probability:} &&\text{The photon is registered in } D_2: \\
&&&\text{No information on grain status;} \\
&25\%\text{ probability:} &&\text{The photon triggers } D_1: \\
&&&\text{The grain is certified live, } \textit{without destroying it.}
\end{aligned}
$$
(5.1)

Therefore, by launching a single photon, we are able to certify live grains in 25% of the cases. In another 25% of the cases we are unable to certify the grain. In this case we may process it further by launching another photon. The second photon will divide the "undecided" batch of grains into 50% destroyed, 25% undecided, and 25% certified. Therefore, after the second photon, the certification yield is $1/4 + 1/16$. If we continue to launch photons at the resulting undecided batch of grains, the upper limit of the certification yield is:

$$C_\infty = \sum_{n=1}^{\infty} \left(\frac{1}{4}\right)^n = \frac{1}{3}. \qquad (5.2)$$

The certification yield C is one way of communicating the efficiency of a grain-testing scheme.

In summary, we have the following result. If we have a live grain, we can *certify* that it is live with at most 33% efficiency. In 2/3 of the cases the grain will be destroyed. The certification efficiency seems low. We need to remember, however, that this certification yield was achieved "without looking," i.e., according to a completely new scheme of measurement: No photon ever struck the grains. We will see in Section 5.4 that with some ingenuity the certification yield can be increased to levels close to one.

So far we assumed that the two beam splitters in Figure 5.3 are 50% reflective. By varying the reflectivities of the mirrors it is possible to improve the certification yield. Since we have to be able to zero out the detector D_1, we need to make sure that the reflectivities of the two beam splitters are the same. Let us call this reflectivity R and the transmissivity T. Because $R + T = 1$, we have $T = 1 - R$. Then, the probability of destroying the grain by absorbing the photon is T, the probability of a successful interaction-free measurement, i.e., the one-photon certification yield, is $C_1 = RT$, and the probability of an undecided outcome is R^2. Thus, the n-photon certification yield is

$$C_n = RT \sum_{\nu=1}^{n} (R^2)^{\nu-1} = RT \left(\frac{1-R^{2n}}{1-R^2}\right) = R\left(\frac{1-R^{2n}}{1+R}\right). \qquad (5.3)$$

Thus, in the limit of infinitely many photons, we obtain:

$$C_\infty = \frac{R}{1+R}. \qquad (5.4)$$

Since the derivative of C_∞ with respect to R, $dC_\infty(R)/dR = 1/(1+R)^2$, is always positive, C_∞ is a monotonically increasing function of R. For $R = 0$ ($T = 1$) we obtain the worst result, $C_\infty = 0$. Physically this is so, since in this case all injected photons transmit through the beam splitter and destroy the grain. For $R = 1/2$ we obtain the previous result of $C_\infty = 1/3$, but already for $R = 2/3$, for example, we obtain $C_\infty = 0.4$, which exceeds the previous result. We obtain the maximum efficiency of the Elitzur-Vaidman Mach-Zehnder scheme for $R \to 1$. In this limit the certification yield approaches 50%. This is the best result we can obtain with the Elitzur-Vaidman Mach-Zehnder setup. In Section 5.4 we will discuss a different experimental setup that allows us to exceed $C_\infty = 1/2$ and obtain a certification yield of close to 100%.

Exercises:

5.3.1 The first beam splitter, B_1, of the Elitzur-Vaidman grain-testing setup shown in Figure 5.3(b) has a reflectivity of 30%. A single photon is launched into the interferometer. What is the probability of an interaction-free measurement?

5.3.2 A factory produces 1000 photonic nanograins per day. It is known that 20% are duds. The customer buys only certified-live grains. The factory decides to use the Elitzur-Vaidman grain-testing scheme, using only a single test-photon per grain. How many grains, on average, can be certified live per day?

5.3.3 You are using the Elitzur-Vaidman scheme to test a batch of photonic nanograins that are only 30% effective. This means that a photon striking the grain has a 30% chance of being absorbed and thus blackening the grain, and has a 70% chance of being transmitted without blackening the grain. Assuming that the beam splitters have a reflectivity of 50%, what is the probability of an interaction-free measurement?

5.4 Optimal Interaction-Free Measurements

In 1995 Kwiat and collaborators published an experimental demonstration of interaction-free measurement. Instead of using a Mach-Zehnder interferometer, as suggested by Elitzur and Vaidman, Kwiat and collaborators used a Michelson interferometer. A schematic sketch of their experimental setup is shown in Figure 5.4. Using a Michelson interferometer instead of a Mach-Zehnder interferometer has experimental advantages. For instance, as shown in Figure 5.4, only one beam splitter is required. Part (a) of Figure 5.4 shows the situation with no object present. Photons are injected along arm a of the Michelson interferometer. At the beam splitter the photons may either transmit and proceed along arm c toward mirror M_2, or they reflect off of the beam splitter and proceed along arm b toward mirror M_1. The optical path lengths of both arms b and c are adjusted such that no photons are recorded in detector D terminating arm d of the Michelson interferometer. Thus, without an object present, all photons exit the interferometer through

160   CHAPTER 5 • INTERACTION-FREE MEASUREMENTS

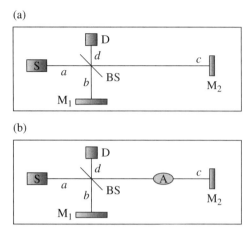

Figure 5.4   Sketch of a Michelson interferometer similar to the one used by Kwiat, Weinfurter, Herzog, Zeilinger, and Kasevich in 1995 for an experimental demonstration of the Elitzur-Vaidman scheme for interaction-free measurement. (a) Sketch of the interferometer without an absorber present in the test arm c. (b) Absorber present in the test arm c of the interferometer.

arm a, the same port through which they were injected originally into the interferometer.

Part (b) of Figure 5.4 shows the situation with a completely absorbing object A present in arm c of the interferometer. This is the configuration that allows interaction-free detection of the absorber A. The logic is very similar to the logic we applied in connection with the Elitzur-Vaidman setup in Section 5.3. A photon is injected along arm a and encounters the beam splitter. It has two possibilities to proceed: (1) Transmit through the beam splitter or (2) reflect off the beam splitter. Possibility (1) immediately leads to the absorption of the photon by A. This is the undesirable result, since no interaction-free measurement has taken place. Alternative (2) may lead to an interaction-free measurement of the presence of A according to the following reasoning. If the photon reflects off the beam splitter BS, retro-reflects off the mirror M_1, and then arrives at BS again, it has a choice of transmitting through the beam splitter or reflecting off of it. The photon has a choice, since due to the presence of the absorber the coherence of the interferometer is broken and unlike in situation (a) of Figure 5.4, the photon is no longer forced to exit the interferometer through port a. If the photon nevertheless decides to reflect off of BS and exit via a, no information on the presence of the absorber is obtained. If the photon, however, decides to transmit through BS, proceed along arm d of the interferometer, and trigger the detector D, we know with certainty, and without the photon ever striking A, that the absorber is present in arm c of the interferometer. If we vary the reflectivity R of the beam splitter we obtain the same certification yield (see Equation 5.4) as for the Elitzur-Vaidman Mach-Zehnder interferometer with a maximum of 50% detection efficiency for $R \to 1$.

Following their experimental demonstration of interaction-free measurements, Kwiat and collaborators demonstrated theoretically and experimentally that the 50% "efficiency barrier" can be broken and close to 100% efficiency of interaction-free detection can be achieved. Their scheme combines the Michelson interferometer with an optical version of the quantum Zeno effect.

A polarization rotator is an optical component that rotates the polarization direction of linearly polarized photons by a fixed amount. Consider the setup shown in Figure 5.5(a). Six polarization rotators are assembled in sequence. If each rotator rotates the polarization direction by 15°, the total rotation angle of the polarization direction after passing all six rotators is 90°. Therefore, with the help of the six rotators in Figure 5.5(a) we may turn horizontally polarized light into vertically polarized light.

Now consider the setup shown in Figure 5.5(b). Between each pair of polarization rotators we insert a horizontal polarization filter. If we feed horizontally polarized light into the first polarization rotator, the light emerges 15° rotated with respect to the horizontal direction. Passing

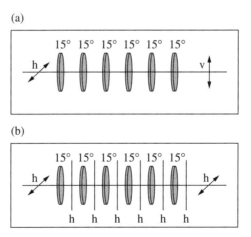

Figure 5.5   (a) With the help of six polarization rotators, each rotating the polarization by 15°, the polarization of a photon, initially polarized horizontally, is rotated to vertical polarization. (b) A horizontal polarization filter is inserted after each polarization rotator. This forces the photon's polarization to stay horizontal but nevertheless allows the photon to traverse the rotator array with appreciable probability.

the first horizontal polarization filter, the light emerges, again polarized horizontally, before entering the second polarization rotator. Emerging from the last horizontal polarizer the light is horizontally polarized, but the question is: What is the transmission probability through the entire cascade of six polarization rotators with horizontal polarizers inserted between them? The answer is $T = \cos^{12}(15°) \approx 0.66$. Therefore, although the successive polarization rotators keep introducing a vertical component of the polarization, which in turn gets erased by the interspersed horizontal polarizers, a large fraction of the originally horizontally polarized light is able to pass through the six stages of rotation and polarization.

Suppose we use N polarization rotators, each rotating the polarization direction by $\pi/(2N)$. Then, if no horizontal filters are present, i.e., a situation similar to the one shown in Figure 5.5(a), the polarization direction will have rotated by 90° when the light emerges to the right of the cascade. In the presence of the linear polarizers, a situation similar to the one shown in Figure 5.5(b), the light will emerge horizontally polarized with transmission probability

$$T = \cos^{2N}[\pi/(2N)] \to 1, \quad \text{for } N \to \infty, \tag{5.5}$$

where we used the quantum Zeno Equations 4.61 and 4.64 to determine the $N \to \infty$ limit of Equation 5.5. As a result of Equation 5.5 we obtain that the full light intensity is transmitted for $N \to \infty$. This is the essence of the optical Zeno effect.

Now consider the experimental setup shown in Figure 5.6. It is essentially a Michelson interferometer with several additional optical elements. The optical component labeled SM in Figure 5.6 is a *switchable mirror*, i.e., a device that can be switched at will, and very rapidly, from "transmission" to "reflection." In the "transmission" setting the device is completely transparent to photons, whereas in the "reflection"

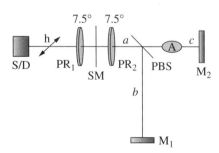

Figure 5.6 Schematic sketch of an experimental setup similar to the one used by Kwiat and collaborators for achieving larger than 50% interaction-free detection efficiency.

5.4 • OPTIMAL INTERACTION-FREE MEASUREMENTS 163

setting the device acts like a 100% reflective mirror. The switchable mirror in Figure 5.6 is used to inject a photon into the interferometer ("transmission" setting), but then to trap it for several passes inside the interferometer ("reflection" setting) before resetting the switchable mirror to "transmission" and letting the photon escape. The component labeled PBS is a *polarizing beam splitter*. It reflects horizontally polarized light with 100% efficiency and lets vertically polarized light pass with 100% efficiency. There is no "arm d" emerging from the polarizing beam splitter since horizontally polarized light reflecting off mirror M_1 stays horizontally polarized and thus is reflected to the left with 100% efficiency when reaching the polarizing beam splitter; vertically polarized light transmitted through the polarizing beam splitter and reflected off mirror M_2 stays vertically polarized and transmits to the left with 100% efficiency through the polarizing beam splitter upon reaching it. The interferometer also contains two polarization rotators labeled PR_1 and PR_2 in Figure 5.6. Each of the rotators rotates the polarization of a photon by 7.5° no matter whether the rotator is traversed from left to right or from right to left. The component labeled S/D in Figure 5.6 serves a dual function. It acts as a single-photon source when it launches single photons toward the interferometer and it acts as a polarization-sensitive detector when a photon exits the interferometer and returns to S/D.

We are now ready to follow a photon through the setup shown in Figure 5.6. Let us first assume that the absorber shown in arm c of the interferometer is absent. A horizontally polarized photon is launched by S/D and sent on its way toward the interferometer. The photon first encounters the polarization rotator PR_1 where its polarization is rotated 7.5°. Next, the photon flies toward the switchable mirror SM, which is set in "transmission" mode. Once the photon has passed SM, the switchable mirror is immediately set to "reflection." Next, the photon transmits through polarization rotator PR_2, where its polarization is rotated an additional 7.5° such that its polarization is now rotated a total of 15° with respect to the horizontal direction. Next, the photon encounters the polarizing beam splitter PBS. Since, due to the actions of PR_1 and PR_2, the photon now has a vertical as well as a horizontal component, it may transmit or reflect at the polarizing beam splitter PBS. The important point, however, is the following: Since we assumed that no absorber is present, the two possible photon paths are indistinguishable and fully coherent such that upon returning to PBS the photon "reassembles" completely and emerges from PBS in the same state as it entered, the only difference being that its momentum is reversed. The photon then transmits through the polarization rotator PR_2, reflects off SM and transmits once more through PR_2 for a total of an additional

15° rotation of its polarization direction. Now with a 30° angle of its polarization direction with respect to the horizontal, the photon is ready for its second pass through the interferometer. It will experience the same effects as discussed in connection with its first pass through the interferometer. Nothing will happen to the photon except that with each pass through the interferometer its angle with respect to the horizontal will increase by 15°. After six passes through the interferometer, SM is set to "transmission" and the photon emerges to the left with a *vertical* polarization.

Let us now assume that an absorber is present in arm c of the interferometer. The photon is injected into the interferometer and its polarization direction is rotated 15°. Again, the photon has a choice when arriving at the polarizing beam splitter: reflect or transmit. If the photon is transmitted, it gets absorbed; its journey is over and no interaction-free measurement has taken place. If, however, the photon decides to reflect off of the polarizing beam splitter, it will be directed into arm b of the interferometer, reflect off of M_1, return to PBS, reflect off of PBS with 100% efficiency and fly to the left along arm a of the interferometer. The important point to notice is that at this point the photon is 100% horizontally polarized. This is so, since because of the presence of the absorber in arm c of the interferometer, we can tell precisely where the photon has or has not been. It could not possibly have visited arm c of the interferometer, since it would have been absorbed and never have reappeared in arm a of the interferometer. The only other possibility is that it took arm b of the interferometer, which means that the photon was horizontally polarized. No matter how often the photon gets re-injected into the interferometer, if it reappears in arm a of the interferometer, it is horizontally polarized. After a time corresponding to six round trips of the photon inside of the interferometer, we switch SM to "transmit" and the photon escapes to the left. If, at this time, we measure a photon at all, the photon will be *horizontally* polarized.

Reviewing the two situations corresponding to "absorber present" and "absorber absent," we notice one decisive difference: If we measure an exit photon and the absorber is *absent*, the exit photon is *vertically* polarized. If the absorber is *present*, and we measure an exit photon, the exit photon is *horizontally* polarized. Therefore, if the exit photon is horizontally polarized, an interaction-free measurement has taken place. If an absorber is present, the probability that the photon will not be absorbed by the absorber, and an exit photon is received, is:

$$P_{\text{ifm}} = \cos^{12}(15°) \approx 0.66. \tag{5.6}$$

This means that, given an absorber is present, we can certify its presence, interaction-free, in 66% of the cases. This exceeds the 50% success rate of the original Elitzur-Vaidman scheme, and the success rate of the single-pass Michelson interferometer. According to Equation 5.5, increasing the number of round trips will increase the success rate even further. There is no fundamental limit that would preclude success rates approaching 100%.

Exercises:

5.4.1 What is the smallest number $N_{\min}$ of round trips of a photon in the multipass Michelson interferometer that results in an interaction-free detection probability P_{ifm} of $P_{\text{ifm}} > 0.9$? Remember that you have to adjust the rotation angle of the polarization rotators to the number of round trips!

5.4.2 Verify Equation 5.5

5.5 Summary

Interaction-free measurement? This sounds like a contradiction in terms. And yet it works. A first hint of how to perform interaction-free measurements was provided by Renninger in the 1950s. He noticed that the *absence* of detection is sometimes a measurement, too, which may provide valuable information on location and momentum of a particle. Renninger called this type of measurement *negative-result measurements*. In Section 5.2 we illustrated this idea in connection with the familiar double-slit experiment. While this setup provides the conceptual foundation for interaction-free measurements, it is not of much use in practice. A first practical scheme for interaction-free measurement was provided by Elitzur and Vaidman who proposed using a Mach-Zehnder interferometer to perform interaction-free measurements. We studied their scheme in detail in Section 5.3. The Elitzur-Vaidman scheme illustrates that interaction-free measurement plays on the wave-particle duality. The wave aspect of light allows us to zero out the signal in one of two detectors. The particle aspect of light forces photons to take one of two paths in case an absorber is present in one of the interferometer arms and destroys wave coherence. The ideas of Elitzur and Vaidman were implemented experimentally by Kwiat and collaborators with the help of a Michelson interferometer. We studied their experimental scheme in Section 5.4. Once the idea of interaction-free measurements had been confirmed experimentally, Kwiat and collaborators went further. With the help of an ingenious optical implementation of the quantum Zeno effect they obtained interaction-free observations exceeding 50% efficiency. Meanwhile interaction-free measurements evolved into interaction-free imaging of small objects. Interaction-free imaging may find useful applications whenever an object to be imaged is particularly sensitive to light.

CHAPTER 5 • INTERACTION-FREE MEASUREMENTS

Chapter Review Exercises:

1. A quantum particle is confined to move in a one-dimensional box of width a. At time $t = t_0$ the particle is in the ground state $|\psi_0\rangle$ of the box (see Figure 5.7(a)). At time $t = t_1 > t_0$, an absorber of width $a/2$, extending from $x = a/4$ to $x = 3a/4$ is briefly switched on. The absorber is assumed to be so effective that, if the particle is present in $[a/4, 3a/4]$, it will be absorbed with certainty. As a result of the measurement at time $t = t_1$ we learn that the particle has *not* been absorbed. Clearly, an interaction-free measurement has taken place; the wave function of the particle immediately after the measurement at $t = t_1$ has collapsed into the wave function shown in Figure 5.7(b).

   (a) Determine the ground-state wave function $\psi_0(x) = \langle x|\psi_0\rangle$, i.e., the wave function of the quantum particle before the measurement takes place.

   (b) Compute the probability P_0 of finding the particle at time $t = t_0$ in the interval $[0, a/4]$.

   (c) Immediately after the interaction-free measurement has taken place at $t = t_1$, what is the probability P_1 of finding the particle in the interval $[0, a/4]$? What is the enhancement factor $f = P_1/P_0$?

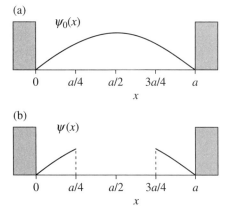

Figure 5.7 Interaction-free measurement of the position of a particle in a one-dimensional box. (a) Wave function before the measurement. (b) Wave function after the interaction-free measurement has occurred.

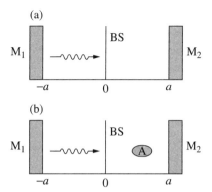

Figure 5.8 Device for making interaction-free measurements, consisting of two planar, parallel mirrors M_1 and M_2, and a beam splitter BS with reflectivity R located between them. (a) Only the photon is present. (b) An absorber is present in the right half-space between the beam splitter and the mirror M_2.

2. Consider the setup shown in Figure 5.8(a). A beam splitter BS of reflectivity

$$R = \cos^2\left(\frac{\pi}{2N}\right)$$

is placed between two planar, parallel mirrors M_1 and M_2, dividing the space between the mirrors into a left and a right cavity. As shown in Figure 5.8(a), a photon starting in the left cavity is launched toward the beam splitter. If N is large, the photon will slowly "slosh" into the right cavity, and will be found with certainty in the right cavity after N bounces. Now consider Figure 5.8(b): A completely absorbing object is present in the right-hand cavity.

(a) Describe qualitatively how the setup shown in Figure 5.8 can be used to make interaction-free measurements.

(b) If an interaction-free measurement has occurred, where is the photon after N bounces, and why?

(c) Compute the probability of an interaction-free measurement for the scheme shown in Figure 5.8.

3. Consider the Elitzur-Vaidman setup for interaction-free measurements. Assume that the beam splitters are 50% reflective. The optical path lengths of the Mach-Zehnder interferometer are adjusted such that without an absorber present one of the detectors is "dark" and the other is "bright."

(a) You are not told whether there is an absorber present in the interferometer or not. To find out, you launch five separate, single photons at the Mach-Zehnder interferometer. Each one of the five photons triggers the bright detector. You conclude that with a high probability no absorber is present in the interferometer. What is the probability of a "false negative," i.e., what is the probability that an absorber was present after all?

(b) How many photons do you have to launch to reduce the probability of a false negative to less than $2^{-20} \approx 10^{-6}$?

(c) Now assume that an absorber is present in the interferometer. What is the maximal probability $P_{\max}(N)$ of interaction-free confirmation of the presence of the absorber if you are allowed to launch N separate, single photons at the Mach-Zehnder interferometer? What is your strategy to obtain $P_{\max}(N)$? What is $P_{\max}(\infty)$?

chapter 6
EPR Paradox

In this chapter:

- ◆ Introduction
- ◆ Hallmarks of Physical Theories
- ◆ EPR and Reality
- ◆ Bell's Theorem
- ◆ Mermin's Reality Machine
- ◆ Summary

6.1 Introduction

Einstein was one of the towering figures who helped usher in the new era of quantum mechanics. In 1905 Einstein formulated the hypothesis of the light quantum, later named the photon, and in 1917 he discovered the principle of stimulated emission that later resulted in the laser. However, it is often claimed that Einstein did not like the new quantum mechanics created in 1925–1926 by Heisenberg, Born, Jordan, and Schrödinger. But this is only partially true. Einstein had no problem with the formalism of quantum mechanics, an amazingly powerful theory that produces correct results wherever applied to real-life problems. What Einstein had trouble with is the interpretation of quantum mechanics. In particular he felt uneasy about the Copenhagen interpretation of quantum mechanics, which claims that our world is fundamentally probabilistic. He longed to return to a deterministic, causal description of nature. All his life he invented objections and counterexamples in the form of thought experiments with the intent to prove that there is something inherently contradictory about the Copenhagen, probabilistic worldview. Therefore, one might get the impression that, after 1925, Einstein no longer played a constructive role in the development of quantum mechanics. Nothing could be further from the truth. All the counterexamples and objections that Einstein threw into the wheels and gears of quantum mechanics served only to clarify the language and interpretation of quantum mechanics. Therefore, Einstein played a pivotal role in the construction of the foundations of quantum mechanics. This is best exemplified with a paper that Einstein published in 1935, co-authored by Podolsky and Rosen. The argument presented in this paper is now known as the "EPR paradox." It went straight to the heart of the relationship between quantum mechanics and reality. The EPR paper gave rise to literally hundreds of papers discussing how much we can know about the world, and how much is forever "hidden from view." And even today the EPR paradox continues to inspire papers on the foundations of quantum mechanics.

This chapter is about the fundamental nature of reality in our world. In Section 6.2 we introduce some basic notions of how physical theories are constructed. In Section 6.3 we learn how Einstein, Podolsky, and Rosen (EPR) defined reality and constructed an ingenious argument that was supposed to show that quantum mechanics is an incomplete theory. Carefully analyzing the collapse of the wave function in the EPR system we show in Section 6.3 that the EPR argument is flawed, but has nevertheless inspired generations of quantum physicists to sharpen the tools and concepts of quantum mechanics. Perhaps the most profound advance toward a resolution of the EPR paradox is the work of Bell published in the 1960s. Bell presents a definitive, quantitative analysis of the EPR argument. He proves conclusively that the local realism favored by EPR does not exist in nature. As a consequence of Bell's proof, presented in Section 6.4, we understand now that nature is fundamentally nonlocal, a property of nature that scientists are only now beginning to exploit scientifically and technologically for the construction of novel quantum devices.

6.2 Hallmarks of Physical Theories

In order to appreciate the arguments advanced by EPR, let us first examine possible connections between "theory" and "objective reality." According to EPR, objective reality is independent of any theory that describes it. EPR are convinced that this is self-evident since objective reality is given by nature and cannot possibly be influenced by how we describe it. Consider Figure 6.1(a). It describes pictorially a possible

6.2 • HALLMARKS OF PHYSICAL THEORIES   171

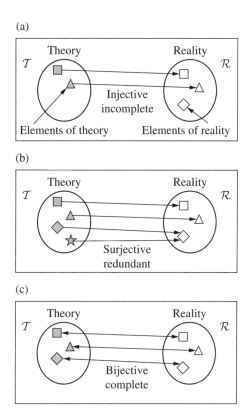

Figure 6.1   Connection between theory and reality. (a) There are more elements of reality than there are elements of theory. Although all elements of theory are in one-to-one correspondence with some elements of reality (injective property), since there are more elements of reality than there are elements of theory, the theory is incomplete. (b) There are more elements of theory than there are elements of reality. Although all elements of reality correspond to at least one element of the theory (surjective property), since there are more elements of theory than there are elements of reality, necessarily, some elements of reality correspond to more than one element of the theory: The theory is "overcomplete" (redundant). (c) There are as many elements of theory as there are elements of reality and both are in one-to-one correspondence with each other (bijective property). The theory is complete. However, although complete, the *quality* of the theory is a different question.

relationship between theory and reality. The theory space $\mathcal{T}$, depicted on the left of Figure 6.1(a), contains *concepts*, usually mathematical entities, that are described and manipulated using the language and procedures of mathematics and logical reasoning. In the case of Figure 6.1(a) the theory space $\mathcal{T}$ contains two concepts, represented by the full plot symbols. The space of objective reality is depicted on the right of Figure 6.1(a). It contains *elements of reality* represented by open plot symbols. Elements of reality and the interactions between them are properties of the world around us. They are discovered with the help of experiments. In the case of Figure 6.1(a) the reality space $\mathcal{R}$ contains three elements of reality. The discipline of theoretical physics concerns itself with establishing a mapping between the theory space and the space of objective reality. There are many ways of establishing mappings between $\mathcal{T}$ and $\mathcal{R}$. Let us examine the case shown in Figure 6.1(a). The concepts, represented by the full plot symbols, are mapped to the elements of reality, represented by the open plot symbols. Each concept in $\mathcal{T}$ corresponds to exactly one element of reality in $\mathcal{R}$. This type of mapping is called *injective*. However, there is one element of reality, represented by the open diamond, which does not correspond to a concept in $\mathcal{T}$. Therefore, the mapping is *incomplete*. In an attempt to complete the mapping, as shown in Figure 6.1(b), let us add additional concepts to $\mathcal{T}$ and map the new concepts to $\mathcal{R}$. Indeed, every element of reality in $\mathcal{R}$ now corresponds to at least one concept in $\mathcal{T}$. Mappings of this kind are called *surjective*. However, two concepts in $\mathcal{T}$ correspond to a single element of reality in $\mathcal{R}$. We call this kind of mapping redundant. It can be made nonredundant by pruning $\mathcal{T}$, i.e., in the case of Figure 6.1(b), by eliminating the concept represented by the full star. The result is shown in Figure 6.1(c). Now, each concept in $\mathcal{T}$ corresponds uniquely to a single element in $\mathcal{R}$, and vice versa. This kind of mapping is called *bijective* and the corresponding theory is *complete*.

Establishing a mapping between $\mathcal{T}$ and $\mathcal{R}$, however, is only a first, but important, step of building a physical theory. We now have to establish rules according to which concepts of $\mathcal{T}$ are manipulated and evolve. Since the space $\mathcal{T}$ is a product of our imagination, we are completely free to choose any set of rules we like. However, the fundamental rules of the time evolution of the elements of reality in $\mathcal{R}$ are not at all subject to our control. They are dictated to us by nature. Therefore, as a second step of theory building, we now have to find a set of rules acting in $\mathcal{T}$ such that an evolution of concepts in $\mathcal{T}$ exactly mimics the actual time evolution or behavior of elements of reality in $\mathcal{R}$. This is illustrated in Figure 6.2(a): Each movement in $\mathcal{R}$ is precisely emulated by a corresponding movement in $\mathcal{T}$. In particular, if an element of reality ρ, corresponding to an element of the theory τ, evolves into ρ', then τ evolves into τ', and a one-to-one

6.2 • HALLMARKS OF PHYSICAL THEORIES 173

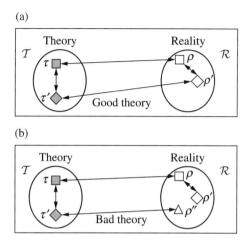

Figure 6.2 Assessment of the quality of a theory. An element of reality evolves from ρ to ρ', while its corresponding element of the theory evolves from τ to τ'. (a) The mapping between the elements of the theory and the elements of reality "tracks" precisely, i.e., the diagram establishing the connection between τ, ρ, ρ', and τ' commutes. This is the hallmark of a good theory. (b) The image of τ' is $\rho'' \neq \rho'$. The mapping does not "track" and the diagram does not commute. This is the hallmark of a bad theory.

mapping exists between τ' and ρ'. In mathematics the illustration in Figure 6.2(a) is known as a *commuting diagram*. It does not matter which way we move along the arrows: If we first move from τ to τ' in $\mathcal{T}$, and then evoke the mapping that exists between τ' and ρ', we can *predict* ρ', an element of reality in $\mathcal{R}$. If we start from $\tau \leftrightarrow \rho$, *observe* ρ evolve into ρ', and then map back to $\mathcal{T}$, we obtain τ', the same result as if we had remained in $\mathcal{T}$ and merely executed the computations that lead from τ to τ' in $\mathcal{T}$. If we can establish a commuting diagram in the sense of Figure 6.2(a), we have an excellent, complete theory of reality.

However, it may also happen that an element of reality ρ in $\mathcal{R}$ evolves into ρ' in $\mathcal{R}$, but the concept τ in $\mathcal{T}$, corresponding initially to ρ, evolves into τ' whose mapping image in $\mathcal{R}$ is ρ'', different from ρ' (see Figure 6.2(b)). This is a frequent situation in theory building, and often of not much concern as long as the theory is complete and each element of reality always has a concept counterpart in $\mathcal{T}$. Discrepancies between ρ'' predicted by theory and the actual result ρ', determined, for instance, by measurement, merely indicate that the theory is not very good. Apparently, the quality of a theory is determined by how well the diagrams in Figure 6.2 commute. Therefore, the best theories are the ones in which the diagrams in Figure 6.2 commute exactly (see Figure 6.2(a)), and where reality (space $\mathcal{R}$) is mapped onto a mathematical "game" (played in space $\mathcal{T}$) whose rules are known. Then, by "working things out" in the theory domain we can *predict reality*. Also, since the rules of the theory domain are usually simpler and more easily controlled and manipulated than reality itself, which usually requires complicated and costly experiments, the theory domain and its rules give us a complete *model of reality*.

In summary, a *complete* theory is one for which each element of reality in $\mathcal{R}$ has a counterpart concept in $\mathcal{T}$, and vice versa. A good theory is one for which the diagrams in Figure 6.2 commute. We also note that a good theory is not necessarily complete, and a complete theory is not necessarily good.

Exercises:

6.2.1 As a theory of nature is classical mechanics (a) good, (b) complete?

6.2.2 Historically, facilitating the computation of products, logarithms provided one of the first applications of "commuting diagrams." Draw a commuting diagram to illustrate how this works.

6.2.3 In 1905 Einstein showed that for large speeds Newtonian mechanics makes inaccurate predictions. Why do we still study (and teach) Newtonian mechanics?

6.3 EPR and Reality

The paper by Einstein, Podolsky, and Rosen (EPR), published in 1935, asks the following question: "Can the quantum mechanical description of physical reality be considered complete?" EPR readily concede that quantum mechanics is a correct theory within its realm of applicability. But they insist that quantum mechanics is an *incomplete* theory. Thus, EPR are convinced that in the case of quantum mechanics we are dealing with the situation portrayed in Figure 6.1(a), i.e., quantum mechanics establishes an injective mapping into $\mathcal{R}$, but not all elements of $\mathcal{R}$ are covered. But in admitting the *correctness* of quantum mechanics they also agree that quantum mechanics, within its reach of elements of reality in $\mathcal{R}$, establishes exactly commuting diagrams in the sense of Figure 6.2(a). It is important to realize that EPR do not mount an all-out attack on quantum mechanics; by 1935, and fully aware of a decade of quantum success, calling quantum mechanics an incorrect theory would not have been prudent. By conceding the success of quantum mechanics in all applications known in 1935, EPR concentrate exclusively on its *completeness*, i.e., on the question of whether quantum mechanics describes all aspects of reality.

EPR deny that quantum mechanics is a complete theory. To make their case, they attempt to construct elements of reality that have no counterpart in quantum mechanics. This way, they try to prove that a bijective mapping between reality and quantum mechanics in the sense of Figure 6.1(c) does not exist. If EPR are right, this would indeed show that quantum mechanics is an *incomplete* theory of nature.

EPR define completeness in the following way:

$$\boxed{\text{Every element of the physical reality must have a counterpart in the physical theory.}} \quad (6.1)$$

This corresponds precisely to our definition of completeness in Section 6.2; we have no problem with it. EPR, in their 1935 publication, also provide a sufficient criterion for "element of reality":

$$\boxed{\begin{array}{l}\text{"If, without in any way disturbing a system,}\\ \text{we can predict with certainty}\\ \text{(i.e., with probability equal to unity)}\\ \text{the value of a physical quantity,}\\ \text{then there exists an element of physical reality}\\ \text{corresponding to this physical quantity."}\end{array}} \quad (6.2)$$

We illustrate this definition with an example taken from classical mechanics. Consider a mass point in free motion. According to Figure 6.3, the velocity at a distance of 1 m from the starting point is *real*, because, within classical mechanics, we can predict it with certainty.

We are now going to set the stage for the EPR argument. The fundamental concept of quantum mechanics is the *state*, i.e.,

$$|\psi\rangle. \quad (6.3)$$

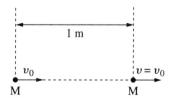

Figure 6.3 Mass point in free motion.

According to conventional quantum mechanics, it represents all we can possibly know about a system. In x representation,

$$\langle x|\psi\rangle = \psi(x), \qquad (6.4)$$

we obtain the *wave function* $\psi(x)$ of the quantum system. In Section 3.3 we established that an operator $\hat{\theta}$ corresponds to each physically observable quantity θ. If $\psi_\vartheta(x)$ is an eigenfunction of $\hat{\theta}$,

$$\hat{\theta}\,\psi_\vartheta(x) = \vartheta\,\psi_\vartheta(x), \qquad (6.5)$$

where ϑ is a real number, then the observable θ has the value ϑ with *certainty* in a state with wavefunction $\psi_\vartheta(x)$. It follows that an element of reality, namely ϑ, corresponds to the observable quantity θ.

As an example, let us examine the momentum p of a particle in one dimension. The physical observable p corresponds to the operator $\hat{p} = -i\hbar\partial/\partial x$. The wave functions $\psi_{p_0}(x)$, where p_0 is a real number, are eigenfunctions of the momentum operator according to:

$$\hat{p}\,\psi_{p_0}(x) = p_0\,\psi_{p_0}(x). \qquad (6.6)$$

It follows that the momentum p of a particle in the wave function $\psi_{p_0}(x)$ has the sharp value p_0 with certainty. Therefore, an element of reality, i.e., p_0, corresponds to $\psi_{p_0}(x)$.

Now consider the position operator $\hat{x}$:

$$\hat{x}\,\psi_{p_0}(x) = x\psi_{p_0}(x) \neq x_0\,\psi_{p_0}(x), \qquad (6.7)$$

where x_0 is some constant. Therefore, the coordinate is *not real* in the wave function $\psi_{p_0}(x)$. Thus, as a general result, we obtain the following: If the momentum of a particle is known exactly, its coordinate has no physical reality.

In Section 3.3 we learned that if the operators $\hat{P}$ and $\hat{Q}$ associated with two observables P and Q commute, i.e., $[\hat{P},\hat{Q}] = 0$, then P and Q may both have sharp values in a state $|\psi\rangle$, i.e., both P and Q are real. Conversely, if $[\hat{P},\hat{Q}] \neq 0$, P and Q are never simultaneously sharp in any state $|\psi\rangle$. Therefore, in this case, P and Q do not have simultaneous reality. For instance, p and x cannot both be simultaneously sharp, and therefore cannot both be simultaneously real in $\psi_{p_0}(x)$ since $[\hat{x},\hat{p}] = i\hbar \neq 0$ prohibits this for *any* state $|\psi\rangle$.

At this point EPR argue in the following way. If the quantum mechanical description of reality by $|\psi\rangle$ is complete, then, if $[\hat{P},\hat{Q}] \neq 0$, P and Q do not have simultaneous reality. If, however, a counterexample can be constructed such that for some system P and Q are real despite the fact that $[\hat{P},\hat{Q}] \neq 0$, then the quantum mechanical description of reality

6.3 • EPR AND REALITY

by $|\psi\rangle$ is incomplete. EPR summarize this argument in the following statement:

> (1) Either the quantum mechanical description of reality by $\psi(x)$ is incomplete, or (2) if $[\hat{P}, \hat{Q}] \neq 0$, P and Q do not have simultaneous reality. (6.8)

Two comments are in order. (i) We should emphasize that the "or" used in Statement 6.8 is an *exclusive or*, i.e., either (1) is true or (2) is true, but not both. (ii) By "$\psi(x)$" EPR mean a wave function with as many variables as are necessary for the proper quantum mechanical description of a system. In the following examples "x" stands for the two variables x_1 and x_2.

In quantum mechanics we assert that $\psi(x)$ <u>is</u> the complete description, i.e., we assert that alternative (2) of Statement 6.8 is the correct one. And indeed, in one-dimensional, single-particle quantum systems there never is a problem.

The EPR *paradox* arises when we consider many-particle systems, or systems consisting of several subsystems. For this case EPR constructed such an ingenious counterexample that we may, at least temporarily, be tempted to question our choice of alternative (2) of Statement 6.8.

Consider the situation depicted in Figure 6.4. It shows the space-time diagram of two quantum systems, System I and System II, scattering off of each other. Initially, at time $t = t_i$, Systems I and II are located so far apart that no interaction exists between them. We know the initial wave functions $\psi_I(x_1)$ and $\psi_{II}(x_2)$ of the separated systems. Therefore,

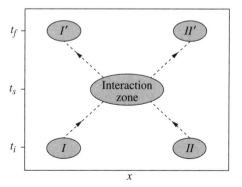

Figure 6.4 The EPR problem. Two systems, *I* and *II*, initially (for $t = t_i$) far apart, scatter off each other in the interaction zone at $t = t_s$. At $t = t_f$ the two subsystems are again spatially well separated. However, I' and II' may now be entangled.

since Systems I and II are initially too far apart to interact, the initial wave function $\psi_i(x_1, x_2)$ of the system is the product

$$\psi_i(x_1, x_2) = \psi_I(x_1)\psi_{II}(x_2). \tag{6.9}$$

The two systems now fly toward each other. At around time t_s, for a brief time interval Δt, the two systems are close enough to interact. Following their interaction, the two systems fly apart. We wait until, at time t_f, the two systems have separated far enough so that no interactions exist any more between them. At time $t = t_f$, in the interaction-free regime after the scattering, the wave function is $\psi_f(x_1, x_2)$. Because of the scattering, $\psi_f(x_1, x_2)$ may be a complicated wave function, certainly more complicated than the simple product in Equation 6.9. However, no matter how complicated the interaction between the two systems had been in the scattering region, we can compute $\psi_f(x_1, x_2)$ exactly, at least in principle, by solving Schrödinger's equation.

We now make a measurement to examine the wave functions of Systems I and II, respectively. The measurement will collapse the wave function of the combined system. Since the EPR argument rests fundamentally on this wave function collapse, we will now study this process in detail.

Let A be an observable pertaining to System I and $\hat{A}$ its associated operator. Then:

$$\hat{A}\, u_n(x_1) = a_n\, u_n(x_1), \tag{6.10}$$

where $u_n(x_1)$ are the normalized eigenfunctions of $\hat{A}$, and a_n are the associated eigenvalues. Since $\hat{A}$ is an operator associated with a physical observable, $\hat{A}$ is a Hermitian operator and therefore $\{u_n(x_1)\}_{n=1}^{\infty}$ is an orthonormal and complete system of functions. We can use this system of functions to expand the final wave function $\psi_f(x_1, x_2)$ of the system of Figure 6.4 in the following way:

$$\psi_f(x_1, x_2) = \sum_{n=1}^{\infty} \varphi_n(x_2)\, u_n(x_1). \tag{6.11}$$

The functions $\varphi_n(x_2)$ are the amplitudes with which the functions $u_n(x_1)$ occur in $\psi_f(x_1, x_2)$.

Now, measure the property A in $\psi_f(x_1, x_2)$. Suppose the result is a_k. Then, we know that after the measurement System II is in the state $\varphi_k(x_2)$ and the wave function of the combined system after the measurement is

$$\psi^{(\text{after})}(x_1, x_2) = \mathcal{N}\varphi_k(x_2)\, u_k(x_1), \tag{6.12}$$

where $\mathcal{N}$ is a normalization constant. This illustrates the process of the collapse of the wave function in the EPR scattering case. The infinite series defined in Equation 6.11 collapses to the single term defined in Equation 6.12. We also notice that measuring one system, we can force the other system into a specific quantum state. This reflects the active role of measurement in quantum mechanics discussed in Chapter 4. This is an important observation for the EPR argument. It is also important as one of the basic principles used in quantum computing as we will see later in Chapters 8 and 10.

But we could as well have chosen an observable B pertaining to System I to make our measurement. In this case we have

$$\hat{B}\, v_m(x_1) \;=\; b_m\, v_m(x_1), \tag{6.13}$$

where $\hat{B}$ is the Hermitian operator corresponding to the observable B, $v_m(x_1)$ are its normalized eigenfunctions, b_m are its eigenvalues, and $\{v_m(x_1)\}_{m=1}^{\infty}$ is an orthonormal system of functions. Using the orthonormal set $\{v_m(x_1)\}_{m=1}^{\infty}$, we obtain the following alternative, but equally valid expansion of the wave function $\psi_f(x_1, x_2)$:

$$\psi_f(x_1, x_2) \;=\; \sum_{m=1}^{\infty} \phi_m(x_2)\, v_m(x_1), \tag{6.14}$$

where $\phi_m(x_2)$ are the expansion coefficients. Suppose that, as a result of measuring the observable B, we obtain the value b_r. Then we know that the System I is in the wave function $v_r(x_1)$, the System II is in the wave function $\phi_r(x_2)$, and the wave function of the combined system after the measurement is

$$\psi^{(\text{after})}(x_1, x_2) \;=\; \mathcal{N}'\, \phi_r(x_2)\, v_r(x_1), \tag{6.15}$$

where $\mathcal{N}'$ is a normalization constant.

From Equations 6.12 and 6.15 we learn that, depending on the type of measurement performed on System I', System II' may be left in two different wave functions, in our case $\varphi_k(x_2)$ or $\phi_r(x_2)$. Now, here is the problem: Since the two systems, I' and II', according to assumption, are so far apart that they no longer interact, how does System II' know how to respond to the measurement done on System I'? To illustrate the severity of the problem: Let System I' reside on Earth, while System II' is located on the star Sirius, 8.6 light years from Earth. Nevertheless, System II' responds *instantaneously* to the choice of measurement (A or B) performed on System I', and finds itself in correspondingly different states. Einstein called this "spooky action at a distance" and strongly disliked this feature of quantum mechanics.

So here is a first paradox: In order to tell System II' which measurement had been performed on Earth, and to ensure *instantaneous* collapse of the wave function on Sirius, some superluminal "message" had to be transmitted. A curious type of message, however, that did not transmit any information, since superluminal information transmission is not permitted by Special Relativity. We call this kind of communication *nonlocality*. This is the part of the EPR paradox that is still a mystery to this very day.

And here is a second paradox: If we assign a *reality* to System II', there is a problem, since the *same* reality (i.e., System II') leads to two different measurement outcomes! The reason why we may want to assign a reality to System II' is this: Given $\psi_i(x_1, x_2)$ at time $t = t_i$, its time evolution, using the *deterministic* Schrödinger equation, is well defined for all times $t > t_i$, and is therefore always the same at $t = t_f$. Therefore, $\psi_f(x_1, x_2)$ is uniquely defined.

At this point EPR make their decisive argument. They construct an example for which $\varphi_k(x_2)$ is an eigenfunction of an operator $\hat{P}$ corresponding to an observable P and $\phi_r(x_2)$ is an eigenfunction of an operator $\hat{Q}$ corresponding to an observable Q, with

$$\hat{P}\varphi_k(x_2) = p_k \varphi_k(x_2),$$
$$\hat{Q}\phi_r(x_2) = q_r \phi_r(x_2), \qquad (6.16)$$

and

$$[\hat{P}, \hat{Q}] \neq 0. \qquad (6.17)$$

But then, EPR argue, they have extracted precise values p_k and q_r from the *same* state for System II', although $[\hat{P}, \hat{Q}] \neq 0$. According to this reasoning alternative (2) of Statement 6.8 is wrong. Therefore, EPR felt compelled to choose alternative (1) of Statement 6.8, i.e., quantum mechanics is not complete.

EPR admit that one could immediately level the following objection against their reasoning: Yes, sharp values for P and Q were extracted, but *not simultaneously*. In fact, two independent measurements were performed, one after another, on two identically prepared systems. EPR do not feel that this fact provides a decisive counter argument against their reasoning. And indeed the modern interpretation and resolution of the EPR paradox does not rely on this line of argument. The EPR argument is flawed, but not because of this reason alone. The EPR argument rests decisively on the assumption that System II' is in a *state*. This is not true. System II' is *entangled* with System I'. Neither one of these two systems, after having interacted with each other, is in a state by itself. Therefore, one of the fundamental assumptions of the EPR argument, namely that II' has independent reality, is flawed. Only the quantum state of the *combined* system, composed of I' and II' has meaning. Thus, EPR did not prove that quantum mechanics is incomplete.

Although the EPR argument is flawed, the historical significance of the EPR "paradox" lies in the fact that it contributed *substantially* to the clarification and sharpening of quantum mechanics itself. Primarily, as we saw previously, to the concept of a system, and the system's wave function. The failure of the EPR argument showed that there are no additional elements of reality that would not already be included in the conventional formalism of quantum mechanics. Therefore, as far as we know now, quantum mechanics is a *complete* theory of nature.

An even sharper defense of quantum mechanics was formulated by Bell in 1964. With the help of his famous inequality, to be discussed in the following section, he rules out all hope of ever discovering more about reality than is already contained in quantum mechanics.

Exercises:

6.3.1 Show that the x and z components of the spin of a spin-1/2 particle do not have simultaneous reality.

6.3.2 Show that the amplitudes $\varphi_n(x_2)$ and $\phi_m(x_2)$ in Equations 6.11 and 6.14 satisfy

$$\sum_{n=1}^{\infty} \int_{-\infty}^{\infty} |\varphi_n(x_2)|^2 \, dx_2 = 1, \quad \sum_{m=1}^{\infty} \int_{-\infty}^{\infty} |\phi_m(x_2)|^2 \, dx_2 = 1.$$

6.3.3 The functions

$$u_n(x) = \sqrt{\frac{2}{a}} \sin\left(\frac{n\pi}{a}x\right), \quad n = 1, 2, \ldots,$$

form a complete, orthonormal set of functions in the space $0 \leq x \leq a$ with the boundary condition that wave functions vanish at $x = 0$ and at $x = a$. Two quantum particles, labeled 1 and 2, with coordinates x_1 and x_2, respectively, are described by the wave function

$$\psi(x_1, x_2) = \mathcal{N} x_1 x_2 \sin\left(\frac{\pi x_1}{a}\right) \sin\left(\frac{\pi x_2}{a}\right),$$

where $\mathcal{N}$ is a normalization constant. Compute the amplitudes $\varphi_n(x_2)$ in the expansion

$$\psi(x_1, x_2) = \sum_{n=1}^{\infty} \varphi_n(x_2) u_n(x_1).$$

6.4 Bell's Theorem

According to EPR, quantum mechanics yields correct results within its range of applicability, but often fails to predict the precise values of observables that can be predicted with classical mechanics without problems. Therefore, according to EPR, quantum mechanics is an incomplete description of nature. Analyzing the EPR argument in depth, we notice that the EPR argument is as much about the nature of reality in our world as it is about completeness or incompleteness of quantum mechanics. EPR have a certain preconceived notion about reality that is based on our experiences with the "classical world." To illustrate: Suppose we have two marbles, a red one and a green one. Without looking, one of our marbles is placed into a box to the left of us, and one is placed in a box to the right of us. The boxes are then sealed so that we have no way of telling which box contains which marble. One thing is for sure, though: Even without opening either of the boxes we know that one of the boxes contains a red marble and one of the boxes contains a green marble. We call this a *preexisting reality*. A preexisting reality does not care whether we know about it or not. It exists anyway; in our example it is the fact that, without looking, one of the boxes *already* contains a red marble and the other box *already* contains a green marble. We open the left box: It contains a green marble. *Instantaneously* we know that the right box contains a red marble. From the point of view of a preexisting reality this is not surprising. It has nothing to do with Einstein's "spooky action at a distance." By opening the left box we merely discover a reality that had existed all along. Deep down this is the argument that EPR make in connection with Systems I and II, which are reminiscent of our two boxes containing the marbles, and where the marbles and their colors are reminiscent of the observables being measured in Systems I and II. But the original EPR argument presented in Section 6.3 is not easy to understand intuitively since it is couched in much technical language and even the EPR system itself (see Figure 6.4) is a complicated scattering system. Therefore, in the 1950s, Bohm recast the original EPR system into a much simpler system that employs a minimal amount of formalism, maximally emphasizing the physics.

Bohm noticed that for the EPR argument to fly, it is not necessary to treat a full scattering system such as the one shown in Figure 6.4. It is enough to consider a decaying system, which in "EPR language" would be equivalent to a "half-scattering system," such as the scattering system shown in Figure 6.4 for $t > t_s$. Bohm chose the system shown in Figure 6.5. Bohm's EPR system contains the same physics as the EPR system discussed in Section 6.3, but it is far simpler to treat formally.

6.4 • BELL'S THEOREM   183

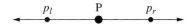

Figure 6.5   Bohm's EPR setup consisting of a particle P, initially at rest, decaying into two spin-1/2 particles p_l and p_r. Particle p_l flies off to the left while particle p_r flies off to the right.

A spin-0 particle P, originally at rest, decays into two spin-1/2 particles, p_l and p_r. Because of momentum conservation, the two particles fly away from each other, back to back, with the same absolute value of linear momentum. Since, before the decay, the total angular momentum S is zero, and since $S_z = 0$, too, quantum mechanics dictates that the two particles must be in the singlet state

$$|\psi\rangle = \frac{1}{\sqrt{2}} \left[|\uparrow_l\rangle|\downarrow_r\rangle - |\downarrow_l\rangle|\uparrow_r\rangle \right] \tag{6.18}$$

of spin angular momentum, where the subscripts l and r refer to p_l and p_r, respectively. Since Bohm's EPR system works with spin-1/2 particles, the Hilbert space of each one of the particles is only two-dimensional (spin-up and spin-down) compared with the infinite-dimensional Hilbert space of the EPR example ($\{u_n(x_1)\}_{n=1}^{\infty}$, $\{v_m(x_1)\}_{m=1}^{\infty}$). Bohm's EPR system is also a direct analogue of our classical "box-and-marble" system, where the marbles stand for the two particles, p_l and p_r, the two colors, "green" and "red," stand for spin-up and spin-down, respectively, and the sealed boxes stand for our ignorance of the spins of the two particles before a measurement occurs.

With the help of Bohm's EPR system and our classical box-and-marble system, we can now bring the EPR argument to the point: When a marble is placed into the left box, it already has its assigned color. We may not know this color, but it is there. In Bohm's decay system, the left particle flies away from the point of decay and the question is: While still in flight and not yet measured, does p_l—in analogy to a classical marble—already possess the spin that will be measured when p_l finally arrives at the measuring instrument? The EPR school of thought emphatically agrees: The particle p_l, although we have not yet measured it, already possesses a definite spin value while still in flight. The quantum measurement simply discovers this preexisting reality. Since according to this philosophy the particle already possesses a definite value of its spin at every point of its journey to the detector, we call this train of thought the philosophy of *local realism*. If it is true that p_l immediately after the decay already possesses a definite value of its spin, then quantum mechanics certainly does not know about it, since it describes the system consisting of p_l and p_r with the wave function defined in Equation 6.18, which is a superposition of two situations, one in which p_l has spin-up and one in which p_l has spin-down. Therefore, if local realism is correct, then p_l has a definite value of its spin, but quantum mechanics, according to Equation 6.18, is ambiguous about this issue, and cannot predict it. This is why EPR call quantum mechanics incomplete.

At this point, thanks to Bohm's EPR system, we have a much more intuitive understanding of what EPR actually meant. And since Einstein was involved in the argument, no wonder that the points raised go straight to the foundations of how our universe works at its most fundamental level. Following EPR we are actually exploring the nature of reality in our world. In particular, does our world work in terms of local realism, or is the way nature "does business" more closely related to nonlocal action at a distance, the notion so reviled by Einstein? To resolve these questions seems more like a task for an analytical philosopher than a physicist. It therefore came as a complete surprise when John Bell, in 1964, was able to *prove* that local realism is wrong and quantum mechanics—and therefore our world—is fundamentally nonlocal. Many scientists and historians of science agree that Bell's discovery ranks as one of the most profound discoveries of 20th century physics.

The basic idea of Bell's proof is the following. We analyze Bohm's EPR system shown in Figure 6.5 in two different ways: assuming the validity of local realism in one way, and assuming the validity of quantum mechanics in another. Surprisingly, there are situations in which the two theories make contradictory predictions, i.e., Bell showed that local realism and quantum mechanics are incompatible with each other! At this point, by turning the question over to experiment, we let nature decide which of the two theories is correct. It turns out that nature favors quantum mechanics and is in disagreement with the predictions of theories based on local realism. This way Bell proved that quantum mechanics is correct and complete and that local realism is an incorrect description of nature. We should not hesitate to point out that the demise of local realism is a revolution of the same order of magnitude as, say, Special Relativity or quantum mechanics itself. It changes our fundamental understanding of nature and flies squarely into the face of our "classical intuition" illustrated with the box-and-marble system.

We are now ready for the formal proof of Bell's theorem. Bell's proof starts by assuming the correctness of local realism. As a consequence of this assumption, we have to accept that quantum mechanics is incomplete. In order to arrive at a more complete quantum theory, we introduce additional variables λ into the wave function, whose values immediately before performing a measurement allow us to predict the outcome of this measurement with certainty. Since physicists have not yet discovered these additional variables, they are called *hidden variables*, and quantum theories that contain them are called *hidden variable theories*. There are many ways of extending quantum mechanics with the help of hidden variables. We are interested in constructing theories that implement local realism. Therefore, we call these theories *local hidden variable theories*. There are countless ways to construct local hidden variable theories and it seems like a daunting job to disprove them all, one-by-one. Amazingly, according to Bell, this is not necessary. By showing that the very idea of introducing additional variables into the wave function is fundamentally flawed and yields results that are inconsistent with experiment, Bell ruled out the whole class of local hidden variable theories, without the necessity of constructing a single one explicitly.

Bell's argument (1964) is the following: Instead of measuring only the z components of the spins of p_l and p_r, let us measure their spins along some arbitrary directions $\boldsymbol{a}$ (for p_l) and $\boldsymbol{b}$ (for p_r), respectively. The corresponding measurement operators are $\hat{\boldsymbol{\sigma}} \cdot \boldsymbol{a}$ and $\hat{\boldsymbol{\sigma}} \cdot \boldsymbol{b}$, where $\boldsymbol{a}$ and $\boldsymbol{b}$ are unit vectors. Since both p_l and p_r are spin-1/2 particles, no matter what the directions of $\boldsymbol{a}$ and $\boldsymbol{b}$, we always obtain ± 1 as the eigenvalues of $\hat{\boldsymbol{\sigma}} \cdot \boldsymbol{a}$ and $\hat{\boldsymbol{\sigma}} \cdot \boldsymbol{b}$, and therefore we obtain only ± 1 as a result of the spin measurements. We denote the result of a measurement along $\boldsymbol{a}$ by $\sigma_{\boldsymbol{a}}$ and the result of a measurement along $\boldsymbol{b}$ by $\sigma_{\boldsymbol{b}}$. Then, the only possible values for $\sigma_{\boldsymbol{a}}$ and $\sigma_{\boldsymbol{b}}$ are ± 1. Since $\boldsymbol{a}$ and $\boldsymbol{b}$ are no longer parallel, there is no reason why the spins in the $\boldsymbol{a}$ and $\boldsymbol{b}$ direction should be perfectly correlated. For generic orientations of $\boldsymbol{a}$ and $\boldsymbol{b}$ all four combinations $(\sigma_{\boldsymbol{a}}, \sigma_{\boldsymbol{b}}) = (1,1), (1,-1), (-1,1),$ and $(-1,-1)$ occur. In order to characterize the correlations between $\sigma_{\boldsymbol{a}}$ and $\sigma_{\boldsymbol{b}}$ we compute the expectation value of the product $\sigma_{\boldsymbol{a}} \cdot \sigma_{\boldsymbol{b}}$, i.e., the average

$$P(\boldsymbol{a},\boldsymbol{b}) = \frac{1}{N} \sum_{j=1}^{N} \sigma_{\boldsymbol{a},j} \sigma_{\boldsymbol{b},j} \tag{6.19}$$

over N measurements for $N \to \infty$. Quantum mechanically, $P(\boldsymbol{a},\boldsymbol{b})$ defined in Equation 6.19 is the expectation value of the operator $(\hat{\boldsymbol{\sigma}}^{(l)} \cdot \boldsymbol{a})(\hat{\boldsymbol{\sigma}}^{(r)} \cdot \boldsymbol{b})$ in the state defined in Equation 6.18, i.e.,

$$\begin{aligned} P(\boldsymbol{a},\boldsymbol{b}) &= \langle \psi | (\hat{\boldsymbol{\sigma}}^{(l)} \cdot \boldsymbol{a})(\hat{\boldsymbol{\sigma}}^{(r)} \cdot \boldsymbol{b}) | \psi \rangle \\ &= \frac{1}{2} \{ \langle \uparrow \downarrow | - \langle \downarrow \uparrow | \} (\hat{\boldsymbol{\sigma}}^{(l)} \cdot \boldsymbol{a})(\hat{\boldsymbol{\sigma}}^{(r)} \cdot \boldsymbol{b}) \{ | \uparrow \downarrow \rangle - | \downarrow \uparrow \rangle \} \\ &= \frac{1}{2}(A + B + C + D), \end{aligned} \tag{6.20}$$

where the first arrow in the bra and ket states as well as the operator $\hat{\boldsymbol{\sigma}}^{(l)}$ refer to p_l, the second arrow in the bra and ket states as well as the operator $\hat{\boldsymbol{\sigma}}^{(r)}$ refer to p_r, and

$$A = \langle \uparrow \downarrow | (\hat{\boldsymbol{\sigma}}^{(l)} \cdot \boldsymbol{a})(\hat{\boldsymbol{\sigma}}^{(r)} \cdot \boldsymbol{b}) | \uparrow \downarrow \rangle = -a_z b_z, \tag{6.21}$$

$$B = -\langle\uparrow\downarrow|(\hat{\boldsymbol{\sigma}}^{(l)}\cdot\boldsymbol{a})(\hat{\boldsymbol{\sigma}}^{(r)}\cdot\boldsymbol{b})|\downarrow\uparrow\rangle$$
$$= -(a_x b_x + a_y b_y) - i(a_x b_y - a_y b_x), \qquad (6.22)$$

$$C = -\langle\downarrow\uparrow|(\hat{\boldsymbol{\sigma}}^{(l)}\cdot\boldsymbol{a})(\hat{\boldsymbol{\sigma}}^{(r)}\cdot\boldsymbol{b})|\uparrow\downarrow\rangle$$
$$= -(a_x b_x + a_y b_y) + i(a_x b_y - a_y b_x), \qquad (6.23)$$

and

$$D = \langle\downarrow\uparrow|(\hat{\boldsymbol{\sigma}}^{(l)}\cdot\boldsymbol{a})(\hat{\boldsymbol{\sigma}}^{(r)}\cdot\boldsymbol{b})|\downarrow\uparrow\rangle = -a_z b_z. \qquad (6.24)$$

Together:

$$P(\boldsymbol{a},\boldsymbol{b}) = \frac{1}{2}(A + B + C + D) = -\boldsymbol{a}\cdot\boldsymbol{b}. \qquad (6.25)$$

Is the resulting Equation 6.25 consistent with local hidden-variable theories? Bell's answer is "no." We see this in the following way. In the spirit of a local hidden-variable theory we assume that for each particle decay event some variable λ determines whether we measure $+1$ or -1 for the spin. We define a function $A(\boldsymbol{a},\lambda)$, which predicts the outcome of the spin measurement of p_l for a given measurement direction $\boldsymbol{a}$ and a given value of the variable λ. In the same way we define a function $B(\boldsymbol{b},\lambda)$, which predicts the outcome of a spin-measurement performed on p_r. The assumption of local reality enters in the fact that once λ is chosen, the spins of each of the two particles are predetermined and a measurement of the spin of p_l, predicted by the function A, then depends only on the state of the left measuring device, characterized by $\boldsymbol{a}$, and does not depend on the state of the right measuring device, characterized by $\boldsymbol{b}$, which may be assumed to be far away from the place of measurement of p_l. In the same way B depends only on $\boldsymbol{b}$, and does not depend on $\boldsymbol{a}$. After all, if the measuring instruments that measure the spins of p_l and p_r are sufficiently far apart from each other, and the particles, immediately after being created in the decay of the particle P, have *already* "made up their minds" which spins they are going to reveal in a measurement (here, remember the box-and-marbles example), the setting of $\boldsymbol{a}$ cannot influence the spin direction of p_r, and the setting of $\boldsymbol{b}$ cannot influence the spin direction of p_l.

Since there are only two possible outcomes for the spin measurements, namely ± 1, the allowed values for A and B are ± 1. Since the values ± 1 for the spins of p_l and p_r occur randomly, and since λ determines this outcome uniquely, we have to assume that the values of λ occur randomly, too. We characterize the probability of occurrence of specific λ values with the probability density function $\rho(\lambda)$, normalized such that

$$\int_{-\infty}^{\infty} \rho(\lambda)\, d\lambda = 1. \qquad (6.26)$$

Since $\rho(\lambda)$ is a probability density, and probabilities are never negative, we also require

$$\rho(\lambda) \geq 0 \quad \text{for all } \lambda. \tag{6.27}$$

With the help of the functions $A(\boldsymbol{a}, \lambda)$ and $B(\boldsymbol{b}, \lambda)$ we are now in a position to predict the expectation value, defined in Equation 6.19, of the product of the spins:

$$P(\boldsymbol{a}, \boldsymbol{b}) = \int_{-\infty}^{\infty} \rho(\lambda) A(\boldsymbol{a}, \lambda) B(\boldsymbol{b}, \lambda) \, d\lambda. \tag{6.28}$$

Since the spins of p_l and p_r are in an $S = 0$ state, it follows that if $A(\boldsymbol{b}, \lambda)$ is $+1$, then $B(\boldsymbol{b}, \lambda)$ is -1, and vice versa. In general:

$$B(\boldsymbol{b}, \lambda) = -A(\boldsymbol{b}, \lambda). \tag{6.29}$$

We use this result to write

$$P(\boldsymbol{a}, \boldsymbol{b}) = \int_{-\infty}^{\infty} \rho(\lambda) A(\boldsymbol{a}, \lambda) [-A(\boldsymbol{b}, \lambda)] \, d\lambda$$

$$= -\int_{-\infty}^{\infty} \rho(\lambda) A(\boldsymbol{a}, \lambda) A(\boldsymbol{b}, \lambda) \, d\lambda. \tag{6.30}$$

Now let $\boldsymbol{c}$ be some additional, arbitrary unit vector. Since $\boldsymbol{b}$ is arbitrary, Equation 6.30 holds for the vector $\boldsymbol{c}$, too, and we obtain

$$P(\boldsymbol{a}, \boldsymbol{c}) = -\int_{-\infty}^{\infty} \rho(\lambda) A(\boldsymbol{a}, \lambda) A(\boldsymbol{c}, \lambda) \, d\lambda. \tag{6.31}$$

Taking the difference between Equations 6.30 and 6.31, we obtain:

$$P(\boldsymbol{a}, \boldsymbol{b}) - P(\boldsymbol{a}, \boldsymbol{c}) = -\int_{-\infty}^{\infty} \rho(\lambda) [A(\boldsymbol{a}, \lambda) A(\boldsymbol{b}, \lambda) - A(\boldsymbol{a}, \lambda) A(\boldsymbol{c}, \lambda)] \, d\lambda. \tag{6.32}$$

Since A can only take the values ± 1, its square is always equal to 1, i.e.,

$$A^2(\boldsymbol{b}, \lambda) = 1. \tag{6.33}$$

With Equation 6.33 we may write the integrand of Equation 6.32 in the following way:

$$P(\boldsymbol{a}, \boldsymbol{b}) - P(\boldsymbol{a}, \boldsymbol{c}) = -\int_{-\infty}^{\infty} \rho(\lambda) \left[1 - A(\boldsymbol{b}, \lambda) A(\boldsymbol{c}, \lambda)\right] A(\boldsymbol{a}, \lambda) A(\boldsymbol{b}, \lambda) \, d\lambda. \tag{6.34}$$

Since, independently of its arguments, the function A can only take the values ± 1, we have immediately

$$A(\boldsymbol{b},\lambda)A(\boldsymbol{c},\lambda) = \pm 1. \tag{6.35}$$

This means that the expression

$$[1 - A(\boldsymbol{b},\lambda)A(\boldsymbol{c},\lambda)] \tag{6.36}$$

can only take the values 0 or 2. Therefore:

$$1 - A(\boldsymbol{b},\lambda)A(\boldsymbol{c},\lambda) \geq 0. \tag{6.37}$$

Because of Equation 6.27 we may write Equation 6.37 in the form:

$$\rho(\lambda)[1 - A(\boldsymbol{b},\lambda)A(\boldsymbol{c},\lambda)] \geq 0. \tag{6.38}$$

Returning to Equation 6.34, we may now write:

$$|P(\boldsymbol{a},\boldsymbol{b}) - P(\boldsymbol{a},\boldsymbol{c})| = \left|\int_{-\infty}^{\infty} \rho(\lambda)[1 - A(\boldsymbol{b},\lambda)A(\boldsymbol{c},\lambda)]A(\boldsymbol{a},\lambda)A(\boldsymbol{b},\lambda)\,d\lambda\right|$$

$$\leq \int_{-\infty}^{\infty} |\rho(\lambda)[1 - A(\boldsymbol{b},\lambda)A(\boldsymbol{c},\lambda)]||A(\boldsymbol{a},\lambda)A(\boldsymbol{b},\lambda)|\,d\lambda$$

$$= \int_{-\infty}^{\infty} \rho(\lambda)[1 - A(\boldsymbol{b},\lambda)A(\boldsymbol{c},\lambda)]\,d\lambda. \tag{6.39}$$

To derive the preceding result we used Equation 6.35, from which we obtain

$$|A(\boldsymbol{b},\lambda)A(\boldsymbol{c},\lambda)| = |\pm 1| = 1, \tag{6.40}$$

and the inequality

$$\left|\int_{-\infty}^{\infty} f(x)\,g(x)\,dx\right| \leq \int_{-\infty}^{\infty} |f(x)|\,|g(x)|\,dx, \tag{6.41}$$

which holds for any arbitrary functions f and g. Using the normalization defined in Equation 6.26 together with the definition, Equation 6.28, and its equivalent formulation, Equation 6.30, we may write Equation 6.39 as:

$$|P(\boldsymbol{a},\boldsymbol{b}) - P(\boldsymbol{a},\boldsymbol{c})| \leq 1 + P(\boldsymbol{b},\boldsymbol{c}). \tag{6.42}$$

This equation is known as *Bell's inequality*. Since $\rho(\lambda)$, and, indeed, the parameter λ itself, do not occur in Equation 6.42, Bell's inequality is a universal prediction of any local hidden-variable theory. If quantum mechanics confirms the prediction in Equation 6.42, nothing interesting can be concluded. If, however, we find a situation in which the quantum prediction is inconsistent with Bell's inequality (Equation 6.42), we have a most powerful result: We have a quantitative handle on distinguishing between local hidden variable theories and quantum mechanics! At this point only *one* of these competing theories may be correct, not both!

A quantum prediction that violates Equation 6.42 is indeed not difficult to construct. Let us choose

$$\boldsymbol{a} = \boldsymbol{e}_x, \quad \boldsymbol{b} = \boldsymbol{e}_y, \quad \boldsymbol{c} = \frac{1}{\sqrt{2}}(\boldsymbol{e}_x + \boldsymbol{e}_y), \tag{6.43}$$

where $\boldsymbol{e}_x$ and $\boldsymbol{e}_y$ are the unit vectors in the x- and y-directions. Using Equation 6.25, which is the quantum prediction for $P(\boldsymbol{a},\boldsymbol{b})$, we obtain:

$$P(\boldsymbol{a},\boldsymbol{b}) = -\boldsymbol{a}\cdot\boldsymbol{b} = 0,$$
$$P(\boldsymbol{a},\boldsymbol{c}) = -\boldsymbol{a}\cdot\boldsymbol{c} = -\frac{1}{\sqrt{2}},$$
$$P(\boldsymbol{b},\boldsymbol{c}) = -\boldsymbol{b}\cdot\boldsymbol{c} = -\frac{1}{\sqrt{2}}.$$
$$\tag{6.44}$$

Then:

$$|P(\boldsymbol{a},\boldsymbol{b}) - P(\boldsymbol{a},\boldsymbol{c})| = |0 + \frac{1}{\sqrt{2}}| = \frac{1}{\sqrt{2}} = 0.707\ldots, \tag{6.45}$$

whereas

$$1 + P(\boldsymbol{b},\boldsymbol{c}) = 1 - \frac{1}{\sqrt{2}} = 0.292\ldots. \tag{6.46}$$

Therefore, since, not by any stretch of the imagination, is 0.707 smaller than 0.292, quantum mechanics flatly contradicts Bell's inequality in Equation 6.42 for the choice of measuring directions defined in Equation 6.43, and therefore contradicts the prediction of any local hidden-variable theory.

We now have two results that are incompatible with each other. On the one hand we have Bell's inequality, Equation 6.42, a valid prediction of any local hidden-variable theory, while, on the other hand, we have the prediction of quantum mechanics, which is inconsistent with Bell's inequality, Equation 6.42. Which one is correct? Since both predictions are formally valid predictions of their respective theoretical frameworks (local hidden-variable theories and quantum mechanics, respectively), only experiment can decide! Experiments were indeed performed. It turned out that the results of these experiments are consistent with quantum mechanics and inconsistent with the prediction in Equation 6.42 of local hidden-variable theories. Thus, local hidden-variable theories are proved to be incorrect theories of nature and are ruled out once and for all.

Thus we have the final verdict on EPR's quest for introducing local realism into the framework of quantum mechanics: Local realism is incompatible with how nature works. Events at one point in space may be influenced instantaneously by events happening far away. This is called *nonlocality*. As a result of Bell's inequalities we have to accept that nature is fundamentally nonlocal, a feature of our world correctly portrayed by the theory of quantum mechanics. It is precisely these nonlocalities that are at the foundations of quantum information and quantum computing (see Chapters 7, 8, and 10). Thus, far more than idle metaphysical and philosophical speculation, nonlocality has reached the stage of technical applicability and currently drives many new developments and applications in science and technology.

Exercises:

6.4.1 The four Bell states

$$|\Phi^+\rangle = \frac{1}{\sqrt{2}}\left(|\downarrow\downarrow\rangle + |\uparrow\uparrow\rangle\right), \quad |\Phi^-\rangle = \frac{1}{\sqrt{2}}\left(|\downarrow\downarrow\rangle - |\uparrow\uparrow\rangle\right),$$
$$|\Psi^+\rangle = \frac{1}{\sqrt{2}}\left(|\downarrow\uparrow\rangle + |\uparrow\downarrow\rangle\right), \quad |\Psi^-\rangle = \frac{1}{\sqrt{2}}\left(|\downarrow\uparrow\rangle - |\uparrow\downarrow\rangle\right)$$
(6.47)

are the four maximally entangled states that can be constructed for two spin-1/2 particles. Show that the four Bell states form a complete orthonormal basis in the combined spin space of two spin-1/2 particles. Then show that only $|\Psi^-\rangle$ is odd under exchange of the two particles, while the remaining three states are even under exchange of the two particles.

6.4.2 In the text we outlined the calculations necessary to show that

$$P(\boldsymbol{a},\boldsymbol{b}) = \langle\psi|(\hat{\boldsymbol{\sigma}}^{(l)}\cdot\boldsymbol{a})(\hat{\boldsymbol{\sigma}}^{(r)}\cdot\boldsymbol{b})|\psi\rangle = -\boldsymbol{a}\cdot\boldsymbol{b},$$

where $|\psi\rangle$ is defined in Equation 6.18 and $\hat{\boldsymbol{\sigma}}$ is the vector of Pauli operators. Fill in the details.

6.4.3 For the derivation of Bell's inequality we used the singlet wave function $|\psi\rangle$ defined in Equation 6.18. Show that indeed

$$\hat{S}^2|\psi\rangle = \hbar^2 S(S+1)|\psi\rangle,$$
$$\hat{S}_z|\psi\rangle = \hbar M|\psi\rangle,$$

where $S = M = 0$. Recall that $\hat{\boldsymbol{S}} = \hat{\boldsymbol{s}}^{(1)} + \hat{\boldsymbol{s}}^{(2)}$ is the total spin of two particles "1" and "2", and that $\hat{S}_z = \hat{s}_z^{(1)} + \hat{s}_z^{(2)}$ is its z component.

6.4.4 Prove Bell's inequality

$$|P(\boldsymbol{a},\boldsymbol{b}) + P(\boldsymbol{a},\boldsymbol{c})| \leq 1 - P(\boldsymbol{b},\boldsymbol{c}). \tag{6.48}$$

Then, find a set of unit vectors $\boldsymbol{a}$, $\boldsymbol{b}$, and $\boldsymbol{c}$ for which Bell's inequality as defined in Equation 6.48 is violated when P is computed quantum mechanically.

6.5 Mermin's Reality Machine

Mermin's reality machine is an illustration of the EPR puzzle. The setup is shown in Figure 6.6. In the spirit of Bohm's EPR system, a decay event at $x = z = 0$ generates a pair of particles, p_l and p_r, in the spin singlet state (see Equation 6.18)

$$|\psi\rangle = \frac{|\uparrow_l\rangle|\downarrow_r\rangle - |\downarrow_l\rangle|\uparrow_r\rangle}{\sqrt{2}}. \tag{6.49}$$

Following the decay, particle p_l flies to the left and particle p_r flies to the right toward detectors D_l and D_r, respectively. As shown in Figure 6.6, each of the detectors is equipped with a switch, capable of three settings. In setting 1 the particle's spin is measured with respect to the direction

$$\boldsymbol{n}_1 = \boldsymbol{e}_z, \tag{6.50}$$

where $\boldsymbol{e}_z$ is the unit vector in the z-direction, in setting 2 the particle's spin is measured with respect to the direction

$$\boldsymbol{n}_2 = \frac{\sqrt{3}}{2}\boldsymbol{e}_x - \frac{1}{2}\boldsymbol{e}_z, \tag{6.51}$$

and in setting 3 the particle's spin is measured with respect to the direction

$$\boldsymbol{n}_3 = -\frac{\sqrt{3}}{2}\boldsymbol{e}_x - \frac{1}{2}\boldsymbol{e}_z. \tag{6.52}$$

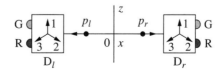

Figure 6.6 Sketch of Mermin's reality machine.

If, as a result of the measurement, the particle's spin is found to be aligned with the measurement direction chosen by a detector's switch setting, the green light ("G" in Figure 6.6) of this detector flashes; if the particle's spin points in the opposite direction, the red light ("R" in Figure 6.6) flashes. If the switches of both detectors are in the same position, we always find a perfect anticorrelation, i.e., if G of D_l flashes, then R of D_r flashes, and vice versa. Since the detectors' switch settings may be changed while the two particles are still in flight, but we always measure the same perfect anticorrelation, classical thinking dictates that the particles must have "agreed" immediately after the decay how to respond to the different switch settings. This is most conveniently expressed by assuming that each particle carries a "gene" that tells the particle how to respond to the different switch settings. If, for instance, particle p_l carries the gene RGR, particle p_l makes D_l flash red if D_l is in switch setting 1, it makes D_l flash green if D_l is in switch setting 2, and it makes D_l flash red if D_l is in switch setting 3. Particle p_r carries the complementary gene GRG to account for the perfect anticorrelation observed if the switches of D_l and D_r are set to the same positions. Table 6.1 lists the eight possible genes:

Since there is nothing special about any of these genes, each of these eight genes occurs with probability 1/8 in any given decay event. We assume that the detectors are 100% efficient such that each decay event always triggers both detectors. Therefore, each decay event leads to one of the following four possible measurement outcomes: GG, GR, RG, RR. The first symbol in this notation refers to the color of the light that detector D_l flashes, and the second symbol refers to the color of the light that detector D_r flashes.

Assuming that the switch of D_l is set to 2 and the switch of D_r is set to 1, we are interested in computing the probability of the measurement outcome GR. According to commonsense reasoning, this probability is

Table 6.1 Gene combinations carried by p_l and p_r.

Gene carried by p_l	Complementary gene carried by p_r
GGG	RRR
GGR	RRG
GRG	RGR
GRR	RGG
RGG	GRR
RGR	GRG
RRG	GGR
RRR	GGG

not difficult to compute. Since the eight gene combinations in Table 6.1 occur with equal probability, all we have to do is determine the number of gene combinations that lead to the measurement outcome GR and divide this number by the total number of gene combinations, i.e., eight. According to Table 6.1, the gene combinations that lead to GR, given that the switch of D_l is in position 2 and the switch of D_r is in position 1, are GGG-RRR and GGR-RRG. Thus, two gene combinations lead to the measurement outcome GR, which means that the commonsense probability for the occurrence of GR is

$$P^{(C)}_{\text{GR}} = \frac{2}{8} = \frac{1}{4}. \tag{6.53}$$

In order to determine the quantum mechanical probability for the measurement outcome GR, we reason in the following way. The measurement operator for p_l is

$$\hat{\boldsymbol{\sigma}} \cdot \boldsymbol{n}_2 = \frac{\sqrt{3}}{2}\hat{\sigma}_x - \frac{1}{2}\hat{\sigma}_z. \tag{6.54}$$

Using

$$\hat{\sigma}_z|\uparrow\rangle = |\uparrow\rangle, \quad \hat{\sigma}_z|\downarrow\rangle = -|\downarrow\rangle,$$
$$\hat{\sigma}_x|\uparrow\rangle = |\downarrow\rangle, \quad \hat{\sigma}_x|\downarrow\rangle = |\uparrow\rangle, \tag{6.55}$$

we verify immediately that

$$|+\rangle = \frac{1}{2}|\uparrow\rangle + \frac{\sqrt{3}}{2}|\downarrow\rangle,$$
$$|-\rangle = \frac{\sqrt{3}}{2}|\uparrow\rangle - \frac{1}{2}|\downarrow\rangle \tag{6.56}$$

are the eigenstates of the measurement operator defined in Equation 6.54 with eigenvalues ± 1, respectively. The measurement operator for p_r is

$$\hat{\boldsymbol{\sigma}} \cdot \boldsymbol{n}_1 = \hat{\sigma}_z, \tag{6.57}$$

whose eigenstates, trivially, are $|\uparrow\rangle$ and $|\downarrow\rangle$ with eigenvalues $+1$ and -1, respectively. In order to measure GR, particle p_l has to be in state $|+\rangle$ and particle p_r has to be in state $|\downarrow\rangle$. Thus, the quantum mechanical probability of measuring GR is

$$P^{(Q)}_{\text{GR}} = |\langle +_l|\langle \downarrow_r|\psi\rangle|^2 = \frac{1}{2}\left|\langle +_l|\langle \downarrow_r|\left(|\uparrow_l\rangle|\downarrow_r\rangle - |\downarrow_l\rangle|\uparrow_r\rangle\right)\right|^2$$
$$= \frac{1}{2}|\langle +|\uparrow\rangle|^2 = \frac{1}{8}. \tag{6.58}$$

We obtain the stunning result that our commonsense prediction for the probability of occurrence of a GR measurement outcome is twice as large as what quantum mechanics predicts. Since our commonsense reasoning was based on the assumption of local reality (the "genes"), Mermin's reality machine provides a simple and direct illustration of the fact that on the fundamental, quantum mechanical level, our world does not operate according to the principle of local reality, but, on the contrary, is fundamentally nonlocal.

Exercises:

6.5.1 The switch of detector D_l of Mermin's Reality Machine is set to position 2; the switch of D_r is set to position 1. In the text we computed $P_{\text{GR}}^{(C,Q)}$ for this switch setting.

(a) Compute the remaining six probabilities $P_{\text{GG}}^{(C,Q)}$, $P_{\text{RG}}^{(C,Q)}$, $P_{\text{RR}}^{(C,Q)}$.

(b) Check your results by confirming that

$$P_{\text{GG}}^{(C,Q)} + P_{\text{GR}}^{(C,Q)} + P_{\text{RG}}^{(C,Q)} + P_{\text{RR}}^{(C,Q)} = 1.$$

(c) You will find that $P_{jk}^{(C)} \neq P_{jk}^{(Q)}$ for all $j, k \in \{G, R\}$. What does this mean?

6.5.2 In the text we determined $P_{\text{GR}}^{(Q)}$ by making a simultaneous measurement of the (p_l, p_r) two-particle state. Alternatively, we could make a sequential measurement by first measuring the spin state of one particle and then, later, the spin state of the other particle. For the switch settings in the text, compute $P_{\text{GR}}^{(Q)}$ by assuming that you first measure the spin state of p_r and then, some time later, the spin state of p_l. Compare your result for $P_{\text{GR}}^{(Q)}$ with the result we obtained in the text.

6.6 Summary

Although Einstein helped create quantum mechanics in the years before 1925, he never accepted the "final product," the quantum mechanics of Heisenberg and Schrödinger, created in 1925–1926. Einstein was particularly displeased with Born's probabilistic interpretation of quantum mechanics, and Heisenberg's Uncertainty Principle, which both imply that quantum theory, unlike classical mechanics, is unable to predict simultaneously exact values of the positions and momenta of particles. In addition, Einstein thought it untenable that reality should be tied to measurements, and that, according to quantum mechanics, a local, objective reality may not even exist between quantum events and measurements. Einstein's dissatisfaction with these "deficiencies" of quantum theory is only natural. Einstein was a classical physicist at heart. He lived in the Newtonian tradition where the attributes of particles, such as positions and momenta, are always objectively real, i.e., existing independently of observers and measuring devices. In fact, with his special and general theories of relativity, he completed the classical worldview. Therefore, fully accepting the successes of quantum mechanics within its range or applicability, he was dissatisfied by what he considered to be essential limitations of quantum mechanics, namely its probabilistic interpretation, and the inability of quantum mechanics to predict sharp results for physical observables that correspond to noncommuting quantum operators. As a consequence, Einstein thought that quantum mechanics is an incomplete theory that is badly in need of improvement. In order to understand what EPR mean by an "incomplete theory," we studied theory building and the properties of physical theories in Section 6.2. Einstein's best stab at proving the incompleteness of quantum mechanics occurred in 1935 when he published a paper co-authored by Podolsky and Rosen, the "EPR paper." In it, EPR present an ingeniously constructed scattering system that seemingly leads to the paradox of being able to predict sharp values of noncommuting physical observables. This is referred to as the "EPR paradox." We reviewed this part of the EPR paper in Section 6.3. However, analyzing in detail the collapse of the EPR wave function, we concluded that EPR's argument is flawed. EPR assigned a definite quantum mechanical state to subsystems of the EPR system that are entangled, and therefore not in a definite state at all. Understanding this part of quantum mechanics resolves the EPR paradox, and leads to a better understanding of quantum systems, subsystems, and entanglement. Thus, although ultimately flawed, the EPR argument contributed decisively to sharpening the concepts and the language of quantum mechanics. In Section 6.4 we presented Bell's analysis of the EPR paradox. Reaching far beyond qualitative, philosophical arguments, Bell proved quantitatively, using mathematical derivations, that the local realism that EPR favored, simply does not exist as a universally valid concept in nature. Bell showed this by proving that local hidden-variable theories, possible implementations of Einstein's "local reality program," are incorrect descriptions of nature. More than an elaboration of an obscure, technical point in quantum mechanics, by proving that local reality is not a fundamental feature of nature, Bell's work is as revolutionary as Einstein's relativity and quantum mechanics themselves, and marks one of the greatest scientific revolutions of 20th century physics. An illustration of the EPR "paradox" is Mermin's

Reality Machine, presented in Section 6.5. It allows a more intuitive and more direct appreciation of the problem of local reality in nature.

Chapter Review Exercises:

1. Prove Bell's inequality:

$$|P(\boldsymbol{a},\boldsymbol{b}) - P(\boldsymbol{a},\boldsymbol{c})| + |P(\boldsymbol{a},\boldsymbol{b}) + P(\boldsymbol{a},\boldsymbol{c})| \leq 2.$$

Is this a useful inequality? Explain why or why not.

2. Show that Bell's inequality in Equation 6.42 is violated for the choice:

$$\boldsymbol{a} = \frac{1}{\sqrt{2}}(\boldsymbol{e}_x - \boldsymbol{e}_y), \quad \boldsymbol{b} = \frac{1}{\sqrt{2}}(\boldsymbol{e}_x + \boldsymbol{e}_y), \quad \boldsymbol{c} = \boldsymbol{e}_y.$$

3. The unit vectors $\boldsymbol{a}$, $\boldsymbol{b}$, and $\boldsymbol{c}$ form an equilateral triangle such that $\boldsymbol{c} = \boldsymbol{a} - \boldsymbol{b}$. Show that, quantum mechanically,

   (a) $P(\boldsymbol{a},\boldsymbol{c}) + P(\boldsymbol{b},\boldsymbol{c}) = 0$,
   (b) $P(\boldsymbol{a},\boldsymbol{b}) + P(\boldsymbol{a},\boldsymbol{c}) = -1$.

4. The switch of detector D_l of Mermin's reality machine is set to position 2; the switch of D_r is set to position 3.

   (a) Compute the eight probabilities $P_{GG}^{(C,Q)}, P_{GR}^{(C,Q)}, P_{RG}^{(C,Q)}, P_{RR}^{(C,Q)}$.
   (b) Check that

   $$P_{GG}^{(C,Q)} + P_{GR}^{(C,Q)} + P_{RG}^{(C,Q)} + P_{RR}^{(C,Q)} = 1.$$

   (c) Check that $P_{jk}^{(C)} \neq P_{jk}^{(Q)}$ for all $j,k \in \{G,R\}$. Briefly explain what this means.

chapter 7

Classical and Quantum Information

In this chapter:

- ◆ Introduction
- ◆ Bits and Qubits
- ◆ Classical Gates
- ◆ Quantum Gates
- ◆ Classical and Quantum Circuits
- ◆ Teleportation
- ◆ Summary

7.1 Introduction

We all are in awe of the incredible power of modern-day computers and information processing systems. But a closer look reveals that these systems are nothing but an enormous number of on-off switches strung together into a logic network. Therefore, conceptually, the working principle of these machines is straightforward to understand. In the span of only a few decades, information and information processing based on these machines has pervaded nearly every aspect of our lives to the extent that following the industrial age, we have entered a new era of human development, the *information age*. Nowadays, all modern consumer electronics contain at least some form of logic circuits, if not an entire microprocessor. The secret of the currently observed rapid advances in microelectronics is miniaturization. More and more transistors are packed onto tiny silicon wafers, which means that more and more logic circuits are realized in the same tiny spaces, and more logic circuits translate into more computer and information processing power. As an example, current microprocessors typically pack anywhere from about 50 million to one billion transistors onto a silicon chip the size of a fingernail. The question is: Can this miniaturization continue forever?

According to Moore's law, which held with astonishing accuracy over the past five decades, the number of transistors that, at affordable cost, can reasonably be packed onto a microchip, increases by about a factor of two every two years. Since current microchips are essentially two-dimensional structures of approximately constant size, this means that the linear dimensions of electronic elements are reduced by a factor of $\sqrt{2}$ every two years. In currently available commercial chips the electronic elements have a linear length scale of about 50 nm. Obviously the reduction of the linear dimensions of electronic components cannot continue forever, since at some point atomic dimensions are reached. Thus, the quantization of matter into atoms and molecules sets a fundamental limit for the size of on-off switches that operate according to the principles of conventional, classical electronics. Since the diameter of a silicon atom is about 2 Å, and assuming that Moore's law continues to hold, we may compute the number of years N that the miniaturization process can continue before the fundamental atomic limit is reached. We obtain:

$$\frac{50\,\mathrm{nm}}{(\sqrt{2})^{N/2}} = 2\,\mathrm{\AA} \quad \Longrightarrow \quad 2^{N/4} = 250 \quad \Longrightarrow \quad N = 32. \quad (7.1)$$

Therefore, the fundamental miniaturization limit is reached in about 30 years. But this, of course, does not mean the end of the road for classical logic circuits and processors. There are many ways out, for instance the fabrication of three-dimensional chip structures or the utilization of parallel computing by having many processors work on a problem simultaneously. Still, when approaching the atomic limit, quantum effects in electronic circuitry will become increasingly important. This is not to say that the transistors employed in current electronic devices are not already quantum devices. They manifestly are: We need quantum mechanics to understand the inner workings of a transistor. However, when employed in electronic circuitry, transistors are operating in "classical mode," i.e., to the outside world, the transistors act like simple classical on-off switches. When approaching the atomic limit, however, the ultimate quantum nature of transistors will become more and more apparent, even to the circuitry that employs them as switching elements, and their quantum nature will have to be taken into account. All this sounds like an unwelcome complication shortly before reaching the fundamental atomic limit. But instead of fearing the "quantum barrier," why not embrace it and investigate if quantum mechanics has qualitatively new features to offer for computing and data processing? This is indeed the case and in this chapter we will learn how to make the transition from classical to quantum computing.

We start in Section 7.2 by introducing the fundamental classical element of information, the *bit*, and its quantum mechanical generalization, the *qubit*. In Sections 7.3 and 7.4 we study classical and quantum logic gates that allow us to make primitive classical and quantum computations. In Section 7.5 we combine gates into circuits and compare their classical and quantum versions. In Section 7.5 we will also encounter our first meaningful quantum circuit that allows us to perform a quantum operation that is so powerful that even for a relatively modest number of qubits it cannot be simulated by any conceivable classical computer. In Section 7.6 we will learn how to transmit quantum information via teleportation. On the fundamental, conceptual level teleportation elucidates the interplay between classical and quantum information and confirms Einstein's cosmic speed limit that prohibits information to travel faster than the speed of light.

7.2 Bits and Qubits

Any kind of classical information can be coded in the form of "yes or no" decisions. Whenever we obtain an answer to a "yes or no" question, we obtain one bit of information. The two possible values of a bit may be represented by any two symbols; the most common are "0" and "1." Physically a bit may be implemented and represented by any device that is capable of exhibiting two distinct states. An example is a switch that is able to open and close an electric circuit. If the switch is open, no current flows, and we identify this situation with the logical "0." If the switch is closed, current flows, which represents the logical "1." In fact, historically, the logical circuits of some of the first digital computers were indeed constructed with the help of hundreds of electromechanical switches. Soon, however, the mechanical switches were replaced by electronic vacuum tubes, and the tubes, subsequently, by transistors, a technology still in use today.

Classical two-state systems have definite technological advantages. They are rugged and relatively robust against error-causing perturbations such as stray electric and magnetic fields. Current rapid progress in microelectronics based on classical two-state systems attests to this fact.

Quantum bits, known as *qubits*, are much more versatile, but also much more delicate. Qubits are not so forgiving when exposed to noise, such as stray electromagnetic fields. But despite this shortcoming they offer such enormous computing power and qualitatively new possibilities for information processing that, for the time being, their fragility takes a back seat and researchers are much more interested in exploring the vast new possibilities of a new quantum information age based on the boundless possibilities of the qubit.

As we saw in Section 3.7, any quantum mechanical two-level system is capable of exhibiting two states, $|0\rangle$ and $|1\rangle$. Therefore, we may represent a classical bit of information with the help of a quantum mechanical two-level system. All that is required is a simple identification such as

$$|0\rangle \equiv 0, \quad |1\rangle \equiv 1. \tag{7.2}$$

But, as we saw in Section 3.7, a quantum mechanical two-level system can do much more. A quantum two-level system may be in a superposition state

$$|\psi\rangle = a|1\rangle + b|0\rangle, \quad |a|^2 + |b|^2 = 1. \tag{7.3}$$

This means that the quantum system may be in the two logical states 0 and 1 at the same time! This realization is at the root of the incredible power of quantum computing and the new field of quantum information. Let us explore this new possibility in some more detail.

Since a and b in Equation 7.3 may be complex numbers, we write

$$a = |a|e^{i\alpha}, \quad b = |b|e^{i\beta} \tag{7.4}$$

with $0 \leq \alpha, \beta < 2\pi$. Also, since $|a|^2 + |b|^2 = 1$, we may parametrize

$$|a| = \cos(\theta), \quad |b| = \sin(\theta), \quad 0 \leq \theta \leq \pi/2. \tag{7.5}$$

The limitation of θ to the interval $0 \leq \theta \leq \pi/2$ is necessary since for $\pi/2 < \theta < 2\pi$ either $\cos(\theta)$ or $\sin(\theta)$ or both are negative, which is incompatible with the fact that they represent the positive quantities $|a|$ and $|b|$. With Equations 7.4 and 7.5, the state $|\psi\rangle$ in Equation 7.3 may be written as

$$|\psi\rangle = e^{i\alpha}[\cos(\theta)|1\rangle + \sin(\theta)e^{i\varphi}|0\rangle], \tag{7.6}$$

where $\varphi = \beta - \alpha$. Since the global phase $\exp(i\alpha)$ in Equation 7.6 cannot be measured, we suppress it and obtain

$$|\psi\rangle = \cos(\theta)|1\rangle + \sin(\theta)e^{i\varphi}|0\rangle. \tag{7.7}$$

This shows that unlike a classical bit, which has only two discrete states, a quantum mechanical qubit has two continuous degrees of freedom, i.e., the two angles θ and φ. Since a qubit depends on two angles, it is tempting to represent a qubit as a point on the surface of a sphere. This is indeed possible. Defining

$$\begin{align} x &= \sin(2\theta)\cos(\varphi), \\ y &= \sin(2\theta)\sin(\varphi), \\ z &= \cos(2\theta), \end{align} \tag{7.8}$$

a qubit with angles θ and φ is represented as a point with coordinates (x, y, z) on the surface of a sphere of radius $x^2 + y^2 + z^2 = 1$ with polar angle 2θ and azimuthal angle φ. This sphere is known as the *Bloch sphere*. It is shown in Figure 7.1. As shown in Figure 7.1, the north pole ($2\theta = 0$) of the Bloch sphere corresponds to the state $|1\rangle$, while the south pole of the Bloch sphere ($2\theta = \pi$) corresponds to the state $|0\rangle$. Had we chosen θ in the transformation defined in Equation 7.8 instead of 2θ, the state $|0\rangle$ would have corresponded to the equator of the Bloch sphere, and the lower half of the Bloch sphere would have gone unused.

Technically, qubits may be implemented by any quantum system that is capable of exhibiting two different quantum states. Examples are the polarization state of a photon ("horizontal"/"vertical"), the spin state of an electron ("spin-up"/"spin-down"), the state of an electronic two-level system ("ground state"/"excited state"), or even the linear momentum state of a photon ("photon going to the left"/"photon going to the right").

Logically, a collection of bits is called a *word*; a collection of qubits is called a *quantum word*. As part of a device, such as, say, a (quantum) computer, a collection of bits is a *register* and a collection of qubits is a *quantum register*. An *n*-bit word, register, or quantum register consists of n bits or qubits. Conceptually, a *computation* is a mapping of an input (quantum) register to an output (quantum) register. The mapping itself is accomplished by means of networks of *logic gates* discussed in Section 7.3 for classical computations and in Section 7.4 for quantum

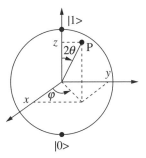

Figure 7.1 The Bloch sphere.

computations. We will find that there are profound conceptual differences between classical and quantum gates.

Exercises:

7.2.1 Bits may be strung together to form n-bit words. An example is the 5-bit word 01100.

  (a) What is the total number of all possible n-bit words?

  (b) How many "yes or no" decisions are necessary to identify a given n-bit word in the set of all possible n-bit words?

7.2.2 Define the operator

$$\hat{\rho} = |\psi\rangle \langle\psi|,$$

where $|\psi\rangle$ is the qubit state defined in Equation 7.7.

  (a) Compute the matrix elements ρ_{00}, ρ_{01}, ρ_{10}, and ρ_{11} of the operator $\hat{\rho}$.

  (b) Show that the Bloch coordinates x, y, and z may be expressed as

$$x = \rho_{10} + \rho_{01},$$
$$y = i(\rho_{10} - \rho_{01}),$$
$$z = \rho_{11} - \rho_{00}.$$

7.2.3 In a certain quantum device a qubit is realized with the help of an electron spin. We identify $|\downarrow\rangle \equiv |0\rangle$ and $|\uparrow\rangle \equiv |1\rangle$. Initially the spin state of the electron is $|\uparrow\rangle$, corresponding to the north pole of the Bloch sphere. At time $t = 0$ a magnetic field is switched on, resulting in an interaction potential $\hat{V} = V_0 \boldsymbol{a} \cdot \hat{\boldsymbol{\sigma}}$ of the electron spin with the magnetic field, where V_0 is a real constant and $\boldsymbol{a}$ is a real unit vector. After time $t = T$ the magnetic field is switched off and the qubit state ends up on the equator of the Bloch sphere with coordinates $x = 1/\sqrt{2}$, $y = 1/\sqrt{2}$, $z = 0$. For this to occur, determine the product $V_0 T$ and the components a_x, a_y, a_z of the unit vector $\boldsymbol{a}$.

7.3 Classical Gates

Bits combined into registers may be used to store information, but by themselves they are incapable of performing computations. In order to perform computations and data processing tasks, we need to combine one or more input bits in order to produce one or more output bits. This is accomplished with the help of logic operations that are technically realized in the form of logic gates. The simplest logic gates are one-bit gates that transform one input bit into one output bit. There are four possibilities:

$$\begin{aligned} \text{I}: \quad & 0 \to 0, \quad 1 \to 1, \\ \text{NOT}: \quad & 0 \to 1, \quad 1 \to 0, \\ \text{ZERO}: \quad & 0 \to 0, \quad 1 \to 0, \\ \text{ONE}: \quad & 0 \to 1, \quad 1 \to 1. \end{aligned} \quad (7.9)$$

Graphically, we represent these gates by boxes as shown in Figure 7.2. The lines leading into and out of the boxes are called *leads*. They remind us of the electrical wires leading into and out of actual electronic realizations of these gates.

The four one-bit gates of Equation 7.9 are not enough to perform meaningful computations. We need to be able to perform logical operations on two or more input bits simultaneously. Conventional two-bit gates that are found in most electronic devices have two input leads and one output lead. This means that a gate G is equivalent to a binary-valued function f_G of two binary variables, i.e.,

$$\text{G}: \quad y = f_G(x_1, x_2), \quad y, x_1, x_2 \in \{0, 1\}. \quad (7.10)$$

There are 16 different two-bit gates G represented by 16 functions f_G. The action of a given gate G is most conveniently represented in the

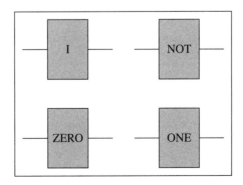

Figure 7.2 Classical one-bit gates.

form of a *truth table* that displays the result, $f_G(x_1, x_2)$, as a function of x_1 and x_2 in the form of a table according to

G	$x_2 = 0$	$x_2 = 1$
$x_1 = 0$	$f_G(0,0)$	$f_G(0,1)$
$x_1 = 1$	$f_G(1,0)$	$f_G(1,1)$

Four of the most important named, conventional two-bit gates are

AND	0	1
0	0	0
1	0	1

OR	0	1
0	0	1
1	1	1

NAND	0	1
0	1	1
1	1	0

NOR	0	1
0	1	0
1	0	0

They are represented graphically as boxes with two input leads and one output lead as shown in Figure 7.3. Notice that gates represented by gate functions of the form defined in Equation 7.10 are irreversible in the sense that, in general, we cannot reconstruct the input bits x_1 and x_2 from knowing only the output bit y.

Irreversible gates destroy information and therefore, necessarily, produce entropy and heat. In order to eliminate this problem, reversible gates with two input leads and two output leads are sometimes investigated. At the present stage of classical computer technology, however, reversible gates do not offer any decisive technological or computational advantages and are therefore, at present, not considered for industrial applications. This is quite different for quantum gates, essential components for building a quantum computer. Since a quantum computer processes quantum information, and, unless a measurement occurs, quantum time evolution is unitary, quantum gates *always* have as many input leads as output leads. Quantum gates are the topic of the following section.

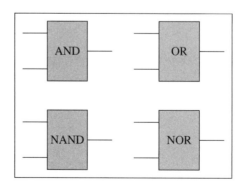

Figure 7.3 Classical two-bit gates.

Exercises:

7.3.1 In the text we claimed that there are 16 two-bit gates. Show that this is indeed the case.

7.3.2 Graphically, a reversible two-bit gate is represented by a box with two input leads and two output leads. Mathematically it is represented by a binary function

$$\boldsymbol{y} = \boldsymbol{f}_G(\boldsymbol{x}), \quad y_1 = f_G^{(1)}(x_1, x_2), \quad y_2 = f_G^{(2)}(x_1, x_2),$$
$$y_1, y_2, x_1, x_2 \in \{0, 1\}.$$

How many different reversible two-bit gates are there?

7.3.3 Connecting its two input leads turns a two-bit gate into a one-bit gate. According to this procedure, what are the one-bit derivatives of the AND, OR, NAND, and NOR gates?

7.3.4 All conventional, classical two-bit gates discussed in this chapter are irreversible in the sense that each application of an irreversible two-bit gate destroys one bit of information. According to Landauer's Principle, each destroyed bit produces an amount $kT\ln(2)$ of heat, where $k = 1.38 \times 10^{-23}$ J/K is Boltzmann's constant and T is the temperature of the gate. Suppose that a state-of-the-art computer chip consisting of 10^9 gates and, running at a temperature of 300 K at a clock speed of 1 GHz, erases 10^{18} bits/second.

(a) Compute the thermal power generated by the chip due to bit erasure.

(b) The Ohmic losses of the chip are 10 W. Compare the thermal power generated due to bit erasure with the heat generated by Ohmic losses. Is the heat generated due to Landauer's Principle a problem for current, state-of-the-art, classical computing technology?

7.4 Quantum Gates

Each of the classical one- and two-bit gates has a quantum version, i.e., a quantum gate. Quantum gates are linear operators. Instead of acting on bits, quantum gates act on qubits. The quantum analogues of the four classical one-bit gates of Equation 7.9 are:

$$\begin{aligned}
\text{QI}: & \quad |0\rangle \to |0\rangle, \quad |1\rangle \to |1\rangle, \\
\text{QNOT}: & \quad |0\rangle \to |1\rangle, \quad |1\rangle \to |0\rangle, \\
\text{QZERO}: & \quad |0\rangle \to |0\rangle, \quad |1\rangle \to |0\rangle, \\
\text{QONE}: & \quad |0\rangle \to |1\rangle, \quad |1\rangle \to |1\rangle.
\end{aligned} \quad (7.11)$$

Only QI and QNOT are useful for quantum computations since only QI and QNOT correspond to invertible, unitary, one-qubit transformations. In addition to QI and QNOT there are two other important one-qubit quantum gates, the Hadamard gate

$$\begin{aligned} \text{H}: \quad |0\rangle &\to \frac{1}{\sqrt{2}}(|0\rangle + |1\rangle), \\ |1\rangle &\to \frac{1}{\sqrt{2}}(|0\rangle - |1\rangle) \end{aligned} \quad (7.12)$$

and the phase rotation gate

$$\begin{aligned} \text{R}(\theta): \quad |0\rangle &\to |0\rangle, \\ |1\rangle &\to e^{i\theta}|1\rangle. \end{aligned} \quad (7.13)$$

Neither the Hadamard gate defined in Equation 7.12 nor the phase rotation gate defined in Equation 7.13 have any classical analogues.

There are many different two-qubit quantum gates. However, only one of them, the controlled-NOT gate, CNOT, is usually considered. It is defined as follows:

$$\begin{aligned} \text{CNOT}: \quad |00\rangle &\to |00\rangle, \\ |01\rangle &\to |01\rangle, \\ |10\rangle &\to |11\rangle, \\ |11\rangle &\to |10\rangle. \end{aligned} \quad (7.14)$$

The CNOT gate is called the "controlled-NOT" gate since it flips the second qubit of a two-qubit state only if the first qubit is $|1\rangle$. Thus, as shown in Equation 7.14, the CNOT gate acts like the identity gate on the second qubit if the first qubit is $|0\rangle$; it acts like the QNOT gate on the second qubit if the first qubit is $|1\rangle$. Therefore, the first qubit acts as a control, "activating" the QNOT feature of the CNOT gate only if the first qubit is $|1\rangle$. The CNOT gate always leaves the first qubit unchanged.

As shown in Figure 7.4, the CNOT gate has two input leads and two output leads. The input and output leads represent the two qubits before and after the gate has acted on them. The top lead, representing the first qubit, is called the *control*. The name is justified since, according to Equation 7.14, the state of the first qubit controls the state of the second qubit of the CNOT gate. The bottom lead is called the *target*, since its state is changed, or targeted, depending on the state of the control qubit.

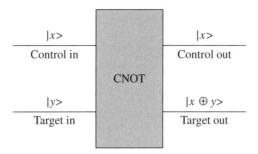

Figure 7.4   Circuit diagram of the CNOT gate.

The action of the CNOT gate may be expressed more concisely in the following way:

$$\text{CNOT}|x\rangle|y\rangle = |x\rangle|x \oplus y\rangle, \quad x, y \in \{0, 1\}, \tag{7.15}$$

where the operation "$\oplus$" is defined as addition modulo 2:

$$x \oplus y = (x + y) \bmod 2 = \begin{cases} 0, & \text{if } x + y \text{ is even,} \\ 1, & \text{if } x + y \text{ is odd.} \end{cases} \tag{7.16}$$

The main difference between classical and quantum gates is that a quantum gate accepts qubits as input whereas a classical gate accepts only bits. This results in the fact that quantum gates have more information processing power and can perform qualitatively new operations not available to classical gates. For instance the CNOT gate can entangle separable states. The following is an example. We start with the separable product state

$$|\psi\rangle = \frac{1}{\sqrt{2}}(|0\rangle + |1\rangle)|0\rangle \tag{7.17}$$

and act with the CNOT gate on it to obtain the entangled state

$$\begin{aligned} \text{CNOT}|\psi\rangle &= \frac{1}{\sqrt{2}} \text{CNOT}(|0\rangle|0\rangle + |1\rangle|0\rangle) \\ &= \frac{1}{\sqrt{2}}(|0\rangle|0\rangle + |1\rangle|1\rangle). \end{aligned} \tag{7.18}$$

The CNOT gate may also be used to disentangle two-qubit states. For instance, if we choose

$$|\varphi\rangle = \frac{1}{\sqrt{2}}(|0\rangle|1\rangle + |1\rangle|0\rangle), \tag{7.19}$$

application of CNOT on $|\varphi\rangle$ yields the product state

$$\begin{aligned}\text{CNOT}\,|\varphi\rangle &= \frac{1}{\sqrt{2}}\,\text{CNOT}\,(|0\rangle|1\rangle + |1\rangle|0\rangle)\\ &= \frac{1}{\sqrt{2}}(|0\rangle|1\rangle + |1\rangle|1\rangle) = \frac{1}{\sqrt{2}}(|0\rangle + |1\rangle)|1\rangle.\end{aligned} \qquad (7.20)$$

Of course our choice of designating the first qubit as the control qubit and the second qubit as the target qubit is arbitrary. We could as well have chosen the second qubit as the control qubit, and the first qubit as the target qubit. This results in the $\overline{\text{CNOT}}$ gate, defined as:

$$\begin{aligned}\overline{\text{CNOT}}\,|0\rangle|0\rangle &= |0\rangle|0\rangle,\\ \overline{\text{CNOT}}\,|0\rangle|1\rangle &= |1\rangle|1\rangle,\\ \overline{\text{CNOT}}\,|1\rangle|0\rangle &= |1\rangle|0\rangle,\\ \overline{\text{CNOT}}\,|1\rangle|1\rangle &= |0\rangle|1\rangle.\end{aligned}$$

Exercises:

7.4.1 The QNOT gate implies the transformation

$$|0\rangle \xrightarrow{\text{QNOT}} |1\rangle, \quad |1\rangle \xrightarrow{\text{QNOT}} |0\rangle.$$

This transformation can be represented by the unitary matrix

$$\text{QNOT} = \begin{pmatrix} 0 & 1 \\ 1 & 0 \end{pmatrix}.$$

There are four different $\sqrt{\text{QNOT}}$ gates, $\text{SQNOT}_0, \ldots, \text{SQNOT}_3$. Show this by computing the four unitary matrices $\text{SQNOT}_0, \ldots, \text{SQNOT}_3$ that satisfy

$$\text{SQNOT}_j^2 = \begin{pmatrix} 0 & 1 \\ 1 & 0 \end{pmatrix}, \quad j = 0, \ldots, 3.$$

7.4.2 State a 4×4 matrix representation of the CNOT gate. Then show that the CNOT gate corresponds to a unitary transformation.

7.4.3 Since the classical NOT gate changes 0 to 1 and 1 to 0, a classical input bit is never invariant under the classical NOT operation. There is, however, a one-qubit state $|\psi\rangle$ that is invariant under the QNOT operation, i.e., $\text{QNOT}|\psi\rangle = |\psi\rangle$. Determine $|\psi\rangle$.

7.4.4 Identify $|1\rangle$ with the column vector $\begin{pmatrix} 1 \\ 0 \end{pmatrix}$ and $|0\rangle$ with the column vector $\begin{pmatrix} 0 \\ 1 \end{pmatrix}$. Then state 2×2 matrix representations for the one-qubit gates QI, QNOT, QZERO, QONE, H, and R defined in Equations 7.11, 7.12, and 7.13, respectively.

7.5 Classical and Quantum Circuits

Feeding the output of a logic gate into the input of another gate, classical and quantum gates may be combined into logic circuits. For instance, feeding the output of an AND gate to a NOT gate, we obtain a logic circuit that is equivalent to the NAND gate:

$$\text{NAND} = \text{NOT AND}. \tag{7.21}$$

This circuit is shown in Figure 7.5(a). A qualitatively different circuit is shown in Figure 7.5(b). It contains a branch point. Branch points copy the content of one lead into another and are a powerful tool of classical logic circuitry. This tool is *not* available for quantum circuits, since copying a qubit violates the quantum no-cloning theorem (see Section 4.4). In addition, since a branch point creates an additional output lead, a branch point violates unitarity. Allowed quantum circuits always have as many input leads (qubits) as output leads to assure a unitary time evolution during quantum processing. Only at the end of a quantum

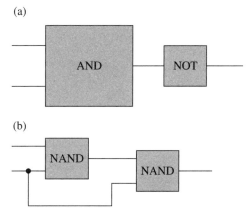

Figure 7.5 (a) Block diagram of a classical circuit equivalent to the NAND gate. (b) A circuit of NAND gates including a branch point.

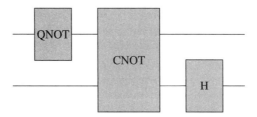

Figure 7.6   Example of a quantum circuit.

circuit is it permitted to make a measurement and terminate quantum leads in order to check the result of the quantum circuit's action.

An example of a quantum circuit is shown in Figure 7.6. Notice that while there are one-qubit gates situated in the leads of this two-qubit quantum circuit, there are no branch points and the number (in this case two) of quantum leads is strictly conserved.

It can be shown that circuits consisting of the two-qubit CNOT gate together with the one-qubit Hadamard gate defined in Equation 7.12 and the phase-rotation gate defined in Equation 7.13 are sufficient to perform all possible quantum operations necessary to build a quantum computer. This means that in terms of quantum logic circuits the CNOT gate is universal in the same sense that the NAND gate is universal for classical logic circuits.

We now discuss the quantum register loading operation as our first meaningful quantum computation. When we do a classical digital computation we start with an N-bit register containing N input bits. An 8-bit register R, for example, might contain the input word $R = 00000000$. The computer acts on R and generates a string of output bits that depend on the initial content of R, in our case $R = 00000000$. It is important to realize that only one register content at a time can be processed into a single stream of output information by a classical computer. If we wanted to load R with different input information, the old information would have to be erased and replaced by the new information, i.e., a new binary 8-bit word. Processing several different input words in parallel is possible, but it requires loading the input registers of several different computers with these input words and then running these computers simultaneously. This is known as *parallel processing* and is a well-known and often practiced procedure in supercomputing laboratories. To compute the results corresponding to all possible 8-bit register contents simultaneously, we would have to run $2^8 = 256$ classical computers in parallel. By modern standards this is not a particularly large number of computers to combine into a parallel-computing cluster. However, exploring all possible outputs corresponding to a 64-bit register requires a cluster consisting of $2^{64} \approx 2 \times 10^{19}$ computers, an impossibly large number.

But we may wonder whether using the quantum superposition principle may in fact allow us to process all 2^{64} inputs in parallel using only a *single* quantum computer. To accomplish this we would have to generate a superposition state of 2^{64} quantum registers, each containing a 64-qubit quantum word according to

$$|\psi\rangle = \Big(|\underbrace{00\cdots 0}_{64 \text{ qubits}}\rangle + |\underbrace{00\cdots 1}_{64 \text{ qubits}}\rangle + \ldots + |\underbrace{11\cdots 1}_{64 \text{ qubits}}\rangle\Big)/2^{32}, \quad (7.22)$$

where we divided by $2^{32} = \sqrt{2^{64}}$ to normalize $|\psi\rangle$ to $\langle\psi|\psi\rangle = 1$. As we know, the sum on the right-hand side of Equation 7.22 consists of $2^{64} = 2 \times 10^{19}$ terms, impossible to generate according to any classical procedure. However, quantum mechanically, there is a way. Let us start with 64 qubits, all initialized to $|0\rangle$. This results in the initial state

$$|\psi_0\rangle = |\underbrace{00\cdots 0}_{64 \text{ qubits}}\rangle. \quad (7.23)$$

Applying an H gate individually to each of the 64 qubits in $|\psi_0\rangle$, we obtain the superposition state $|\psi\rangle$ of Equation 7.22. Thus, 64 applications of an H gate has generated $2^{64} \approx 2 \times 10^{19}$ terms! Here we get our first taste of the awesome power of quantum information processing. A certain task, the generation of 2^{64} terms, according to classical logic impossible to even contemplate, is near trivial to accomplish quantum mechanically.

Let us see how this works in detail. Let us start with a one-qubit quantum register. According to Equation 7.12, application of H to the starting state $|0\rangle$ produces $(|0\rangle+|1\rangle)/\sqrt{2}$, a quantum superposition of two one-qubit input registers. Now let us try a two-qubit quantum register. Starting with $|0\rangle|0\rangle$ and applying an H gate to each of the two qubits, we obtain

$$\text{H}\,|0\rangle\,\text{H}\,|0\rangle = \frac{1}{\sqrt{2}}(|0\rangle+|1\rangle)\frac{1}{\sqrt{2}}(|0\rangle+|1\rangle)$$
$$= (|00\rangle + |01\rangle + |10\rangle + |11\rangle)/2. \quad (7.24)$$

Continuing in this way it is straightforward to see how the state defined in Equation 7.22 comes about if we apply an H gate to each of the 64 qubits of the starting state defined in Equation 7.23.

The application of H gates to a quantum register initialized to $|0\rangle|0\rangle\ldots|0\rangle$ is called *quantum register loading*. The importance of this

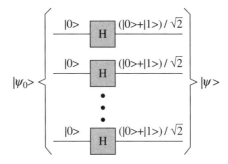

Figure 7.7 Sketch of the quantum register loading circuit.

procedure cannot be overstated. For instance, without the register-loading procedure that produces a superposition of 2^N states with only N applications of a quantum gate, it would be impossible to make use of the new computational power that quantum superposition provides us with. If we imagine that we generated the superposition state "the pedestrian way," one state at a time, for instance with the help of a classical front-end computer, it would be impossible to even initialize a quantum computer, since generating the exponentially many states necessary to do so would in all cases of interest take more than the age of the universe to complete.

A schematic sketch of the quantum register loading circuit is shown in Figure 7.7. This circuit is at the same time extremely simple and extremely powerful. In a sense it is the "enabling circuit" that makes it possible for quantum computers to outclass classical computers.

Exercises:

7.5.1 Show that the eight states

$$|\alpha^{(\pm)}\rangle = \frac{1}{\sqrt{2}}(|\uparrow\uparrow\uparrow\rangle \pm |\downarrow\downarrow\downarrow\rangle),$$

$$|\beta^{(\pm)}\rangle = \frac{1}{\sqrt{2}}(|\uparrow\uparrow\downarrow\rangle \pm |\downarrow\downarrow\uparrow\rangle),$$

$$|\gamma^{(\pm)}\rangle = \frac{1}{\sqrt{2}}(|\uparrow\downarrow\uparrow\rangle \pm |\downarrow\uparrow\downarrow\rangle),$$

$$|\delta^{(\pm)}\rangle = \frac{1}{\sqrt{2}}(|\uparrow\downarrow\downarrow\rangle \pm |\downarrow\uparrow\uparrow\rangle)$$

form a complete basis in the space of three-qubit states.

7.5.2 Express the four classical one-bit gates I, NOT, ZERO, and ONE, defined in Equation 7.9, using exclusively

(a) NOR gates,

(b) NAND gates.

7.5.3 Show that

(a) two successive H gates,

(b) two successive CNOT gates,

produce the identity gate. Therefore, two successive H or CNOT gates may be removed from a quantum circuit.

7.5.4 The quantum circuit in Figure 7.6 may be replaced by a single two-qubit quantum gate. Let us call this gate G.

(a) State the results of $G\,|x\rangle|y\rangle$, where $x, y \in \{0, 1\}$.

(b) What is the input state $|\psi\rangle$ that yields

$$|\psi'\rangle = G\,|\psi\rangle = |1\rangle|0\rangle$$

as an output state?

(c) Compute the eigenstates

$$G\,|\varphi^{\pm}\rangle = \pm|\varphi^{\pm}\rangle$$

of the G gate.

(d) Give an example of an input product state that, on output, is entangled by G.

7.6 Teleportation

Teleportation is a core staple of science fiction. Persons or objects disappear (dematerialize) at a location A and reappear (rematerialize) at another, distant location B. While the net effect of teleportation may be achieved just as well by boarding a train at location A and getting off the train at location B, the point of teleportation is to achieve the transport from A to B quasi-instantaneously in a disembodied form. While this is difficult to achieve technically with persons and macroscopic objects, there is a way to teleport elementary particles such as electrons or photons. In fact, since electrons and photons are identical throughout the universe, it is enough to teleport the quantum state of the elementary particle. To illustrate, suppose Alice, at location A, has an electron e_1^- whose quantum state

$$|\phi_1\rangle = a|\uparrow_1\rangle + b|\downarrow_1\rangle, \quad |a|^2 + |b|^2 = 1, \quad (7.25)$$

she would like to communicate to Bob, miles away at location B. If it is known that $|\phi_1\rangle$ is in an eigenstate of $\hat{s}_z$ (spin-up or spin-down), all Alice has to do is to measure the spin state of e_1^- and communicate this information, with a phone call for example, essentially at the speed of light, to Bob. Depending on the instructions received from Alice ("up" or "down") Bob would then simply create a spin-up or spin-down electron at his location and thus obtain a precise copy of e_1^- at his location. In this trivial case teleportation would not even require destruction of the quantum state of Alice's particle e_1^-; the entire operation would be like making a copy on a fax machine. Moreover, with the information Alice provided, Bob could create any number of copies of e_1^- at his location B.

But now suppose that Alice does not have any prior knowledge of the spin state of her particle. It will most likely be in a superposition state of the form of Equation 7.25 with $a, b \neq 0$. A measurement of the spin state of e_1^- will force the particle to collapse into either spin-up or spin-down with probabilities $|a|^2$ or $|b|^2$, respectively, which destroys the relative phase information between the amplitudes a and b. Even worse: Since after the measurement the quantum state of e_1^- is unpredictably altered, Alice has only one shot at making a single measurement to learn whatever she can about $|\phi_1\rangle$. But since a single measurement cannot even reveal the absolute squares of a and b, absolutely no information is obtained about $|\phi_1\rangle$ with a single von Neumann-type measurement. Obviously, Alice has to be more sophisticated than that to convey the quantum state $|\phi_1\rangle$ of e_1^- to Bob. She may teleport the quantum state $|\phi_1\rangle$ to Bob using the following scheme.

Alice makes use of two auxiliary electrons, e_2^- and e_3^- in the entangled EPR singlet state

$$|\psi_{23}^{(-)}\rangle = \frac{1}{\sqrt{2}} \{|\uparrow_2\rangle|\downarrow_3\rangle - |\downarrow_2\rangle|\uparrow_3\rangle\}. \tag{7.26}$$

She keeps e_2^- for herself, and sends e_3^- to Bob. The complete state of the three-electron system at this point is

$$|\psi_{123}\rangle = |\phi_1\rangle|\psi_{23}^{(-)}\rangle. \tag{7.27}$$

At this point nothing has been achieved. Bob has a particle that contains no information on $|\phi_1\rangle$, and the three-electron state $|\psi_{123}\rangle$ itself is a separable product state. But by performing a suitable measurement on both of her particles, Alice can entangle the quantum state of e_1^- with the quantum state of e_2^- and thus, via the entanglement of e_2^- with e_3^-, communicate information about the quantum state of e_1^- to Bob.

She performs a measurement by projecting the state of the two-particle (e_1^-, e_2^-)-system onto the two-particle Bell states

$$|\psi_{12}^{(\pm)}\rangle = \frac{1}{\sqrt{2}} \{|\uparrow_1\rangle|\downarrow_2\rangle \pm |\downarrow_1\rangle|\uparrow_2\rangle\},$$

$$|\phi_{12}^{(\pm)}\rangle = \frac{1}{\sqrt{2}} \{|\uparrow_1\rangle|\uparrow_2\rangle \pm |\downarrow_1\rangle|\downarrow_2\rangle\}. \quad (7.28)$$

The four Bell states defined in Equation 7.28 are a complete, orthonormal basis set. In order to compute the result of this Bell-state measurement, we first expand the three-particle state $|\psi_{123}\rangle$ into a sum of elementary three-particle spin-up and spin-down product states:

$$\begin{aligned}
|\psi_{123}\rangle &= |\phi_1\rangle|\psi_{23}^{(-)}\rangle \\
&= (a|\uparrow_1\rangle + b|\downarrow_1\rangle) \frac{1}{\sqrt{2}}(|\uparrow_2\rangle|\downarrow_3\rangle - |\downarrow_2\rangle|\uparrow_3\rangle) \\
&= \frac{a}{\sqrt{2}} \{|\uparrow_1\rangle|\uparrow_2\rangle|\downarrow_3\rangle - |\uparrow_1\rangle|\downarrow_2\rangle|\uparrow_3\rangle\} \\
&\quad + \frac{b}{\sqrt{2}} \{|\downarrow_1\rangle|\uparrow_2\rangle|\downarrow_3\rangle - |\downarrow_1\rangle|\downarrow_2\rangle|\uparrow_3\rangle\}. \quad (7.29)
\end{aligned}$$

With this result we compute the scalar products

$$\begin{aligned}
\langle\psi_{12}^{(-)}|\psi_{123}\rangle &= \frac{1}{2}(-a|\uparrow_3\rangle - b|\downarrow_3\rangle), \\
\langle\psi_{12}^{(+)}|\psi_{123}\rangle &= \frac{1}{2}(-a|\uparrow_3\rangle + b|\downarrow_3\rangle), \\
\langle\phi_{12}^{(-)}|\psi_{123}\rangle &= \frac{1}{2}(a|\downarrow_3\rangle + b|\uparrow_3\rangle), \\
\langle\phi_{12}^{(+)}|\psi_{123}\rangle &= \frac{1}{2}(a|\downarrow_3\rangle - b|\uparrow_3\rangle). \quad (7.30)
\end{aligned}$$

Now, because of

$$|\psi_{12}^{(-)}\rangle\langle\psi_{12}^{(-)}| + |\psi_{12}^{(+)}\rangle\langle\psi_{12}^{(+)}| + |\phi_{12}^{(-)}\rangle\langle\phi_{12}^{(-)}| + |\phi_{12}^{(+)}\rangle\langle\phi_{12}^{(+)}| = \hat{1}, \quad (7.31)$$

we have

$$\begin{aligned}
|\psi_{123}\rangle &= \hat{1}|\psi_{123}\rangle \\
&= |\psi_{12}^{(-)}\rangle\langle\psi_{12}^{(-)}|\psi_{123}\rangle + |\psi_{12}^{(+)}\rangle\langle\psi_{12}^{(+)}|\psi_{123}\rangle \\
&\quad + |\phi_{12}^{(-)}\rangle\langle\phi_{12}^{(-)}|\psi_{123}\rangle + |\phi_{12}^{(+)}\rangle\langle\phi_{12}^{(+)}|\psi_{123}\rangle \\
&= \frac{1}{2}\Big[|\psi_{12}^{(-)}\rangle(-a|\uparrow_3\rangle - b|\downarrow_3\rangle) + |\psi_{12}^{(+)}\rangle(-a|\uparrow_3\rangle + b|\downarrow_3\rangle) \\
&\quad + |\phi_{12}^{(-)}\rangle(a|\downarrow_3\rangle + b|\uparrow_3\rangle) + |\phi_{12}^{(+)}\rangle(a|\downarrow_3\rangle - b|\uparrow_3\rangle)\Big]. \quad (7.32)
\end{aligned}$$

According to Equation 7.32, Alice's Bell-state measurement may find the electron system (e_1^-, e_2^-) in any of the four Bell states in Equation 7.28 with equal probability 1/4. Suppose Alice's Bell-state measurement collapses the (e_1^-, e_2^-) two-particle state onto $|\psi_{12}^{(-)}\rangle$. Then, according to Equation 7.32, electron e_3^- is forced into the state $|e_3^-\rangle = -a|\uparrow_3\rangle - b|\downarrow_3\rangle$. Therefore, as soon as Bob obtains the news from Alice that her Bell-state measurement revealed $|\psi_{12}^{(-)}\rangle$, all Bob has to do to force his particle e_3^- into the state $|\phi_1\rangle$ is to apply the unitary transformation $-I$ to his EPR particle e_3^-:

Alice finds $|\psi_{12}^{(-)}\rangle \rightarrow$ Bob applies:

$$-I|e_3^-\rangle = \begin{pmatrix} -1 & 0 \\ 0 & -1 \end{pmatrix} \begin{pmatrix} -a \\ -b \end{pmatrix} = |\phi_1\rangle. \qquad (7.33)$$

Similarly, depending on Alice's results, Bob may apply appropriate unitary transformations in the three remaining cases according to:

Alice finds $|\psi_{12}^{(+)}\rangle \rightarrow$ Bob applies:

$$-\hat{\sigma}_z|e_3^-\rangle = \begin{pmatrix} -1 & 0 \\ 0 & 1 \end{pmatrix} \begin{pmatrix} -a \\ b \end{pmatrix} = |\phi_1\rangle; \qquad (7.34)$$

Alice finds $|\phi_{12}^{(-)}\rangle \rightarrow$ Bob applies:

$$\hat{\sigma}_x|e_3^-\rangle = \begin{pmatrix} 0 & 1 \\ 1 & 0 \end{pmatrix} \begin{pmatrix} b \\ a \end{pmatrix} = |\phi_1\rangle; \qquad (7.35)$$

Alice finds $|\phi_{12}^{(+)}\rangle \rightarrow$ Bob applies:

$$i\hat{\sigma}_y|e_3^-\rangle = \begin{pmatrix} 0 & 1 \\ -1 & 0 \end{pmatrix} \begin{pmatrix} -b \\ a \end{pmatrix} = |\phi_1\rangle. \qquad (7.36)$$

In summary, once Alice communicates the outcome of her Bell-state measurement to Bob using conventional, classical communication channels (for instance via a telephone), Bob uses this information to apply the corresponding unitary transformation to the state of his EPR particle e_3^-, which then acquires the quantum state $|\phi_1\rangle$ of Alice's original electron.

The most economical way for Alice to transmit the result of her Bell-state measurement is to label the Bell states in Equation 7.28 in binary code according to

$$\begin{aligned} |\psi_{12}^{(-)}\rangle &\rightarrow 00, & |\psi_{12}^{(+)}\rangle &\rightarrow 01, \\ |\phi_{12}^{(-)}\rangle &\rightarrow 10, & |\phi_{12}^{(+)}\rangle &\rightarrow 11. \end{aligned} \qquad (7.37)$$

Then, the result of Alice's Bell-state measurement may be transmitted as a brief two-bit message. The message "10," for example, means that Alice has performed her Bell-state measurement and has found the (e_1^-, e_2^-) system in the Bell state $|\phi_{12}^{(-)}\rangle$.

It is important to realize that once Alice has measured the quantum state of the electron pair (e_1^-, e_2^-), the quantum state $|\phi_1\rangle$ of Alice's electron e_1^- is irreparably destroyed. Alice does not learn anything about the quantum state $|\phi_1\rangle$ of her original electron e_1^-. Bob, too, despite being provided with the result of Alice's Bell-state measurement, does not learn anything about $|\phi_1\rangle$. Thus, the quantum no-cloning theorem (see Section 4.4) is not violated: Quantum information has been swapped from e_1^- to e_3^-, but has not been duplicated. Therefore, the net outcome of teleportation is a single quantum state that has been translated in space. This property of quantum information transmission may be formulated equivalently in a different context. Since classically we may make as many copies of a given original as we wish, we may send, or *broadcast*, the copies in all directions. Examples include sending out fax, phone, radio, or TV messages. Quantum mechanically, however, only a single copy of an unknown state $|\phi_1\rangle$ may exist at all times. While this copy can be sent from one person to another, i.e., point-to-point, it cannot be "broadcast" to many persons at the same time, since this would require many copies of the original quantum state $|\phi_1\rangle$, which is forbidden by the no-cloning theorem (see Section 4.4). This is the essence of the *no-broadcasting theorem* of quantum mechanics.

However, broadcasting in a narrower sense is possible and useful in practical applications. Bob does not have to be in a specific location for teleportation to work. It is not even necessary for Alice to know where Bob is, in order to communicate $|\phi_1\rangle$ to Bob. This works in the following way. Once the EPR pair of electrons (e_2^-, e_3^-) is shared between Alice and Bob, Bob simply takes e_3^- with him wherever he travels. Once Alice makes her Bell measurement, she broadcasts the *classical* two-bit information in all directions, for instance using radio, TV, or the internet. This is possible since the broadcasting of *classical* information is *not* forbidden. Once Bob receives the broadcasted classical two-bit information, he can then recreate $|\phi_1\rangle$ on his electron e_3^-, independently of where he is located in space.

Quantum entanglement does not "expire," i.e., the entanglement does not decay in time unless it is corrupted by external influences such as electromagnetic noise. This feature of entanglement can be used by Alice and Bob to share a number of EPR pairs long before the actual communication via teleportation commences. This way, after sharing the pairs, many states may be transmitted simultaneously without the need to wait until the next member of an EPR pair arrives at Bob's location. This is especially useful if Bob decides to travel to unknown destinations before Alice commences teleportation.

Teleportation seems to violate the Uncertainty Principle, since Alice extracts *all* information from $|\phi_1\rangle$ and sends it on to Bob, who can produce an exact copy of $|\phi_1\rangle$. However, since both Alice and Bob are only *manipulating* the state $|\phi_1\rangle$, but do not *learn* anything about $|\phi_1\rangle$, their actions do *not* violate the Uncertainty Principle.

In the beginning of this section we mentioned that the net effect of teleportation may also be achieved if Alice simply sends her electron e_1^- to Bob in a way that preserves its quantum state. But teleportation, the disembodied transport of classical information and quantum EPR correlations may have decisive advantages. For instance, it may not be possible to transport e_1^- from Alice to Bob without destroying its quantum state, for example, if the space between Alice and Bob is not a vacuum, but filled with air, or if the space between Alice and Bob contains strong electromagnetic fields that would destroy the spin-state of e_1^-. Instead, when using teleportation to send the spin state of e_1^-, the "murky" medium between Alice and Bob is usually no great obstacle to sending the classical two-bit message. The necessary prior exchange of EPR pairs is also not impeded, since, as mentioned previously, this can be arranged in a meeting between Alice and Bob, each of whom will then safeguard their EPR particles in a shielded box and take the particles with them wherever they travel, or EPR particle exchange could be arranged in times when the electromagnetic interference or other adverse effects are at a minimum. Another reason that argues for teleportation and against the physical exchange of particles is that teleportation works at speeds close to the speed of light which is usually much faster than the material exchange of electrons, especially if they have to be carried in shielded containers. And finally, since teleportation requires only the exchange of information, which can easily be sent "around corners" (optical fibers) or may be broadcast in all directions, teleportation is far more convenient than sending material particles, which, if sent as a particle beam, requires a clear line of sight between sender and receiver, and if mailed in a shielded container requires the address of the receiving station. The entire process of teleportation is summarized pictorially in Figure 7.8.

7.6 • TELEPORTATION 219

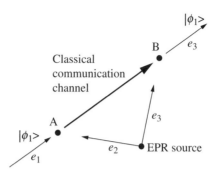

Figure 7.8 Diagram illustrating teleportation of the quantum state of a spin-1/2 particle from Alice (A) to Bob (B) with the help of a correlated spin-1/2 EPR pair.

Exercises:

7.6.1 Given are the matrices

$$s_x = \frac{\hbar}{\sqrt{2}} \begin{pmatrix} 0 & 1 & 0 \\ 1 & 0 & 1 \\ 0 & 1 & 0 \end{pmatrix},$$

$$s_y = \frac{\hbar}{\sqrt{2}} \begin{pmatrix} 0 & -i & 0 \\ i & 0 & -i \\ 0 & i & 0 \end{pmatrix},$$

$$s_z = \hbar \begin{pmatrix} 1 & 0 & 0 \\ 0 & 0 & 0 \\ 0 & 0 & -1 \end{pmatrix},$$

(a) Show that s_x, s_y, and s_z are matrix representations of the spin operators $\hat{s}_x$, $\hat{s}_y$, and $\hat{s}_z$, respectively, of a spin-1 particle.

(b) Show that the state

$$|\text{singlet}\rangle = \frac{1}{\sqrt{3}} (|1,-1\rangle - |00\rangle + |-1,1\rangle)$$

of two spin-1 particles has $S = 0$ and $S_z = 0$, where $\hat{\boldsymbol{S}} = \hat{\boldsymbol{s}}^{(1)} + \hat{\boldsymbol{s}}^{(2)}$ and $\hat{S}_z = \hat{s}_z^{(1)} + \hat{s}_z^{(2)}$.

7.6.2 Alice wants to teleport the quantum state

$$|\Omega\rangle = A|1\rangle + B|0\rangle + C|-1\rangle$$

220 CHAPTER 7 • CLASSICAL AND QUANTUM INFORMATION

of a spin-1 particle P_1 to Bob. Alice and Bob share an EPR pair of two spin-1 particles, P_2 and P_3, in the singlet state

$$|\text{singlet}\rangle_{23} = \frac{1}{\sqrt{3}}\{|1,-1\rangle_{23} - |00\rangle_{23} + |-1,1\rangle_{23}\}.$$

Alice has particle P_2; Bob has particle P_3.

(a) Alice makes a measurement on her two particles P_1 and P_2 and reports to Bob that she found her particle pair in the state

$$|\psi\rangle_{12} = \frac{1}{\sqrt{3}}\{|11\rangle_{12} + |00\rangle_{12} + |-1,-1\rangle_{12}\}.$$

Determine the state $|\phi\rangle$ of Bob's particle P_3.

(b) Determine the unitary operator $\hat{U}$ that Bob has to apply to $|\phi\rangle$ to put his particle P_3 into the state $|\Omega\rangle$ and thus complete the teleportation transaction.

7.6.3 How many classical bits of information have to be transmitted to teleport the quantum state $|\psi\rangle = A|1\rangle + B|0\rangle + C|-1\rangle$, $|A|^2 + |B|^2 + |C|^2 = 1$, of a massive spin-1 particle?

7.7 Summary

In this chapter we laid the conceptual foundations of quantum computing and quantum information processing. Although the computational power of classical computers is impressive, increasing their performance relies on continued miniaturization of classical electronic circuitry. In Section 7.1 we saw that due to the atomistic nature of matter this miniaturization cannot go on forever, setting distinct fundamental limits to classical computing. Therefore, a further increase in computing and information processing power can be achieved only if we change the computing paradigm. Quantum computing and information processing provides an example of such a paradigm shift. It promises unprecedented computer power, computer power so tremendous that it easily exceeds the computer power of a classical computer the size of the universe! While classical computers are based on binary switches, i.e., bits, that can take only the logical values 0 and 1, quantum computers are based on qubits that may be in a quantum superposition of $|0\rangle$ and $|1\rangle$, which accounts for one of the "secrets" of quantum computing. We studied classical bits and quantum qubits in Section 7.2. In order to perform a computation, no matter whether classical or quantum mechanical, we need gates that allow us to perform logic operations on bits (classical gates) or qubits (quantum gates). We studied classical and quantum gates together with their main differences in Sections 7.3 and 7.4, respectively. In Section 7.5 we combined classical and quantum logic gates into classical and quantum circuits. In this section we also encountered our first meaningful quantum circuit, the quantum register loading circuit. This circuit is a common element in quantum algorithms. In fact, without it, it is hard to imagine how we could possibly make effective use of quantum parallel processing and obtain the exponential speed-ups that quantum algorithms afford us. In Section 7.6 we teleported the quantum state of a single electron. Although teleportation of the quantum state of a macroscopic object is impossible to do with present-day technology, there are no physical laws that would forbid it. While teleportation has obvious technical applications that even include the possibility of a quantum internet (see Chapter 12), its main scientific value is of a conceptual nature. Teleportation shows how to resolve the quantum information of a given single-qubit state $|\phi_1\rangle$ into two components, two bits of classical information and a purely quantum EPR correlation.

Chapter Review Exercises:

1. There are 16 possible two-bit gates G_j, $j = 0, \ldots, 15$. The action of any one of these gates is to map two input bits to an output bit according to

$$(0,0) \to \rho_0, \ (0,1) \to \rho_1, \ (1,0) \to \rho_2, \ (1,1) \to \rho_3,$$

where $\rho_0, \rho_1, \rho_2, \rho_3 \in \{0,1\}$. We may use the outputs $\rho_0, \ldots, \rho_3$ to assign a unique binary sequence number, $j = \rho_3\rho_2\rho_1\rho_0$, to each of the 16 possible gates. Example: $\rho_0 = 1$, $\rho_1 = 0$, $\rho_2 = 1$, $\rho_3 = 1$ corresponds to the gate with sequence number $j = 1101$ in binary notation, i.e., gate number $j = 13$ in decimal notation.

   (a) What are the gate sequence numbers, expressed in decimal notation, of the (i) AND, (ii) NAND, (iii) OR, and (iv) NOR gates?

   (b) Show that each of the 16 gates may be represented by an equivalent logical circuit that involves only NAND gates. This proves that the NAND gate is universal.

2. Express the logic functions (i) f_{NOT}, (ii) f_{AND}, and (iii) f_{OR} in terms of the logic function f_{NAND}.

3. All 16 two-bit gates can be realized with circuits consisting only of the NAND gate. On the basis of this result show that all 16 two-bit

gates can be realized with circuits consisting only of NOR gates. This shows that the NOR gate is a universal two-bit gate.

4. The two-qubit analogues of the four Bell states defined in Equation 6.47 are

$$|\Phi^+\rangle = \frac{1}{\sqrt{2}}\Big(|00\rangle + |11\rangle\Big), \quad |\Phi^-\rangle = \frac{1}{\sqrt{2}}\Big(|00\rangle - |11\rangle\Big),$$
$$|\Psi^+\rangle = \frac{1}{\sqrt{2}}\Big(|01\rangle + |10\rangle\Big), \quad |\Psi^-\rangle = \frac{1}{\sqrt{2}}\Big(|01\rangle - |10\rangle\Big). \quad (7.38)$$

Show that application of the CNOT gate transforms each of the four Bell states into a product state.

chapter 8
Quantum Computing

In this chapter:

- Introduction
- Our First Quantum Computer
- Deutsch's Algorithm
- Deutsch-Jozsa Algorithm
- Grover's Search Algorithm
- Summary

8.1 Introduction

Although a standard PC provides more computer power than the average person needs for everyday use, such as using the Internet, writing letters, and paying bills, there are plenty of problems for which even the computer power of all the world's PCs combined is not enough. It is therefore not surprising that scientists have long searched for a new computing paradigm and found it in quantum computers. Quantum computers are unlike any classical computer. Making use of quantum superposition and entanglement, quantum computers are orders of magnitude faster than classical computers. In many cases, a task that takes exponentially long to perform on a classical computer, can be performed in polynomial time on a quantum computer. Since algorithms with exponential execution time are technically intractable, quantum algorithms shift a problem from the realm of impossibilities into the realm of possibilities: a profoundly important qualitative advance. Classical computers are limited in two ways: (1) Algorithms (software) and (2) processors (hardware). Because of the discreteness of matter at the atomic scale, it is fundamentally impossible to increase the density of transistors in the processor chips of classical computers beyond about one transistor per cubic Å. This places a fundamental limit on the hardware of classical computers that cannot be overcome in principle (see Section 7.1). Because it is subject to classical reasoning based on the discreteness of classical bits, classical software is limited, too.

In this chapter we will study important examples of quantum software and show how this software outperforms software based on classical reasoning. In Section 8.2 we encounter our first quantum computer. This computer uses quantum superposition and illustrates quantum parallel processing, but is not superior to a corresponding classical computer. In Section 8.3 we encounter Deutsch's algorithm. It was the first quantum algorithm that outperforms any conceivable classical algorithm. Still, although a conceptual advance over any classical algorithm, the actual computational task performed by Deutsch's algorithm is modest enough so that it can be simulated on any classical computer. In Section 8.4, however, when advancing from Deutsch's algorithm to the quantum algorithm by Deutsch and Jozsa, we leave classical computers in the dust. The Deutsch-Jozsa algorithm shows an exponential speed-up with respect to any conceivable classical algorithm. Finally, in Section 8.5, we discuss a quantum algorithm that is of practical relevance: Grover's quantum search algorithm. It addresses the universally important problem of locating an item in an unsorted database. Grover's algorithm provides a quadratic speed-up compared with any existing, or conceivable, classical search algorithm.

8.2 Our First Quantum Computer

Recalling the action of the CNOT gate,

$$|x\rangle|y\rangle \stackrel{\text{CNOT}}{\longrightarrow} |x\rangle|x \oplus y\rangle, \qquad (8.1)$$

we notice that the CNOT gate itself already performs a computation: Addition modulo 2 of two input qubits. Therefore, the CNOT gate itself is already a primitive quantum computer. A circuit diagram of the CNOT quantum computer is shown in Figure 8.1(a). If both inputs $|x\rangle$ and $|y\rangle$ are pure states, then the result $x \oplus y$, available on the target-out line of the CNOT gate, is a pure state, too, and can be obtained with a single measurement. In fact, supplied with pure states, our quantum computer works just like a classical computer. We say that supplied with pure states we operate the quantum computer in *classical mode*. In classical mode our quantum computer is equivalent to a classical computer

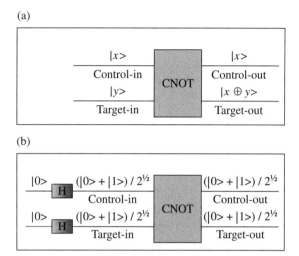

Figure 8.1 Circuit diagram of our first quantum computer.

and shows no indications of being in any way superior to a classical computer. But this changes immediately when we equip our quantum computer with two Hadamard gates as shown in Figure 8.1(b). Applying the states $|0\rangle$ on the input leads of both Hadamard gates turns the states $|0\rangle$ into the superposition $(|0\rangle+|1\rangle)/\sqrt{2}$, which is then fed to the control-in and target-in leads of the CNOT gate. Therefore, the two-qubit input state is

$$\frac{1}{\sqrt{2}}(|0\rangle + |1\rangle)\frac{1}{\sqrt{2}}(|0\rangle + |1\rangle) = \frac{1}{2}(|0\rangle|0\rangle + |0\rangle|1\rangle + |1\rangle|0\rangle + |1\rangle|1\rangle). \tag{8.2}$$

Applying the CNOT gate, this input state is transformed into the output state

$$\frac{1}{2}(|0\rangle|0\rangle + |0\rangle|1\rangle + |1\rangle|1\rangle + |1\rangle|0\rangle), \tag{8.3}$$

which can be written as the product state

$$\frac{1}{\sqrt{2}}(|0\rangle + |1\rangle)\frac{1}{\sqrt{2}}(|0\rangle + |1\rangle). \tag{8.4}$$

Therefore, as shown in Figure 8.1(b), both control-out and target-out are in the state $(|0\rangle + |1\rangle)/\sqrt{2}$.

What happened in this quantum computation is the following: We provided a single two-qubit input state, $|0\rangle|0\rangle$. This state got "fanned out" into a coherent superposition of all four possible input states, $|0\rangle|0\rangle$, $|0\rangle|1\rangle$, $|1\rangle|0\rangle$, and $|1\rangle|1\rangle$. The CNOT gate then performed the four computations $0 \oplus 0$, $0 \oplus 1$, $1 \oplus 0$, and $1 \oplus 1$ *in parallel*, and returned all four results as a single two-qubit state. This process reveals one of the central strengths of a quantum computer: *Parallel processing* by making use of the superposition of states. This is something a *single* classical processor cannot do. Only four classical processors working in parallel would be able to accomplish the same net result. However, there is one essential drawback. The output of our quantum computer is in a superposition of the four possible "results," i.e., $|0\rangle|0\rangle$, $|0\rangle|1\rangle$, $|1\rangle|0\rangle$, and $|1\rangle|1\rangle$. In order to obtain an individual result, we have to perform a measurement. But this will yield any one of the four results randomly. Therefore, apart from illustrating quantum parallel processing, our CNOT quantum computer does not perform a computation with a useful result. Another drawback is that in order to obtain each of the four results at least once, we have to run the quantum computer at least four times to be able to perform four independent measurements. But even if we follow this protocol and run the quantum computer four times, we are not guaranteed to obtain each of the four different results. We might just as well be unlucky and obtain $|0\rangle|0\rangle$ four times in a row.

So far this is disappointing. On the one hand, yes, our quantum computer exhibits parallelism, on the other hand, there is no speed-up with respect to a classical computer when it comes to actually reading out the results. In the best-case scenario we have to run our quantum computer four times, just like a classical computer, to obtain each of the four results. If we are unlucky, we need more runs, since we might obtain one or more of the results several times in a row. Thus, it looks like our quantum computer, on average, is even *slower* than a classical computer. It appears that because of the quantum superposition principle, quantum computers are very good at generating massive amounts of results in parallel, but are not very good at discriminating between the computed results and locating the desired result. Thus, quantum computers do not seem to have any advantage over classical computers. This was indeed a problem until, in 1985, David Deutsch presented the first quantum algorithm that (1) performed a useful computation and (2) showed a factor-2 speed-up with respect to *any* classical computer whatsoever. This speed-up is achieved by using *interference*, another distinct quantum phenomenon *in addition to* superposition. Interference is the process by which quantum computers select the desired results from the host of results they are capable of computing in parallel. Using superposition *as well as* quantum interference, Deutsch presented a quantum algorithm that was superior to any conceivable classical algorithm. Thus, Deutsch's quantum algorithm was the first unequivocal demonstration of the conceptual superiority of quantum computing versus classical computing. A detailed discussion of Deutsch's algorithm is the subject of the following section.

Exercises:

8.2.1 Multiplication by 2 is a common operation for both classical and quantum computers. On a classical two-bit computer we can perform exactly two such computations:

$$00 \xrightarrow{\times 2} 00, \quad 01 \xrightarrow{\times 2} 10.$$

Quantum mechanically, this corresponds to the transformation

$$|0\rangle|0\rangle \xrightarrow{\times 2} |0\rangle|0\rangle, \quad |0\rangle|1\rangle \xrightarrow{\times 2} |1\rangle|0\rangle. \tag{8.5}$$

Use a CNOT and a $\overline{\text{CNOT}}$ gate to realize the quantum computation defined in Equation 8.5.

8.2.2 Use a CNOT and a $\overline{\text{CNOT}}$ gate to construct a quantum computer that realizes division by 2, i.e.,

$$|0\rangle|0\rangle \xrightarrow{/2} |0\rangle|0\rangle, \quad |1\rangle|0\rangle \xrightarrow{/2} |0\rangle|1\rangle.$$

8.2.3 Show that two-qubit multiplication by 3,

$$|0\rangle|0\rangle \xrightarrow{\times 3} |0\rangle|0\rangle, \quad |0\rangle|1\rangle \xrightarrow{\times 3} |1\rangle|1\rangle, \tag{8.6}$$

can be implemented with a single $\overline{\text{CNOT}}$ gate.

8.3 Deutsch's Algorithm

Historically, Deutsch's algorithm was the first quantum algorithm that demonstrated the superiority of quantum algorithms over classical algorithms. This is sometimes referred to as *quantum speed-up*. Deutsch's algorithm deals with a class of functions f that map a single binary argument to the two values 0 or 1, i.e., $f(x) \in \{0, 1\}$ for $x \in \{0, 1\}$. Since f can assume only two possible values, the class of functions f is known as the class of binary-valued functions of a single binary argument. There are exactly four such functions:

$$\begin{aligned} f_0(0) &= 0, & f_0(1) &= 0, \\ f_1(0) &= 0, & f_1(1) &= 1, \\ f_2(0) &= 1, & f_2(1) &= 0, \\ f_3(0) &= 1, & f_3(1) &= 1. \end{aligned} \tag{8.7}$$

Since f_0 and f_3 return a constant value (0 and 1, respectively) independent of the value of their arguments, these two functions are *constant* functions. Since f_1 and f_2 return exactly one 0 and one 1, f_1 and f_3 are *balanced* functions.

We are now ready to state Deutsch's problem: We are given one of the four functions $f_0, \ldots, f_3$ defined in Equation 8.7, i.e., we are given $f \in \{f_0, f_1, f_2, f_3\}$. Question: Is f constant? In other words: Is $f(0) = f(1)$?

Obviously, a classical solution of this problem, using classical reasoning, requires *two* evaluations of the function f. Only by evaluating f for both arguments, 0 and 1, can we be sure that f is either constant or not.

Our goal is the following: Devise a quantum algorithm that beats the classical paradigm, i.e., devise a quantum algorithm that needs only a *single* evaluation of the function f. The solution is Deutsch's algorithm.

As preparation for Deutsch's algorithm we recall our notation for addition modulo 2:

$$(p+q) \mod 2 \equiv p \oplus q. \tag{8.8}$$

With this notation, we have, for example,

$$(-1)^{a+b} = (-1)^{(a+b) \mod 2} = (-1)^{a \oplus b}. \tag{8.9}$$

We now prove that, if $x, y \in \{0, 1\}$ and $f(x) \in \{0, 1\}$, then:

$$f(x) = f(y) \iff f(x) \oplus f(y) = 0. \tag{8.10}$$

The proof is straightforward.

"$\Longrightarrow$": Since f can only assume the two values 0 and 1, $f(x) = f(y)$ implies that either $f(x) = f(y) = 0$ or $f(x) = f(y) = 1$. There are no other possibilities. In both cases we have $f(x) \oplus f(y) = 0$.

"$\Longleftarrow$": On the other hand, if $f(x) \oplus f(y) = 0$, this can happen only if both $f(x)$ and $f(y)$ are zero or both $f(x)$ and $f(y)$ are one, i.e., $f(x) = f(y)$ in both cases.

Having shown both directions, "$\Longrightarrow$" and "$\Longleftarrow$" of the equivalence defined in Equation 8.10, we proved "$\iff$", i.e., Equation 8.10.

Deutsch's algorithm is a two-qubit quantum algorithm. It works in the following way. We choose the starting state $|0\rangle|1\rangle$. Then, we apply a Hadamard transformation to both qubits of the starting state to obtain the state

$$|\psi\rangle = H|0\rangle \, H|1\rangle = \frac{1}{\sqrt{2}}(|0\rangle + |1\rangle)\frac{1}{\sqrt{2}}(|0\rangle - |1\rangle)$$

$$= \frac{1}{2}(|0\rangle|0\rangle - |0\rangle|1\rangle + |1\rangle|0\rangle - |1\rangle|1\rangle). \tag{8.11}$$

8.3 • DEUTSCH'S ALGORITHM

We define the f-controlled CNOT gate, f-CNOT, a linear operator, in the following way:

$$f\text{-CNOT}(|x\rangle|y\rangle) = |x\rangle|f(x) \oplus y\rangle. \tag{8.12}$$

The realization of f-CNOT may be hard-wired in the quantum computer. Now, we apply f-CNOT on $|\psi\rangle$. Using the linearity of f-CNOT together with its definition in Equation 8.12, we obtain:

$$\begin{aligned} f\text{-CNOT}(|\psi\rangle) &= \frac{1}{2} f\text{-CNOT}(|0\rangle|0\rangle - |0\rangle|1\rangle + |1\rangle|0\rangle - |1\rangle|1\rangle) \\ &= \frac{1}{2} \{|0\rangle|f(0) \oplus 0\rangle - |0\rangle|f(0) \oplus 1\rangle + |1\rangle|f(1) \oplus 0\rangle - |1\rangle|f(1) \oplus 1\rangle\} \\ &= \frac{1}{2} \{|0\rangle \left[|f(0)\rangle - |f(0) \oplus 1\rangle\right] + |1\rangle \left[|f(1)\rangle - |f(1) \oplus 1\rangle\right]\}. \end{aligned} \tag{8.13}$$

There are two possibilities for $f(0)$:

$$\begin{aligned} (i) \quad & f(0) = 0 : |f(0)\rangle - |f(0) \oplus 1\rangle = |0\rangle - |1\rangle, \\ (ii) \quad & f(0) = 1 : |f(0)\rangle - |f(0) \oplus 1\rangle = |1\rangle - |0\rangle \\ & \implies |f(0)\rangle - |f(0) \oplus 1\rangle = (-1)^{f(0)}(|0\rangle - |1\rangle). \end{aligned} \tag{8.14}$$

Similarly we derive:

$$|f(1)\rangle - |f(1) \oplus 1\rangle = (-1)^{f(1)}(|0\rangle - |1\rangle). \tag{8.15}$$

Using the previous results we obtain:

$$\begin{aligned} f\text{-CNOT}(|\psi\rangle) &= \frac{1}{2}\left[(-1)^{f(0)}|0\rangle(|0\rangle - |1\rangle) + (-1)^{f(1)}|1\rangle(|0\rangle - |1\rangle)\right] \\ &= \frac{1}{2}\left[(-1)^{f(0)}|0\rangle + (-1)^{f(1)}|1\rangle\right](|0\rangle - |1\rangle) \\ &= \frac{1}{2}(-1)^{f(0)}\left[|0\rangle + (-1)^{f(0)+f(1)}|1\rangle\right](|0\rangle - |1\rangle). \end{aligned} \tag{8.16}$$

Using Equation 8.9 in Equation 8.16 we obtain:

$$f\text{-CNOT}(|\psi\rangle) = \frac{1}{2}(-1)^{f(0)}\left[|0\rangle + (-1)^{f(0) \oplus f(1)}|1\rangle\right](|0\rangle - |1\rangle). \tag{8.17}$$

Recalling that $H|0\rangle = (|0\rangle + |1\rangle)/\sqrt{2}$ and $H|1\rangle = (|0\rangle - |1\rangle)/\sqrt{2}$, we apply a Hadamard transformation to the first and second qubit of Equation 8.17:

$$\begin{aligned}
|\psi'\rangle &= H_1 H_2 f\text{-CNOT}(|\psi\rangle) \\
&= \frac{1}{2}(-1)^{f(0)}\left[|0\rangle + |1\rangle + (-1)^{f(0)\oplus f(1)}|0\rangle - (-1)^{f(0)\oplus f(1)}|1\rangle\right]|1\rangle \\
&= \frac{1}{2}(-1)^{f(0)}\left\{|0\rangle\left[1 + (-1)^{f(0)\oplus f(1)}\right] + |1\rangle\left[1 - (-1)^{f(0)\oplus f(1)}\right]\right\}|1\rangle \\
&= (-1)^{f(0)}\begin{cases}|0\rangle|1\rangle, & \text{for } f(0) \oplus f(1) = 0 \ (f \text{ constant}), \\ |1\rangle|1\rangle, & \text{for } f(0) \oplus f(1) = 1 \ (f \text{ balanced}).\end{cases}
\end{aligned} \quad (8.18)$$

Inspecting Equation 8.18, we find an astonishing result. Depending on whether f is constant ($f(0) \oplus f(1) = 0$) or balanced ($f(0) \oplus f(1) = 1$), our quantum computer (up to a global phase) is left in the pure states $|0\rangle|1\rangle$ and $|1\rangle|1\rangle$, respectively. Therefore, a single measurement of the first qubit of $|\psi'\rangle$ reveals with certainty whether f is constant or balanced: If we find $|0\rangle$ as a result of the measurement of the first qubit, we know that f is constant, *with certainty*. If we obtain $|1\rangle$, we know that f is balanced *with certainty*. Note that the global phase in Equation 8.18 is of no consequence, since it is eliminated or altered in the process of measurement. A circuit diagram of Deutsch's quantum algorithm is shown in Figure 8.2.

It is remarkable that contrary to any conceivable classical algorithm, only *a single* evaluation of f is required; we need to run our quantum computer only *once* to obtain a complete solution of Deutsch's problem. The following is a partial explanation for how quantum mechanics achieves this "trick." Recalling the classical solution of the problem, i.e., evaluating both $f(0)$ and $f(1)$, we notice that the classical solution of

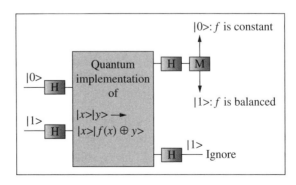

Figure 8.2 Circuit diagram of Deutsch's quantum algorithm.

Deutsch's problem gives us too much information. Apart from solving the problem of whether f is balanced or not, it also tells us *which one* of the four functions defined in Equation 8.7 we were given. This is extraneous information, not required by the statement of Deutsch's problem. Generating too much information is one of the reasons why the classical algorithm is less effective than the corresponding quantum algorithm. Classically there seems to be no way around the conundrum of generating too much information. The quantum solution is more economical in that it solves the problem of whether f is balanced or constant without revealing precisely which one of the f's we were given. Interestingly, if the quantum measurement would give us the phase $(-1)^{f(0)}$, too, we could tell precisely which one of the four functions f we were given. But since the measurement obliterates the phase, we cannot.

Deutsch's algorithm may not be very useful in practice, but it proves the point that quantum computation is a *qualitative* advance over classical computation: The quantum algorithm can do something that no classical algorithm can do in principle. In addition, historically, Deutsch's algorithm provided a template for the construction of Grover's algorithm (see Section 8.5) and Shor's algorithm (see Section 10.4), two of the most useful and most powerful quantum algorithms ever designed. All that is required now is to deliver on the promise that a quantum computer is exponentially faster than any conceivable classical computer. This is shown by means of the Deutsch-Jozsa algorithm discussed in the following section.

Exercises:

8.3.1 One way of debugging a classical computer program is to run it in step-by-step mode. This means that the computer slowly steps through a program line-by-line and displays the contents of all variables addressed in the current line of code. Let us use this technique to run Deutsch's quantum algorithm, step-by-step, in "debug mode."

(a) Find a matrix implementation of $|x\rangle|y\rangle \rightarrow |x\rangle|f(x) \oplus y\rangle$ for the special case $f(0) = 1$, $f(1) = 0$.

(b) Represent all quantum gates of Deutsch's algorithm by matrices.

(c) Now run Deutsch's algorithm, step-by-step, in "debug mode." Do this in the following way. For each tick of the quantum computer's clock, i.e., after each application of a quantum gate, record the current quantum state. Do you obtain a pure state after application of the final gate? Is the state what you expect it to be?

8.3.2 In the text we used $|\alpha_1\rangle = |0\rangle|1\rangle$ as the starting state for Deutsch's algorithm. Explore what happens if you use (i) $|\alpha_0\rangle = |0\rangle|0\rangle$, (ii) $|\alpha_2\rangle = |1\rangle|0\rangle$, and (iii) $|\alpha_3\rangle = |1\rangle|1\rangle$ as starting states. Do the four states $|\alpha_0\rangle, \ldots, |\alpha_3\rangle$ exhaust all possibilities for starting states?

8.3.3 For each one of the four functions defined in Equation 8.7 find a 4×4 matrix implementation of the f-controlled CNOT gate defined in Equation 8.12.

8.4 Deutsch-Jozsa Algorithm

Although Deutsch's algorithm performs a qualitative task that cannot be duplicated with a single classical processor, we may use a classical parallel computer with two classical processors to find the solution of Deutsch's problem in a *single* run of the classical, two-processor machine. Thus, while awed by the conceptual, intellectual advance in computing displayed by Deutsch's algorithm, we are not really impressed, since, in practice, the same result can easily be achieved using multiprocessor, classical computing strategies. However, with the Deutsch-Jozsa problem, defined in the following, we switch gears to a new regime of computation, where classical computers simply cannot follow—in principle. One of the reasons is that classically so many computing steps are required to solve the Deutsch-Jozsa problem that a single supercomputer would demonstrably take longer than the age of the universe to produce the result. Alternatively, in order to cut the execution time to a barely tolerable level of, say, a year, we would need to build a parallel computer that would be larger than the size of the known universe. The Deutsch-Jozsa algorithm executed on a quantum computer, however, shows that making use of the "quantum advantage," already displayed in nascent form by Deutsch's algorithm, we may solve the Deutsch-Jozsa problem—inaccessible to any conceivable classical computer—on a quantum computer of reasonable size. We are now going to present the Deutsch-Jozsa problem, a generalization of Deutsch's problem, and then proceed to solve the Deutsch-Jozsa problem with a quantum algorithm.

The Deutsch-Jozsa problem, a generalization of Deutsch's problem, considers binary-valued functions f of n binary variables $x_1, x_2, \ldots, x_n$ according to

$$f : \{0,1\} \times \{0,1\} \times \ldots \times \{0,1\} = \{0,1\}^{\otimes n} \to \{0,1\},$$
$$f(x_1, x_2, \ldots, x_n) \in \{0,1\}, \quad (x_1, x_2, \ldots, x_n) \in \{0,1\}^{\otimes n}. \quad (8.19)$$

8.4 • DEUTSCH-JOZSA ALGORITHM

We introduced the short-hand notation

$$A^{\otimes n} = \underbrace{A \times A \times \ldots \times A}_{n \text{ times}}, \qquad (8.20)$$

where the symbol A may stand for any object, for instance a set, or an operator.

According to the Deutsch-Jozsa problem, we are given a binary-valued function of n binary arguments as defined in Equation 8.19, and we are told that the function f is either constant, i.e., all values of f on the domain $\{0,1\}^{\otimes n}$ are either all 0 or all 1, or f is balanced, i.e., exactly 50% of its values on $\{0,1\}^{\otimes n}$ are 0 and 50% are 1.

In order to get a feeling for the difficulty and the complexity of the Deutsch-Jozsa problem, we ask how many binary-valued functions f there are for given n. We start with $n = 1$. In this case we already know the answer: There are four possible binary-valued functions: f_0, f_1, f_2, and f_3. They are stated explicitly in Equation 8.7. How many binary-valued functions f are there for $n = 2$? In this case we consider binary-valued functions f defined on the domain $\{0,1\} \times \{0,1\}$, i.e., functions f with two binary arguments. There are 16 such functions:

$$f_0(0,0) = 0, \; f_0(0,1) = 0, \; f_0(1,0) = 0, \; f_0(1,1) = 0,$$
$$f_1(0,0) = 0, \; f_1(0,1) = 0, \; f_1(1,0) = 0, \; f_1(1,1) = 1,$$
$$f_2(0,0) = 0, \; f_2(0,1) = 0, \; f_2(1,0) = 1, \; f_2(1,1) = 0,$$

$$\cdot$$
$$\cdot$$
$$\cdot$$

$$f_{15}(0,0) = 1, \; f_{15}(0,1) = 1, \; f_{15}(1,0) = 1, \; f_{15}(1,1) = 1. \quad (8.21)$$

Examining the two cases, we notice the exponential proliferation of function arguments and the hyperexponential proliferation of functions with increasing n. As a function of n the number of possible arguments is 2^n; the number of possible functions is 2^{2^n}.

The severity of hyperexponential growth is illustrated by the following example. For $n = 9$ there are $2^{2^9} = 2^{512} \approx 10^{154}$ functions f. This amounts to more functions than there are electrons in the universe! This example makes it plain that if we had to answer any question concerning the properties of a given binary-valued function f defined on $\{0,1\}^{\otimes n}$ that requires us to examine all possible functions on $\{0,1\}^{\otimes n}$, where n is larger than 8, we would either need a classical computer the size of the universe, or we would have to wait a *very* long time for the answer.

According to the examples above, and because of the exponential proliferation of function arguments, in order for a classical algorithm to decide *with certainty* whether any *given* f out of the set of 2^{2^n} possible f is constant or balanced, we need, in the worst case, $(2^n/2)+1 = 2^{n-1}+1$ function evaluations. Thus, the classical effort is exponential in n.

While classically, in the worst case, exponentially many function evaluations are necessary to decide whether a given f is constant or balanced, the Deutsch-Jozsa algorithm shows that quantum mechanically only a *single* function evaluation is necessary. This is an astonishing result. Before presenting the Deutsch-Jozsa algorithm in detail we have to provide some preliminary material. First, we note that in order to solve the $\{0,1\}^{\otimes n}$ problem, we need $n+1$ qubits. This requires a generalization of the f-controlled CNOT gate to $n+1$ qubits. Let $|x\rangle$ be an n-qubit state and $|y\rangle$ be a one-qubit state. Then:

$$f\text{-CNOT}\,|x\rangle|y\rangle \;=\; |x\rangle|f(x)\oplus y\rangle, \tag{8.22}$$

where x is an n-bit string,

$$x \;=\; b^{(x)}_{n-1}b^{(x)}_{n-2}\ldots b^{(x)}_0, \quad b^{(x)}_j \in \{0,1\}, \quad j=0,\ldots,n-1, \tag{8.23}$$

and the associated quantum state is

$$|x\rangle \;=\; |b^{(x)}_{n-1}b^{(x)}_{n-2}\ldots b^{(x)}_0\rangle. \tag{8.24}$$

Next, we state a formula for the application of n Hadamard gates on the n qubits of the state $|x\rangle$:

$$\mathrm{H}^{\otimes n}\,|x\rangle \;=\; \mathrm{H}\,|b^{(x)}_{n-1}\rangle\,\mathrm{H}\,|b^{(x)}_{n-2}\rangle\ldots \mathrm{H}\,|b^{(x)}_0\rangle \;=\; \frac{1}{2^{n/2}}\sum_{y=0}^{2^n-1}(-1)^{x\odot y}|y\rangle, \tag{8.25}$$

where

$$x\odot y \;=\; \left[\sum_{j=0}^{n-1} b^{(x)}_j b^{(y)}_j\right] \;\mathrm{mod}\;2 \tag{8.26}$$

is a bit-by-bit "scalar product" modulo 2. The formula defined in Equation 8.25 is known as Hadamard's formula.

We are now ready for a detailed presentation of the Deutsch-Jozsa algorithm.

(i) We start with the input state

$$|x\rangle|y\rangle \;=\; |0\rangle^{\otimes n}|1\rangle \;=\; \underbrace{|0\rangle|0\rangle\ldots|0\rangle}_{n\text{ times}}|1\rangle \;=\; |\underbrace{00\ldots0}_{n\text{ times}}\rangle|1\rangle. \tag{8.27}$$

(ii) Apply an H gate to each of the n qubits of the state $|x\rangle$, (which, at this point, are all in state $|0\rangle$) and to the $|y\rangle$-qubit (which is in the state $|1\rangle$):

$$H|0\rangle H|0\rangle \ldots H|0\rangle H|1\rangle = \frac{1}{\sqrt{2^{n+1}}}(|0\rangle + |1\rangle) \ldots (|0\rangle + |1\rangle)(|0\rangle - |1\rangle)$$

$$= \frac{1}{\sqrt{2^{n+1}}} \sum_{j=0}^{2^n-1} |j\rangle(|0\rangle - |1\rangle), \quad (8.28)$$

where

$$|j\rangle = |b_{n-1}^{(j)} b_{n-2}^{(j)} \ldots b_0^{(j)}\rangle \quad (8.29)$$

and

$$j = b_{n-1}^{(j)} b_{n-2}^{(j)} \ldots b_0^{(j)} \quad (8.30)$$

is the binary representation of j. Equation 8.28 provides an excellent opportunity to point out another strength of quantum computing. Classical thinking suggests that in order to compute the superposition state on the right-hand side of Equation 8.28 we have to sum 2^n terms, a task that quickly becomes impossible to perform even for relatively modest values of n. Quantum mechanically, however, as shown by Equation 8.28, application of only $n+1$ quantum gates is sufficient to generate the required superposition state. Therefore, while classically the effort is exponential in n, quantum mechanically the effort is only linear in n (see the quantum register loading operation in Section 7.5).

(iii) Apply f-CNOT:

$$f\text{-CNOT}|j\rangle(|0\rangle - |1\rangle) = f\text{-CNOT}|j\rangle|0\rangle - f\text{-CNOT}|j\rangle|1\rangle$$
$$= |j\rangle|f(j) \oplus 0\rangle - |j\rangle|f(j) \oplus 1\rangle = |j\rangle|f(j)\rangle - |j\rangle|f(j) \oplus 1\rangle. \quad (8.31)$$

(iv) Although j is an n-bit string, $f(j)$ can only take the values 0 or 1. If $f(j) = 0$, then:

$$f\text{-CNOT}|j\rangle(|0\rangle - |1\rangle) = |j\rangle(|0\rangle - |1\rangle). \quad (8.32)$$

If $f(j) = 1$, then:

$$f\text{-CNOT}|j\rangle(|0\rangle - |1\rangle) = -|j\rangle(|0\rangle - |1\rangle). \quad (8.33)$$

Just like in the case of Deutsch's algorithm, these two results may be combined into:

$$f\text{-CNOT}\,|j\rangle(|0\rangle - |1\rangle) = (-1)^{f(j)}\,|j\rangle(|0\rangle - |1\rangle). \qquad (8.34)$$

At this point the state of the quantum computer is:

$$|\psi\rangle = \frac{1}{\sqrt{2^{n+1}}} \sum_{j=0}^{2^n-1} (-1)^{f(j)}\,|j\rangle(|0\rangle - |1\rangle). \qquad (8.35)$$

(v) Apply $n+1$ Hadamard gates to the $n+1$ qubits of $|\psi\rangle$:

$$H^{\otimes(n+1)}|\psi\rangle = \frac{1}{2^{n/2}} \sum_{j=0}^{2^n-1} (-1)^{f(j)} H^{\otimes n}|j\rangle H\left(\frac{|0\rangle - |1\rangle}{\sqrt{2}}\right). \qquad (8.36)$$

Using Hadamard's formula defined in Equation 8.25 to evaluate $H^{\otimes n}|j\rangle$, we obtain:

$$H^{\otimes(n+1)}|\psi\rangle = |\varphi\rangle|1\rangle, \qquad (8.37)$$

where

$$|\varphi\rangle = \frac{1}{2^n} \sum_{y=0}^{2^n-1} \left[\sum_{j=0}^{2^n-1} (-1)^{f(j)} (-1)^{j\odot y} \right] |y\rangle. \qquad (8.38)$$

(vi) Now let us measure the probability

$$P_0(\varphi) = |\langle 0|^{\otimes n}|\varphi\rangle|^2 = \left| \frac{1}{2^n} \sum_{j=0}^{2^n-1} (-1)^{f(j)} (-1)^{j\odot 0} \right|^2$$

$$= \left| \frac{1}{2^n} \sum_{j=0}^{2^n-1} (-1)^{f(j)} \right|^2 \qquad (8.39)$$

of finding the state $|0\rangle^{\otimes n}$ in $|\varphi\rangle$. There are two possibilities:

(a) f is constant, i.e., $f(j) = 0$ for all j, or $f(j) = 1$ for all j. Then, according to Equation 8.39:

$$P_0(\varphi) = \left| \frac{\pm 1}{2^n} \sum_{j=0}^{2^n-1} 1 \right|^2 = 1. \qquad (8.40)$$

Therefore, if f is constant, the state $|0\rangle^{\otimes n}$ occurs in the state $|\varphi\rangle$ *with certainty*. No other state occurs.

8.4 • DEUTSCH-JOZSA ALGORITHM

(b) f is balanced. In this case $f(j) = 0$ for 50% of its arguments and $f(j) = 1$ for the other 50% of its arguments such that

$$\sum_{j=0}^{2^n-1} (-1)^{f(j)} = 0. \qquad (8.41)$$

Therefore,

$$P_0(\varphi) = \left| \frac{1}{2^n} \sum_{j=0}^{2^n-1} (-1)^{f(j)} \right|^2 = 0. \qquad (8.42)$$

Therefore, in the balanced case, the state $|0\rangle^{\otimes n}$ *never* occurs.

This algorithm solves the Deutsch-Jozsa problem. By running the Deutsch-Jozsa algorithm described above just *once*, and measuring the probability of occurrence of the state $|0\rangle^{\otimes n}$ in the state $|\varphi\rangle$, we obtain the following answer:

If $|0\rangle^{\otimes n}$ occurs in $|\varphi\rangle$, f is constant.
If $|0\rangle^{\otimes n}$ does *not* occur in $|\varphi\rangle$, f is balanced.

Exercises:

8.4.1 Let $|n_1 n_2 \cdots n_M\rangle$, $n_k = -1, +1$, $k = 1, \ldots, M$, denote a spin state of an M electron system, where we made the identification $|\downarrow\rangle \equiv |-1\rangle$ and $|\uparrow\rangle \equiv |+1\rangle$.

(a) What is the number ν_M of different spin states you can construct with M electrons?

(b) There are approximately $N = 10^{80}$ electrons in the universe. How many electrons do you need to satisfy $\nu_M > N$?

8.4.2 In the final step of the Deutsch-Jozsa algorithm we need to perform a measurement to determine the probability $P_0(\varphi)$ for the occurrence of the state $|0\rangle^{\otimes n}$ in the n-qubit state $|\varphi\rangle$. Assuming that the measurement is 100% effective, a single run of the quantum computer is sufficient to solve the Deutsch-Jozsa problem. In practice, however, measurement of $P_0(\varphi)$ requires n individual measurements to determine the quantum state of each of the n qubits of $|\varphi\rangle$. Assume that the probability of correctly identifying an individual qubit's quantum state is 0.99; the probability of misidentifying the qubit's quantum state is 0.01. For $n = 30$:

(a) Determine the probability of a correct measurement of $P_0(\varphi)$.

(b) Devise and justify a measurement protocol that allows you to increase the probability of a correct identification of $|0\rangle^{\otimes n}$ by multiple measurements of $|\varphi\rangle$ generated by multiple runs of the quantum computer.

(c) On the basis of your strategy defined in (b), what is the minimum number of times you have to run the quantum computer to measure $P_0(\varphi)$ with larger than 90% confidence?

8.4.3 Consider the class F_n of binary-valued functions of n binary arguments. A function $f \in F_n$ satisfies:

$$f: \{0,1\}^{\otimes n} \to \{0,1\},$$

$$f(x_1, x_2, \ldots, x_n) \in \{0,1\}, \quad x_j \in \{0,1\}, \quad j = 1, \ldots, n.$$

(a) Show that the number B_n of balanced functions in F_n is given by:

$$B_n = \frac{(2^n)!}{[(2^{n-1})!]^2}. \tag{8.43}$$

(b) Using Stirling's formula,

$$N! \approx \sqrt{2\pi N} \left(\frac{N}{e}\right)^N,$$

which holds to an excellent approximation even for relatively small N, show that B_n rises hyperexponentially (i.e., faster than exponentially) in n.

8.4.4 Prove Hadamard's formula as defined in Equation 8.25

8.5 Grover's Search Algorithm

The success of the Internet search engine GOOGLE demonstrates convincingly that searching is one of the most important tasks of everyday life. Apart from searching the Internet we search for words in dictionaries; we search for our keys and gloves; but most importantly, we search databases for information. Suppose we have an unsorted database with one million files, each containing information on a specific animal. We would like to find the file with information about "elephants." If our database is not alphabetically ordered, we have to check each file, starting from the beginning, until we find the file on elephants. We might be lucky and the file on elephants is the first file we access. But, equally

8.5 • GROVER'S SEARCH ALGORITHM

likely, we might be unlucky and the file on elephants is the last file we access. Therefore, on average, we have to look at 500,000 files before we find the file on "elephants." This is the best any classical search algorithm can do in this case. In general, if an unsorted database consists of N entries, a classical search algorithm has to access the database $N/2$ times, on average, in order to locate the requested item. No classical search algorithm can improve on these odds in principle. But quantum algorithms can! This was demonstrated by Grover who showed that a quantum search algorithm can find the requested item with only about $\sqrt{N}$ accesses of the database. Although we have already studied several quantum algorithms, for instance the Deutsch-Jozsa algorithm, we study Grover's algorithm because there is a decisive qualitative difference between the algorithms of Deutsch-Jozsa and Grover: The Deutsch-Jozsa algorithm is *deterministic* while Grover's algorithm is *probabilistic*. This means the following: Running the Deutsch-Jozsa algorithm once on an ideal quantum computer, we obtain the correct result with probability 1 with a single measurement. Grover's algorithm, however, yields only a *high probability*, i.e., a probability very close to 1, to obtain the desired result after a single measurement. To improve the odds, we have to run Grover's algorithm several times. Still, even taking this handicap into account, Grover's algorithm far outperforms any conceivable classical algorithm that searches for items in an unsorted database.

Suppose we have a database with N objects and we would like to find an object φ in this database. We first assign an orthonormal set of quantum states $|0\rangle$, $|1\rangle$, ..., $|N-1\rangle$, $\langle n|m\rangle = \delta_{nm}$, to each of the N objects of the database. The N quantum states may be realized, for example, with M qubits, where M is the smallest integer larger than $\log_2(N)$. The state $|\varphi\rangle$ assigned to the object φ corresponds to one of the states $|n\rangle$, $n = 0, 1, \ldots, N-1$ of the database. Since all the states of the database are normalized, we have

$$\langle \varphi | \varphi \rangle = 1. \tag{8.44}$$

Although we know that $|\varphi\rangle$ is one of the states of the database, we do not know which one, because we do not know the sequence number n_φ of the object φ. If we knew, a single access of the database would be sufficient to locate φ. But if all we are given is the state $|\varphi\rangle$, and not its sequence number, the naive way of locating φ in an unsorted database is to blindly take scalar products $\langle n|\varphi\rangle$ with the states of the database. According to "classical thinking," since the database is unsorted, there is no better way than starting with the first state of the database and to compute the scalar product $\langle 0|\varphi\rangle$. If the result is 1, we found the object φ in the database. But more likely we will obtain $\langle 0|\varphi\rangle = 0$ and have to try the next state, $\langle 1|\varphi\rangle$. We continue this procedure until we find a state $|n_\varphi\rangle$ in the database with $\langle n_\varphi|\varphi\rangle = 1$. This determines the sequence number n_φ of the object φ and thus locates the object φ in the database. This procedure requires $N/2$ trials on average.

But making use of the quantum nature of the states $|n\rangle$, we can do much better than that. In order to search for the item φ we define a unitary operator

$$\hat{U}_\varphi |n\rangle = |n\rangle, \quad \text{for } |n\rangle \neq |\varphi\rangle,$$
$$\hat{U}_\varphi |\varphi\rangle = -|\varphi\rangle. \qquad (8.45)$$

Equation 8.45 shows that the operator $\hat{U}_\varphi$ is constructed such that it "tags" (with a minus sign) the state $|\varphi\rangle$ we want to find, but leaves all other states invariant. The tagging operator $\hat{U}_\varphi$ is constructed explicitly in the following way:

$$\hat{U}_\varphi = \hat{\mathbf{1}} - 2|\varphi\rangle\langle\varphi|, \qquad (8.46)$$

where $\hat{\mathbf{1}}$ is the unit operator. We can easily check that, because of the mutual orthogonality of the database states $|n\rangle$, $\langle n|m\rangle = \delta_{nm}$, the construction defined in Equation 8.46 indeed satisfies the requirements in Equation 8.45.

As the starting state of our quantum computer we use

$$|s\rangle = \frac{1}{\sqrt{N}} \sum_{n=0}^{N-1} |n\rangle, \qquad (8.47)$$

which can be generated efficiently with about $\log_2(N)$ Hadamard gates (see the quantum register loading operation in Section 7.5). The starting state defined in Equation 8.47 is a normalized, coherent superposition of all the states of the database. For the following computations it is useful to note that

$$\langle s|s\rangle = 1, \quad \langle \varphi|s\rangle = \langle s|\varphi\rangle = \frac{1}{\sqrt{N}}. \qquad (8.48)$$

We define a second unitary operator $\hat{U}_s$, built with the starting state $|s\rangle$, according to:

$$\hat{U}_s = 2|s\rangle\langle s| - \hat{\mathbf{1}}. \qquad (8.49)$$

With the help of $\hat{U}_\varphi$ and $\hat{U}_s$, we construct the *Grover operator*

$$\hat{G} = \hat{U}_s \hat{U}_\varphi. \qquad (8.50)$$

8.5 • GROVER'S SEARCH ALGORITHM

Repeated applications of the Grover operator $\hat{G}$ on the starting state $|s\rangle$ are called *Grover iterations*. We define the state

$$|\gamma_k\rangle = \hat{G}^k |s\rangle, \qquad (8.51)$$

obtained from $|s\rangle$ after k Grover iterations. The Grover operator $\hat{G}$ is constructed such that each Grover iteration drives the starting state $|s\rangle$ closer to the state $|n_\varphi\rangle$, the desired result of our database search. As discussed above, with n_φ known, we can then locate the object φ with a single access of the database.

Let us see how this works. To perform the first Grover iteration $\hat{G}|s\rangle$ we have to first apply $\hat{U}_\varphi$ on the starting state $|s\rangle$, and then the operator $\hat{U}_s$ on the resulting state. We obtain in succession:

$$\hat{U}_\varphi |s\rangle = \left(\hat{1} - 2|\varphi\rangle\langle\varphi|\right)|s\rangle = |s\rangle - \frac{2}{\sqrt{N}}|\varphi\rangle, \qquad (8.52)$$

and, with Equation 8.48,

$$\begin{aligned}
|\gamma_1\rangle &= \hat{G}|s\rangle = \hat{U}_s \hat{U}_\varphi |s\rangle = \hat{U}_s \left(|s\rangle - \frac{2}{\sqrt{N}}|\varphi\rangle\right) \\
&= \left(2|s\rangle\langle s| - \hat{1}\right)\left(|s\rangle - \frac{2}{\sqrt{N}}|\varphi\rangle\right) \\
&= 2|s\rangle - \frac{4}{\sqrt{N}}|s\rangle\langle s|\varphi\rangle - |s\rangle + \frac{2}{\sqrt{N}}|\varphi\rangle \\
&= \left(\frac{N-4}{N}\right)|s\rangle + \frac{2}{\sqrt{N}}|\varphi\rangle. \qquad (8.53)
\end{aligned}$$

For large N we obtain approximately:

$$|\gamma_1\rangle \approx |s\rangle + \frac{2}{\sqrt{N}}|\varphi\rangle. \qquad (8.54)$$

For $N \gg 1$ and $k \ll N$, we may generalize the result of Equation 8.54 to

$$|\gamma_k\rangle \approx |s\rangle + \frac{2k}{\sqrt{N}}|\varphi\rangle. \qquad (8.55)$$

Indeed, according to Equation 8.54, Equation 8.55 holds for $k = 1$, and if Equation 8.55 holds for some k, it also holds for $k + 1$ according to

$$\begin{aligned}
|\gamma_{k+1}\rangle &= \hat{G}|\gamma_k\rangle \approx \hat{G}\left(|s\rangle + \frac{2k}{\sqrt{N}}|\varphi\rangle\right) \\
&= \hat{U}_s\left(\hat{\mathbf{1}} - 2|\varphi\rangle\langle\varphi|\right)\left(|s\rangle + \frac{2k}{\sqrt{N}}|\varphi\rangle\right) \\
&= \hat{U}_s\left(|s\rangle - \frac{2k+2}{\sqrt{N}}|\varphi\rangle\right) \\
&= |s\rangle - 2\left(\frac{2k+2}{N}\right)|s\rangle + \frac{2(k+1)}{\sqrt{N}}|\varphi\rangle \\
&\approx |s\rangle + \frac{2(k+1)}{\sqrt{N}}|\varphi\rangle,
\end{aligned} \qquad (8.56)$$

which establishes Equation 8.55 inductively. Therefore, according to Equation 8.55, and for $N \gg 1$, $k \ll N$, the amplitude of $|\varphi\rangle$ increases by approximately $2/\sqrt{N}$ in each Grover iteration, thereby evolving $|s\rangle$ into $|\varphi\rangle$. Thus, if we stop the iteration at some optimal $k = n_G$, where n_G is on the order of $\sqrt{N}$, a measurement of $|\gamma_{n_G}\rangle$ will return $|n_\varphi\rangle$ with near certainty, solving our database search problem.

Because, in general, the weight of $|\varphi\rangle$ in any of the states $|\gamma_k\rangle$ is not precisely equal to 1, there is always a small chance to obtain a state different from $|n_\varphi\rangle$ after the measurement of $|\gamma_{n_G}\rangle$. This explains why Grover's algorithm is only *probabilistic* and not *deterministic*. Grover's algorithm would be deterministic only if we could guarantee that there is always a κ such that the amplitude of $|\varphi\rangle$ in $|\gamma_\kappa\rangle$ is precisely equal to 1. Since, in general, we cannot, Grover's algorithm is probabilistic. The most important observation, however, is that we need only about $\sqrt{N}$ Grover iterations to drive $|s\rangle$ toward $|\varphi\rangle$ with a probability very close to 1. Since each Grover iteration requires one access to the (quantum) database, Grover's quantum algorithm solves the database search problem with only about $\sqrt{N}$ accesses, while any classical algorithm, as discussed above, requires $N/2$ database accesses, on average.

To get a more intuitive feeling for how Grover's algorithm works, we cast it into geometrical language. As shown in Figure 8.3, the states $|s\rangle$ and $|\varphi\rangle$ span a two-dimensional plane P. If γ is a vector in P, i.e.,

$$|\gamma\rangle = A|s\rangle + B|\varphi\rangle, \qquad (8.57)$$

with amplitudes A and B, then, applying a Grover iteration to $|\gamma\rangle$, yields an image vector $|\gamma'\rangle$ that, again, is entirely contained in P. Indeed, explicit calculation shows that

$$|\gamma'\rangle = \hat{G}|\gamma\rangle = A'|s\rangle + B'|\varphi\rangle, \qquad (8.58)$$

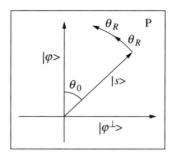

Figure 8.3 Geometric interpretation of Grover's quantum search algorithm in the plane spanned by $|s\rangle$ and $|\varphi\rangle$.

where

$$A' = A\left(\frac{N-4}{N}\right) - B\left(\frac{2}{\sqrt{N}}\right),$$
$$B' = A\left(\frac{2}{\sqrt{N}}\right) + B. \quad (8.59)$$

Because of Equation 8.48, the states $|s\rangle$ and $|\varphi\rangle$ are not orthogonal. Therefore, instead of using $|s\rangle$ and $|\varphi\rangle$ to span P, we choose the states $|\varphi\rangle$ and

$$|\varphi^\perp\rangle = \sqrt{\frac{N}{N-1}}\left(|s\rangle - \frac{1}{\sqrt{N}}|\varphi\rangle\right) \quad (8.60)$$

to span P. Because of Equation 8.48 we have

$$\langle\varphi|\varphi^\perp\rangle = 0, \quad (8.61)$$

i.e., the state $|\varphi^\perp\rangle$, as shown in Figure 8.3, is orthogonal to $|\varphi\rangle$. The states $|\varphi^\perp\rangle$ and $|\varphi\rangle$ form a convenient, orthonormal basis in P. Our next goal is to compute the matrix elements of $\hat{G}$ in the basis $(|\varphi^\perp\rangle, |\varphi\rangle)$. Using

$$\langle\varphi^\perp|s\rangle = \langle s|\varphi^\perp\rangle = \sqrt{\frac{N-1}{N}}, \quad (8.62)$$

which follows from Equations 8.48 and 8.60, together with

$$\hat{G} = \hat{U}_s\hat{U}_\varphi = (2|s\rangle\langle s| - \hat{1})(\hat{1} - 2|\varphi\rangle\langle\varphi|)$$
$$= 2|s\rangle\langle s| - \frac{4}{\sqrt{N}}|s\rangle\langle\varphi| - \hat{1} + 2|\varphi\rangle\langle\varphi|, \quad (8.63)$$

it is not difficult to show that

$$G = \begin{pmatrix} \langle \varphi^\perp | \hat{G} | \varphi^\perp \rangle & \langle \varphi^\perp | \hat{G} | \varphi \rangle \\ \langle \varphi | \hat{G} | \varphi^\perp \rangle & \langle \varphi | \hat{G} | \varphi \rangle \end{pmatrix} = \begin{pmatrix} \frac{N-2}{N} & -\frac{2}{N}\sqrt{N-1} \\ \frac{2}{N}\sqrt{N-1} & \frac{N-2}{N} \end{pmatrix}.$$
(8.64)

Since G is an orthogonal matrix, $G^T G = 1$, with $\det(G) = 1$, we can write the matrix G in the form of a rotation matrix

$$G = \begin{pmatrix} \cos(\theta_R) & -\sin(\theta_R) \\ \sin(\theta_R) & \cos(\theta_R) \end{pmatrix}$$
(8.65)

with rotation angle θ_R, where

$$\cos(\theta_R) = \frac{N-2}{N}, \quad \sin(\theta_R) = \frac{2}{N}\sqrt{N-1}.$$
(8.66)

This reveals the Grover operator $\hat{G}$ as a rotation with rotation angle θ_R in the plane P. Since $\hat{G}$ is repeatedly applied on $|s\rangle$, each Grover iteration rotates the images of $|s\rangle$ closer to $|\varphi\rangle$ by an angle θ_R. This is illustrated in Figure 8.3.

How many Grover iterations do we need to rotate $|s\rangle$ into $|\varphi\rangle$? The answer is the following. The initial angle θ_0 between $|s\rangle$ and $|\varphi\rangle$ is

$$\cos(\theta_0) = \langle s | \varphi \rangle = \frac{1}{\sqrt{N}}, \quad \Longrightarrow \quad \theta_0 = \arccos\left(\frac{1}{\sqrt{N}}\right).$$
(8.67)

Grover's algorithm would work perfectly if an integer number n_G of Grover iterations would align one of the iterates of $|s\rangle$ with $|\varphi\rangle$, i.e., if the Grover ratio

$$\rho_G = \frac{\theta_0}{\theta_R} = \frac{\arccos\left(\frac{1}{\sqrt{N}}\right)}{\arccos\left(\frac{N-2}{N}\right)}$$
(8.68)

is an integer. However, for most N, the Grover ratio ρ_G is not an integer. Therefore, the best we can do is to apply a number n_G of Grover rotations where n_G is the integer closest to ρ_G.

The expression for ρ_G is exact, but not very illuminating. Since Grover's algorithm reveals its power for large N, we look for a large N approximation of ρ_G. For large N the argument of the arccos function in Equation 8.67 is very close to 0. Therefore, the initial angle θ_0 is very close to $\pi/2$. Equation 8.66 tells us that for large N the rotation angle θ_R is very small. Using the second expression of Equation 8.66, we obtain:

$$\theta_R = \arcsin\left(\frac{2}{N}\sqrt{N-1}\right) \approx \frac{2}{N}\sqrt{N-1} \approx \frac{2}{N}\sqrt{N} = \frac{2}{\sqrt{N}}.$$
(8.69)

Therefore, approximately,

$$n_G \approx \rho_G = \frac{\theta_0}{\theta_R} \approx \frac{\pi/2}{2/\sqrt{N}} = \frac{\pi}{4}\sqrt{N}. \qquad (8.70)$$

This result proves that Grover's quantum algorithm indeed locates an item φ in an unsorted database in about $\sqrt{N}$ steps, while any classical algorithm requires $N/2$ steps, on average, to perform the same task.

The closest integer approximation n_G to ρ_G will not rotate $|s\rangle$ exactly into $|\varphi\rangle$. This means that, in general, a measurement of the rotated state will have a probability close to 1, *but not equal to* 1, to yield $|\varphi\rangle$ as the result of the measurement. This means that there is a small chance that Grover's algorithm fails. However, running Grover's algorithm a few times increases the chances for success to near certainty. Running Grover's algorithm a few times does not erase its $\sqrt{N}$ advantage over classical algorithms. Thus, Grover's algorithm is qualitatively superior to any classical algorithm searching an unsorted database.

Finding an item φ in a database is an important task. But equally important is the task of deciding whether a given item φ is in the database at all. Because, if it is not, we might want to add this item to the database. If φ is not in the database, then $\langle\varphi|s\rangle = 0$. This is so, since $|\varphi\rangle$ is not a member of the superposition state $|s\rangle$ defined in Equation 8.47. In this case a Grover iteration applied on $|s\rangle$ yields

$$\begin{aligned}\hat{G}|s\rangle &= \hat{U}_s\left(\hat{\mathbf{1}} - 2|\varphi\rangle\langle\varphi|\right)|s\rangle = \hat{U}_s|s\rangle \\ &= \left(2|s\rangle\langle s| - \hat{\mathbf{1}}\right)|s\rangle = |s\rangle.\end{aligned} \qquad (8.71)$$

We see that if φ is not a member of the database, we remain in the starting state $|s\rangle$; the rotation away from $|s\rangle$ never commences. When we measure the final state after n_G Grover iterations, where n_G is determined according to Equation 8.70, we will find that the results of the measurements jump wildly, pointing to this and that database item according to how the state $|s\rangle$ collapses onto the individual database items as a result of the measurements. Therefore: If several measurements return the same state $|n_\varphi\rangle$ with high probability, the item φ we were asked to locate in the database actually exists in the database and its label is n_φ. If, on the other hand, we get vastly different results for n_φ after running the Grover algorithm several times and measuring the final state after n_G Grover iterations, then φ does not exist in the database.

Exercises:

8.5.1 Show by explicit computation that the operators $\hat{U}_\varphi$ and $\hat{U}_s$ defined in Equations 8.46 and 8.49, respectively, are unitary.

8.5.2 Verify that the operator $\hat{U}_\varphi$ constructed in Equation 8.46 satisfies the requirements 8.45.

8.5.3 Verify Equation 8.48.

8.5.4 Verify the constants A' and B' in Equation 8.59.

8.5.5 Suppose $|\gamma\rangle$ defined in Equation 8.57 is normalized, i.e., $\langle\gamma|\gamma\rangle = 1$. Since $\hat{G}$ is a unitary operator, $|\gamma'\rangle$ defined in Equation 8.58 is normalized, too. Checking the correctness of the constants A' and B' defined in Equation 8.59, verify that $\langle\gamma'|\gamma'\rangle = 1$ if these constants are used in the construction of $|\gamma'\rangle$ according to Equation 8.58.

8.5.6 Using Equations 8.48, 8.62, and 8.63, verify Equation 8.64.

8.5.7 Show that for database sizes $N = 656776$, 12024721, and 17403660 the Grover ratio ρ_G is very close to an integer. This may be of practical interest, since in these cases Grover's algorithm is nearly deterministic.

8.5.8 The Chebyshev polynomials $T_n(x)$ are defined as

$$T_n(x) = \cos[n\arccos(x)].$$

(a) Show that if for some N the Grover ratio ρ_G is an integer, i.e., $\rho_G = n_G$, the size N of the corresponding database satisfies

$$1 = N T_{n_G}^2\left(\frac{N-2}{N}\right). \tag{8.72}$$

(b) Show that for $n_G = 1$ Equation 8.72 has the solution $N = 4$.

8.5.9 Show that for $n_G = 2$ there is no integer N for which $\rho_G = 2$, i.e., there is no database for which Grover's quantum search algorithm is deterministic for $n_G = 2$ Grover iterations.

8.6 Summary

In this chapter we studied the basic ideas behind quantum computing. We started by constructing a primitive quantum computer that illustrates how superposition may be used to achieve parallel processing in quantum computers. We also saw that parallel processing is not enough: It has to be combined with interference in order to select the desired results from the multitude of computed results. Deutsch's algorithm, studied in Section 8.3, was the first to combine both principles. Deutsch's algorithm performs a task that cannot be performed on any classical computer in principle. Thus, Deutsch's algorithm was the first to prove the point that quantum computing is a qualitatively new way of information processing. While Deutsch's algorithm beats any classical algorithm by a factor of two, practically speaking, this is not too impressive given how difficult it is to construct quantum circuitry and keep its coherence. Simply using two off-the-shelf classical processors and running them in parallel, erases the speed advantage of Deutsch's algorithm. This is where the Deutsch-Jozsa algorithm comes in. The Deutsch-Jozsa algorithm demonstrates that quantum computers can perform computations that are out of the league of classical computers, even if we allow for a classical computer the size of the universe! Therefore, when it comes to solving problems of the Deutsch-Jozsa type, classical computers simply are no match for quantum computers. Although quantum computers excel when solving problems of the Deutsch- or Deutsch-Jozsa type, one could argue that these problems and their quantum solution algorithms are rather contrived and do not have any practical applications. To counter this argument, and to show that quantum computers reign supreme even in areas of everyday importance, we studied Grover's quantum algorithm. Grover's algorithm addresses the problem of finding an item in an unsorted database. Not only since the advent of the Internet and its various search engines do we know that searching for items in an unsorted environment is an everyday occurrence. We saw that Grover's algorithm solves the task of locating an object in an unsorted database of size N in approximately $\sqrt{N}$ steps, while any classical algorithm requires $N/2$ steps, on average. This may seem like a marginal advantage. And indeed, for $N = 10$, for instance, there is hardly any difference. For $N = 1,000,000$, however, not even such a large database, the difference between $\sqrt{N} = 1,000$ and $N/2 = 500,000$ is significant. Based on the examples of quantum algorithms presented in this chapter, it is not hard to imagine that quantum computers have the potential to revolutionize the fields of information processing and computing.

Chapter Review Exercises:

1. Compute the matrices

$$U_\varphi = \begin{pmatrix} \langle \varphi^\perp | \hat{U}_\varphi | \varphi^\perp \rangle & \langle \varphi^\perp | \hat{U}_\varphi | \varphi \rangle \\ \langle \varphi | \hat{U}_\varphi | \varphi^\perp \rangle & \langle \varphi | \hat{U}_\varphi | \varphi \rangle \end{pmatrix},$$

$$U_s = \begin{pmatrix} \langle \varphi^\perp | \hat{U}_s | \varphi^\perp \rangle & \langle \varphi^\perp | \hat{U}_s | \varphi \rangle \\ \langle \varphi | \hat{U}_s | \varphi^\perp \rangle & \langle \varphi | \hat{U}_s | \varphi \rangle \end{pmatrix}.$$

Then, by multiplying U_s and U_φ verify the matrix G in Equation 8.64.

248 CHAPTER 8 • QUANTUM COMPUTING

2. In the text we claimed that Grover's algorithm is probabilistic. While this is true in general, there exists at least one N for which Grover's algorithm is deterministic. Show that for $N = 4$ Grover's algorithm is exact and locates an item in the database with a single Grover iteration.

3. Given are three bits $x, y,$ and z that can take the values 0 and 1 each. Use two CNOT gates to construct a quantum computer that adds the three bits modulo 2, i.e, given $x, y,$ and z, the quantum computer returns $x \oplus y \oplus z$.

4. Given is a binary-valued function f of a single binary argument, i.e., $f : \{0,1\} \to \{0,1\}$. Devise a quantum algorithm that is faster than any conceivable classical algorithm and tests whether $\sigma = f(0)+f(1)$ is divisible by 2.

5. A set B of binary-valued functions of two binary arguments contains the following four functions:

$$f_{00}(00) = 1, \quad f_{00}(01) = 0, \quad f_{00}(10) = 0, \quad f_{00}(11) = 0;$$

$$f_{01}(00) = 0, \quad f_{01}(01) = 1, \quad f_{01}(10) = 0, \quad f_{01}(11) = 0;$$

$$f_{10}(00) = 0, \quad f_{10}(01) = 0, \quad f_{10}(10) = 1, \quad f_{10}(11) = 0;$$

$$f_{11}(00) = 0, \quad f_{11}(01) = 0, \quad f_{11}(10) = 0, \quad f_{11}(11) = 1.$$

The function values are zero except if the argument of a function equals its label. We are given a function $f \in B$. Our task is to find out which one of the four functions in B we are given. Quantum mechanically, this task can be accomplished with only one quantum evaluation of f using the three-qubit quantum algorithm whose circuit diagram is shown in Figure 8.4. As indicated in the figure, the first two qubits are control qubits; the third qubit is the target qubit.

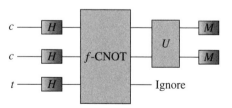

Figure 8.4 Circuit diagram of the quantum computation for identifying a given function out of four.

The f-CNOT gate for this three-qubit circuit works in the following way:
$$f\text{-CNOT}\,|b_2 b_1 b_0\rangle = |b_2 b_1\rangle\,|f(b_2 b_1) \oplus b_0\rangle.$$
The starting state of the quantum computer is $|001\rangle$.

(a) Classically, given f, and in the worst case, how many function evaluations are necessary to reveal the identity of f?

(b) What is the state of the quantum computer immediately after application of the H gates, but before application of the f-CNOT gate?

(c) Show that immediately after application of the f-CNOT gate the state of the quantum computer is:

$$\frac{1}{2\sqrt{2}}\Big[(-1)^{f(00)}|00\rangle + (-1)^{f(01)}|01\rangle \\ + (-1)^{f(10)}|10\rangle + (-1)^{f(11)}|11\rangle\Big]\big[|0\rangle - |1\rangle\big]. \quad (8.73)$$

Notice that application of the f-CNOT gate "tags" the states corresponding to the label of the given function f.

(d) Show that application of the operator

$$\hat{U} = \frac{1}{2} \sum_{a,b,p,q=0}^{1} (-1)^{\delta_{ap}\delta_{bq}} |ab\rangle\langle pq|$$

to the quantum state of the two control qubits in Equation 8.73 results in a pure, two-qubit state whose label is identical with the label of the function f. Thus, as indicated in Figure 8.4, measurement of the two control qubits, following the application of $\hat{U}$, reveals the identity of f. To show this step of the quantum algorithm, you might find it helpful to represent the quantum state of the two control qubits as a vector, and the operator $\hat{U}$ as a matrix in the space of the four basis states $|00\rangle$, $|01\rangle$, $|10\rangle$, and $|11\rangle$.

6. In order to show the enhancement of $|\varphi\rangle$ in the iterates of $|s\rangle$ after two Grover iterations, compute

(a) $\hat{G}^2$,

(b) the enhancement factor $f_2 = |\langle\varphi|\hat{G}^2|s\rangle|^2 / |\langle\varphi|s\rangle|^2$.

7. A database contains $N = 5$ items that are mapped to the five quantum states $|n\rangle$, $n = 1, \ldots, 5$. Your task is to determine the sequence

number n_φ of a given object φ. For $N = 5$ we have $\rho_G \approx 1.2$. Therefore, $n_G = 1$, which means that a single Grover iteration is optimal for determining the sequence number n_φ of φ in the database. Show that, indeed,
$$|\langle n_\varphi|\hat{G}|s\rangle|^2 \gg |\langle n|\hat{G}|s\rangle|^2$$
for $n \neq n_\varphi$.

8. List all balanced binary functions of two binary arguments. Is your result consistent with Equation 8.43?

chapter 9

Classical Cryptology

In this chapter:

- Introduction
- Private-Key Cryptosystems
- RSA Public-Key Cryptosystem
- How Does RSA Work?
- Why Is Integer Factorization So Difficult?
- Summary

9.1 Introduction

The need for secure communication is as old as civilization itself. In ancient times it was mostly the military who relied on the transmission of secret messages; today, it is governments and commerce who rely increasingly on means of secure communication. In fact, both Internet commerce and Internet banking are unthinkable without the possibility of transmitting secure messages. *Cryptology* is the branch of mathematics that concerns itself with secure, and usually secret, communication. Cryptology has two branches: *Cryptography* and *cryptanalysis*. Cryptography concerns itself with converting *plaintext* messages into *ciphers* or *cipher-text* messages that only the sender and the intended receiver can understand. Cryptanalysis provides methods and procedures for unravelling the information concealed in secret messages.

Table 9.1 Illustration of Caesar's substitution code. First row: Roman alphabet around the time of Caesar. The Roman letters are in alphabetical order. Second row: "Caesar shift" $n = 3$. A message is encrypted by replacing any of its letters occurring in the first row with its corresponding letter in the second row.

A	B	C	D	E	F	G	H	I	K	L	M	N	O	P	Q	R	S	T	V	X	Y	Z
D	E	F	G	H	I	K	L	M	N	O	P	Q	R	S	T	V	X	Y	Z	A	B	C

There are two principal means of secure communication: (1) Private-key cryptosystems and (2) public-key cryptosystems. Both have advantages and disadvantages. For modern Internet commerce and Internet banking, however, public-key cryptosystems, in particular the RSA cryptosystem, are of particular importance.

Historically, private-key cryptosystems were developed first. We discuss them in Section 9.2. Public-key cryptosystems are a relatively recent discovery, coinciding with the increased importance of Internet communication. Public-key cryptosystems, with special focus on the RSA cryptosystem, are discussed in Section 9.3. Because of its importance for government, military, and industry, any attack on the security of the RSA cryptosystem is of particular significance. We study the RSA cryptosystem and integer factorization, which it is based on, in Sections 9.4 and 9.5, respectively. While there is a general consensus in the scientific community that RSA is safe from attacks launched by commercially available, classical computers, it was recently demonstrated that RSA is vulnerable to attacks launched by *quantum computers*. In fact, for better or worse, compromising the security of cryptosystems is one of the preeminent strengths of quantum computers. Thus, cryptosystems provide an important playground for quantum computers. This is why we introduce and discuss cryptosystems in detail in this chapter.

9.2 Private-Key Cryptosystems

The Roman military commander Julius Caesar invented one of the first cryptosystems more than 2000 years ago. It is now known as the *Caesar Cipher*. It works in the following way. Suppose Caesar wants to send the message ATTACK to one of his field commanders. Since the content of this communication is sensitive, and should not fall into the wrong hands, Caesar decides to *encrypt* his message. Encryption is performed by an *algorithm* under the control of a *key* that turns a plaintext message into a cipher that only Caesar and the field commander can understand. Caesar's encryption scheme consists of replacing each letter in his message with the letter of the Roman alphabet that occurs three positions shifted to the right (see Table 9.1). This way Caesar turns the plaintext message ATTACK into the cipher-text message DYYDFN. All the field commander has to do is to shift the letters of the cipher-text three positions to the left to recover Caesar's plaintext message ATTACK.

The algorithm of Caesar's encryption scheme is to shift each letter of the plaintext message n positions to the right, in alphabetical order. Caesar chose $n = 3$. This number is the key of Caesar's encryption scheme. Knowing the algorithm is usually not enough to *decipher* a message. Only possession of the key allows instantaneous recovery of the original plaintext message. Since both the sender and the receiver need to know the key, and have to keep this key secret, i.e., private, this type of cryptosystem is known as a *private-key cryptosystem*. Since sender and receiver have the *same* key, private-key cryptosystems are also called *symmetric cryptosystems*. Since Caesar's scheme consists of substituting characters of the plaintext message with other characters, this scheme is also known as a *substitution cipher*. The substitutions do not necessarily have to be characters of the alphabet. They may be arbitrary symbols.

Algorithm and key are the two hallmarks of any cryptosystem. While the algorithm is usually known, or can be discerned with some effort on the basis of the appearance of the cipher-text, it is the task of cryptanalysts to recover the key to be able to *eavesdrop* on intercepted, secret messages. Suppose someone intercepts Caesar's encrypted message. Even without knowledge of the key n, the original message is easily recovered after only a few decryption attempts with randomly chosen keys n. Once a cryptosystem is *broken*, i.e., the key is revealed, a new key has to be chosen.

Compared with modern cryptosystems Caesar's scheme is pathetically insecure and can be broken easily with only a few attempts. This, however, should not distract from its historic significance. Therefore, it still serves us well as an introductory paradigm for cryptosystems.

How does the field commander get Caesar's key? This is known as the *key distribution problem*. Obviously the key cannot be transmitted in encrypted form, since the commander needs the key to decrypt the message. There are only two ways of reasonably safe key distribution: (1) In person at a personal meeting with the commander, or, where this is not possible, (2) via a messenger. Obviously neither of these two methods is satisfactory since (1) someone may overhear Caesar whispering the secret key to the field commander or (2) the messenger may decide that some profit may be made by sharing the secret key with unauthorized parties. To this day key distribution is the most vulnerable phase of symmetric, private-key cryptosystems.

Because of the key distribution problem, private-key cryptosystems are completely inadequate for internet commerce and internet banking. While, for example, one might conceive of exchanging keys by mail,

254 CHAPTER 9 • CLASSICAL CRYPTOLOGY

Table 9.2 The pangram *The quick brown fox jumps over a lazy dog* is turned into a private key by assigning each letter of the English alphabet the first occurrence of a new letter in the pangram.

```
ABC DEFGH IJKLM N O P QRS  T    U V WX Y Z
THE QUICK BROWN FOX JUMPS OVER A LAZY DOG
```

this method is both clumsy, slow, and insecure. Luckily, just in time for internet applications, *public-key cryptosystems* have been invented. The most successful public-key cryptosystem is the RSA scheme discussed in the following section.

Exercises:

9.2.1 Use Caesar's encryption scheme ($n = 3$) to encrypt the plaintext message RETREAT.

9.2.2 Pangrams are an excellent way to construct (and remember) a private key. A pangram is a sentence that uses all the letters of the English alphabet at least once. Consider Table 9.2. It turns the pangram *The quick brown fox jumps over a lazy dog* into an easy-to-remember private key by assigning each letter of the English alphabet the first occurrence of a new letter in the pangram. Use the key constructed in Table 9.2 to encrypt the plaintext message ROSES ARE RED VIOLETS ARE BLUE.

9.2.3 The message that follows was encoded using a simple substitution cipher. Use frequency analysis and the grammatical structure of the English language to decode the message:
zvi tv svzj pjv ibc dvkzapiuvzt vd hbltugt ycmc ibc mcpsf vd hbusvtvhbcmt. ibut gbpzjca ampfpiugpssl ybcz dclzfpz pza vibcm hbltugutit tipmica iv thcgkspic vz ibc hvycm vd qkpzikf gvfhkiuzj.

• 9.3 RSA Public-Key Cryptosystem

The RSA encryption scheme, named after its inventors Rivest, Shamir, and Adleman, is the most important public-key cryptosystem. We illustrate it with the following example.

Suppose Alice wants to set up a communication scheme that allows anybody in the universe to send her a message that can be read only by Alice and the sender. In order to set the stage for this mode of secret communication, Alice performs the following five preparation steps:

Step 1: Choose two prime numbers p and q with $p \neq q$.

Step 2: Compute the *modulus* $n = pq$. Since n is the product of two prime numbers, n is called a *semi-prime*.

Step 3: Compute the *totient* $\phi = (p-1)(q-1)$.

Step 4: Choose a number e, the *public-key exponent*. It has to satisfy two conditions: (i) $1 < e < \phi$ and (ii) e and n have to be relatively prime, i.e. they are not allowed to have any common factors.

Step 5: Compute (if possible; otherwise choose a different e) a number d, the *private-key exponent*, such that $d = (1 + k\phi)/e$ is an integer for some integer $k \geq 1$. This means that $de = 1 \mod \phi$.

Following completion of all five preparation steps, Alice publishes the modulus n and the number e, i.e., anybody who wants to send Alice a message may log in on her website and obtain her *public key* (n, e) consisting of the modulus n, computed in *Step 2*, and the public-key exponent e, chosen in *Step 4*. Alice keeps the two prime numbers p and q, as well as her private-key exponent d, secret.

Following the five-step preparation stage, Bob wants to send Alice a message $\mathcal{M}$, for instance $\mathcal{M} = $ AHOY!. Bob first converts the message $\mathcal{M}$ into an integer N. He does this with the help of a conversion function f. The conversion function f takes character strings as input and turns them into numbers according to $f(\mathcal{M}) = N$. The conversion function f has to be invertible such that N may be unambiguously converted back into the message $\mathcal{M}$. The function f may take the form of a formula or a table, or both. Bob settles on the following conversion function. Given a message $\mathcal{M} = c_1 c_2 \ldots c_m$, where c_j, $j = 1, \ldots, m$, are the individual characters of the message, Bob first converts each character into a number, and then strings the numbers corresponding to individual characters together to form one single, large integer N. Formally, if $\varphi(c)$ is the function that assigns each character an integer, then

$$f(\mathcal{M}) = f(c_1 c_2 \ldots c_m) = \varphi(c_1) \circ \varphi(c_2) \circ \ldots \circ \varphi(c_m) = N, \quad (9.1)$$

where the symbol $\circ$ denotes *concatenation*, i.e., the operation of "stringing together."

Bob constructs the function φ in the form of a look-up table that assigns each letter of the alphabet (and some punctuation marks) a two-digit number as shown in Table 9.3. The two-digit numbers are constructed by adding the number 10 to the numerical position of the corresponding letter in the alphabet; the punctuation marks are assigned arbitrary two-digit numbers not yet used up by other characters. Bob chooses two-digit numbers for encoding characters to ensure that there is no ambiguity when converting numbers back into plain text.

Bob's encoding scheme satisfies the condition of invertibility. This means that given any integer N (with an even number of digits), the original message is easily recovered by back-substitution according to Table 9.3. Bob now publishes Table 9.3 together with the instruction defined in Equation 9.1 on his website.

Table 9.3 Bob's substitution function φ. Characters are converted to numbers, and vice versa, using the following invertible assignment scheme:

A 11	F 16	K 21	P 26	U 31	Z 36
B 12	G 17	L 22	Q 27	V 32	. 37
C 13	H 18	M 23	R 28	W 33	, 38
D 14	I 19	N 24	S 29	X 34	? 39
E 15	J 20	O 25	T 30	Y 35	! 40

We mention an important restriction on the length of a message $\mathcal{M}$. In order to ensure that an RSA-encrypted message can be properly decrypted, the number N generated from $\mathcal{M}$ according to Equation 9.1 has to be smaller than n. This is a serious restriction since most messages $\mathcal{M}$ generate numbers N that are larger than n. This, however, is not a problem. Bob may, for example, chop his message into blocks of lengths $m < n$ and send several blocks instead of a single transmission containing the entire message $\mathcal{M}$. Suppose $N < n$. Then Bob encrypts $\mathcal{M}$ by mapping N into the integer

$$C = N^e \mod n. \tag{9.2}$$

The integer C is the *cipher*, or *cipher text*. Bob then sends the cipher C to Alice. Alice decrypts the message according to the identity

$$N = C^d \mod n. \tag{9.3}$$

She then looks up the function φ on Bob's website and obtains the message $\mathcal{M}$.

Let us illustrate the RSA cryptosystem with the following example. Suppose Alice chooses $p = 5$ and $q = 11$. This produces the modulus $n = pq = 55$ and the totient $\phi = 4 \cdot 10 = 40$. Alice now chooses the public-key exponent $e = 3$. This works, since e and ϕ do not have any factors in common (they are relatively prime) and $1 < e < \phi$. For $e = 3$, Alice realizes that $d = (1 + k\phi)/e$ is an integer already for $k = 2$. She settles on this choice and obtains her private key, $d = 27$. Alice posts the modulus $n = 55$ and the public-key exponent $e = 3$ as public information on her website.

Bob now realizes an important fact: Since the modulus of Alice's key, $n = 55$, is very small, he will be able to encode single letters only, since even a two-letter word would result in $N > n = 55$, which is not permitted by the rules of the RSA scheme. Thus, in order to send his message $\mathcal{M} =$ AHOY!, Bob decides to chop it into five blocks containing one character each.

9.3 • RSA PUBLIC-KEY CRYPTOSYSTEM

Table 9.4 Encryption/decryption table for the example presented in the text.

Character	Numerical value	Encrypted	Decrypted
A	11	11	11
H	18	2	18
O	25	5	25
Y	35	30	35
!	40	35	40

Bob now encodes each character of his message. The letter "A," for example, according to Table 9.3, maps into the number $N_A = \varphi(A) = 11$, which, subsequently, is transformed into the cipher $C_A = 11^3 \mod 55 = 1331 \mod 55 = 11$; the letter "H" maps into the number $N_H = 18$, which subsequently is transformed into the cipher $C_H = 18^3 \mod 55 = 5832 \mod 55 = 2$. Bob continues to perform these operations on the remaining letters of the word "AHOY!," including the exclamation point. The result is shown in Table 9.4. The first column of Table 9.4 shows the characters of the message to be sent, the second column shows their numerical values according to Table 9.3, and the third column shows the result of the encryption step according to Equation 9.2. At this point Bob sends the five numbers $11, 2, 5, 30$, and 35 to Alice. Upon receipt of the numbers Alice decodes the ciphers according to Equation 9.3. Alice starts with the number 11, which has to be taken to the 27th power modulo 55. Alice has only a 10-digit pocket calculator, but the number 11^{27} has 29 digits. Alice solves this problem by using the following elementary result of modular algebra:

$$(ab) \mod n = [(a \mod n)(b \mod n)] \mod n. \qquad (9.4)$$

With Equation 9.4 Alice computes:

$$11^{27} \mod 55 = \left[\left(11^7 \mod 55\right)^3 \left(11^6 \mod 55\right)\right] \mod 55 = 11,$$
$$2^{27} \mod 55 = 134,217,728 \mod 55 = 18,$$
$$5^{27} \mod 55 = \left[(5^{13} \mod 55)^2 \cdot 5\right] \mod 55 = \left[15^2 \cdot 5\right] \mod 55 = 25,$$
$$30^{27} \mod 55 = \left[\left(30^6 \mod 55\right)^4 \cdot \left(30^3 \mod 55\right)\right] \mod 55$$
$$= \left[25^4 \cdot 50\right] \mod 55 = 35,$$
$$35^{27} \mod 55 = \left[\left(35^6 \mod 55\right)^4 \cdot \left(35^3 \mod 55\right)\right] \mod 55$$
$$= \left[20^4 \cdot 30\right] \mod 55 = 40, \qquad (9.5)$$

and obtains the integers listed in column four of Table 9.4. With the help of Bob's substitution table (see Table 9.3) Alice then translates the numbers into the original message: AHOY!.

For our example to be tractable, Alice chose relatively small prime numbers, $p = 5$ and $q = 11$, resulting in the small modulus $n = 55$. A small n, however, results in an insecure encryption scheme. We see this in the following way. For small n, an eavesdropper, Eve, has no problem computing, or guessing, the two prime numbers that the modulus n is composed of. Knowing p and q, Eve may now immediately compute her own private key d' and use it to decrypt Bob's messages. While factoring n into its two prime components is relatively straightforward for small n, it becomes increasingly difficult for larger n, and is nearly impossible to perform with current computer power and algorithmic techniques if n has more than 200 digits. In fact, working from December 2003 until May 2005 it took a team of computer scientists and mathematicians more than a year to factor a number with 200 decimal digits with the help of various computers, including a cluster of 80 state-of-the-art computer workstations. Therefore, it is believed that using a modulus n with more than 250 decimal digits results in secure encryption, at least for the near future. Why it is so difficult to factor large integers is discussed in Section 9.5.

RSA encryption is a deterministic encryption scheme. This means that a given message $\mathcal{M}$ is always mapped into the same number N. Therefore, RSA encryption is vulnerable to *chosen plain-text attacks*. This technique works best when eavesdropping on a sender, who routinely sends short messages containing predictable contents. A bank is a good example. In this case an eavesdropper would compose a large dictionary of millions of plausible banking-related text messages, and store their corresponding ciphers C_j, $j = 1, \ldots, N_D$, where N_D is the number of dictionary entries. Then, when intercepting a particular cipher C_p, it is compared with the dictionary entries C_j, $j = 1, \ldots, N_D$. If a match is found, i.e., if $C_p \in \{C_j, j = 1, \ldots, N_D\}$, the whole message corresponding to C_p is revealed because the plaintext messages corresponding to each C_j, $j = 1, \ldots, N_D$, are known.

As shown by our example, chosen plain-text attacks are mostly successful for small n and short messages. In our example, because n is so small, Bob is only able to encrypt single letters. Therefore, a dictionary with less than 55 entries is enough to decrypt, with ease, any intercepted message. This, again, points to the need for choosing a large modulus n to ensure secure RSA encryption.

The discussion above shows that RSA encryption is secure only if it is true that factoring large integers n (the modulus) is "hard." To date it has not been proven mathematically that factorization of large integers is difficult to do. We only know that no fast algorithm for integer factorization exists to date, although scores of mathematicians and computer scientists have tried to construct one. This, however, is not proof that no fast algorithm exists in principle. Therefore, it is still possible that an algorithm may be found that efficiently factors large integers. In this case RSA encryption would be useless. Also, it is still not known whether, besides factoring n, there isn't an alternative method for breaking RSA. In the absence of any known "RSA code-breaking schemes," however, RSA is generally believed to be safe and convenient. It is used for government and business transactions worldwide.

While currently available personal computers and workstations, and even research-grade supercomputers have difficulties cracking RSA ciphers, this task is easily accomplished by a *quantum computer*. Thus, the prospect of breaking RSA is one of the most compelling arguments for the construction of quantum computers.

Exercises:

9.3.1 Prove Equation 9.4.

9.3.2 You intercept the RSA encrypted cipher

$$C = 2504899119404993239674057596844751135950.$$

You know that the original plaintext message was encrypted using Bob's encoding scheme (see Table 9.3), you know the RSA modulus,

$$n = 2643252984012685297031642898555551661973,$$

and you know the public-key exponent $e = 17$. Reveal the original plaintext message by performing the following four steps.

Step 1: Factor the modulus n using the factoring function provided with any of the now widely available mathematical software packages to reveal the two prime factors p and q of n. Any standard personal computer will accomplish this task in a matter of seconds.

Step 2: With p and q known, compute the totient ϕ.

Step 3: With ϕ known, compute your own private key d'.

Step 4: Use your key d' to decrypt the message hidden in C.

9.3.3 Show that $p = 13$, $q = 17$, $n = 221$, $e = 23$, and $d = 167$ are valid choices for encryption and decryption using the RSA public-key cryptosystem.

9.3.4 Euler's totient function $\varphi(N)$ returns the number of positive integers smaller than N and co-prime to N. Examples are: $\varphi(6) = 2$ and $\varphi(7) = 6$. Suppose $N = pq$, where p and q are prime numbers with $p \neq q$. Show that $\varphi(N) = \phi = (p-1)(q-1)$, where ϕ is the totient computed in *Step 3* of the five-step RSA preparation scheme.

9.4 How Does RSA Work?

RSA encryption and decryption is based on two simple mathematical operations. According to Equation 9.2 encryption is performed by raising the integer N containing the message to the public-key exponent e and taking this number modulo n. This operation yields the cipher C to be transmitted. Decryption of the cipher C, according to Equation 9.3, is performed by raising the cipher C to the private-key exponent d and taking this result modulo n. This operation recovers the original message contained in N. But why does this cipher scheme work? Apart from the simple identity of modular arithmetic stated in Equation 9.4, the answer draws on two theorems of elementary number theory: Fermat's Little Theorem and a theorem related to the Chinese Remainder Theorem.

Fermat's Little Theorem states the following: Let p be a prime number that does not divide a. Then:

$$a^{p-1} = 1 \mod p. \tag{9.6}$$

If p divides a, then

$$a^{p-1} = 0 \mod p. \tag{9.7}$$

It is straightforward to see why Equation 9.7 holds. In this case, since p divides a and $p - 1 \geq 1$, the left-hand side of Equation 9.7 is a multiple of p and Equation 9.7 holds trivially. Proving the case stated in Equation 9.6 is left as an exercise.

The second theorem we need is the following: Suppose p and q are two prime numbers with $p \neq q$, then:

$$x = a \mod p, \quad x = a \mod q \implies x = a \mod (pq). \tag{9.8}$$

The proof of this theorem is not difficult. The first identity of Equation 9.8 implies

$$x = a + Kp, \tag{9.9}$$

while the second identity of Equation 9.8 implies

$$x = a + Lq \tag{9.10}$$

with suitable integers K and L. Subtracting Equation 9.10 from Equation 9.9, we obtain

$$0 = Kp - Lq. \tag{9.11}$$

But since p and q are prime with $p \neq q$, Equation 9.11 holds only if $K = Mq$ and $L = Mp$, where M is an integer. Using $K = Mq$ in Equation 9.9, we obtain immediately $x = a + Mpq$, i.e., the theorem in Equation 9.8.

We are now ready to explain how RSA works. According to Equation 9.2, we encrypt the number N representing the plaintext message by raising it to the eth power modulo n to obtain the cipher C. We decrypt C by raising it to the power d modulo n. Using the identity in Equation 9.4, the combined operation performed on N is:

$$C^d \mod n = N^{ed} \mod n. \tag{9.12}$$

Instead of computing $N^{ed} \mod n$ directly, we first compute $N^{ed} \mod p$ and $N^{ed} \mod q$. Because of *Step 5* of the five-step RSA scheme presented in Section 9.3, we have $de = 1 + k(p-1)(q-1)$ for some integer k. Therefore,

$$\begin{aligned} N^{ed} \mod p &= N^{1+k(p-1)(q-1)} \mod p \\ &= N \left[N^{k(q-1)} \right]^{p-1} \mod p. \end{aligned} \tag{9.13}$$

At this point there are two cases: (1) p and N are co-prime and (2) p divides N. In Case (1) Fermat's Little Theorem (see Equation 9.6) tells us that $N^{k(q-1)}$, some integer co-prime to p, raised to the power $p-1$, is 1 modulo p. Therefore, with Equation 9.4,

$$N^{ed} \mod p = N \mod p. \tag{9.14}$$

In Case (2), i.e., p divides N, the left-hand side of Equation 9.14 is zero, and so is the right-hand side. This means that Equation 9.14 remains true. In the same way we obtain

$$N^{ed} \mod q = N \mod q. \tag{9.15}$$

We now use Equation 9.8 to conclude that

$$N^{ed} \mod n = N \mod n. \tag{9.16}$$

But this result now implies

$$C^d \mod n = N^{ed} \mod n = N \mod n, \qquad (9.17)$$

i.e., the decryption step yields the number N representing the original plaintext message. We also see now why the restriction $N < n$, mentioned in Section 9.3, is so important. According to Equation 9.17, the message N can be revealed without scrambling it by the modulo operation only if $N < n$.

Exercises:

9.4.1 Prove Fermat's Little Theorem stated in Equation 9.6. This is the case where p does not divide a.

9.4.2 When choosing e in *Step 4* of the five-step RSA preparation scheme outlined in Section 9.3, it was necessary to choose e such that (i) $1 < e < \phi$ and (ii) e and n are relatively prime. Explain why (i) and (ii) are necessary.

9.4.3 Let m and n be two positive integers with $m = n \mod (p-1)$, where p is prime.

(a) If a and p are co-prime, show that $a^m = a^n \mod p$.

(b) In what way is (a) relevant for how RSA works?

9.5 Why Is Integer Factorization So Difficult?

Suppose we have to factor a large integer N. The most straightforward solution of the problem is to divide N by all integers $1 < m \leq \sqrt{N}$ and determine those m for which m divides N evenly without a remainder. If N is a D-digit number, $\sqrt{N}$ has approximately $D/2$ digits. Therefore, following this factorization algorithm, we would need to perform approximately $10^{D/2}$ divisions to find the factors of N. Assuming that each division can be done in the same amount of time τ, independent of D, the execution time of this algorithm is $T = 10^{D/2}\tau$. Apparently, the execution time T rises exponentially in D for increasing D.

How serious is this problem? Most personal computers these days have a clock speed in the GHz regime, i.e., one clock cycle takes about 10^{-9} seconds. Let us assume that we are working with a fast, special-purpose computer that performs each division, independently of D, in just one clock cycle. How long does it take to factor a 200-digit RSA modulus n? We have $T = 10^{100} \times 10^{-9}\text{s} = 10^{91}\text{s} = 3 \times 10^{83}$ years. This is orders of magnitude longer than the age of the universe (on the order of 10^{10} years) and clearly not an option for factoring a 200-digit RSA modulus n.

Perhaps our algorithm is too simple. After all, we don't have to try *all* integers $1 < m \leq \sqrt{N}$. It is enough to try all *prime numbers* $p \leq \sqrt{N}$. How many prime numbers smaller than or equal to $\sqrt{N}$ are there? This question is answered by the prime number counting function $\pi(M)$. It returns the number of primes smaller than or equal to M. According to the prime number theorem of analytic number theory we have $\pi(M) \approx M/\ln(M)$. With the π function we can now calculate the number ν of primes p smaller than or equal to $\sqrt{N}$. We obtain $\nu \approx \sqrt{N}/\ln(\sqrt{N}) = 2\sqrt{N}/\ln(N)$. For $N \approx 10^{200}$ this means $\nu \approx 4.3 \times 10^{97}$. This translates into an execution time of $T = 4.3 \times 10^{88}$s $\approx 10^{81}$ years. This result is an insignificant improvement, since it is still orders of magnitude longer than the age of the universe.

Obviously our algorithm is still much too simplistic. Let us turn to industry for a better algorithm. The software package MATHEMATICA offers a factoring algorithm. Since MATHEMATICA is an industry standard in algebraic software, we may assume that the latest know-how went into designing MATHEMATICA's integer factoring algorithm. MATHEMATICA's algorithm was tested using MATHEMATICA's FactorInteger function to factor semi-prime numbers with a number of digits D between 45 and 67. Running MATHEMATICA 5.2 on a PowerBook G4 laptop computer, the execution times of the FactorInteger function are graphed in Figure 9.1 (plot symbols in Figure 9.1). The data points

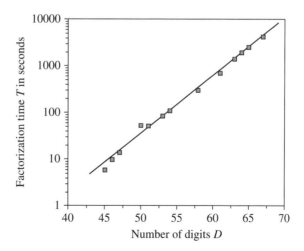

Figure 9.1   Time T in seconds needed on an Apple G4 PowerBook laptop computer to factor D-digit semi-primes using MATHEMATICA 5.2. The straight-line fit to the data points indicates that the time needed to factor a D-digit integer rises exponentially with D. This means that integer factorization is very hard to do on a classical (laptop) computer.

lie approximately on a straight line (linear fit function in Figure 9.1) indicating that MATHEMATICA's algorithm still produces execution times that are exponential in the number of digits D of the integer to be factored. The fit function used in Figure 9.1 is

$$T = 2.57 \times 10^{-5 + 0.123 D} \text{ s}. \qquad (9.18)$$

Thus, for $D = 200$, using MATHEMATICA's FactorInteger function, implemented on a PowerBook G4 laptop computer, it would still take us 3×10^{12} years to factor a 200-digit RSA modulus, more than 200 times longer than the age of the universe. Even a modern supercomputer that is, perhaps, a million times more powerful than a laptop computer, would still require about 3 million years to factor a 200-digit RSA modulus using the algorithm underlying MATHEMATICA's FactorInteger function. This does not mean that we couldn't factor a 200-digit number in a relatively reasonable time using specialized factorization algorithms. Indeed, as mentioned in Section 9.3, a team of mathematicians and computer scientists did just that: They factored a 200-digit RSA modulus, a task that took them a little more than a year to accomplish. But even if some future computer were powerful enough to factor a 200-digit number in a matter of seconds, simply increasing the number of digits from 200 to 2000 is enough to foil this computer as well.

We conclude that based on the current state of number theory, factoring large integers is beyond the power of classical computers, current or future. Large integers can be factored only by (1) a fundamental breakthrough in number theory or (2) a fundamental breakthrough in computer design. In the absence of (1), computer scientists and physicists are currently exploring option (2), using quantum computers to break RSA. We will revisit this topic in Chapter 10.

Exercises:

9.5.1 Suppose you are factoring large D-digit RSA moduli with the MATHEMATICA 5.2 function FactorInteger using a Mac PowerBook G4. Assume that the factorization time for a D-digit semi-prime is given by Equation 9.18. What is the largest semi-prime (i.e., the largest D) you can factor if you budget

(a) a day,

(b) a year

of computer run-time?

9.5.2 Not impressed with the apparent difficulty of factoring large integers, you set out to factor the semi-prime, 309-decimal-digit RSA Challenge Number

$$\begin{aligned}\text{RSA1024} =\ &13506641086599522334960321627880596993888147560566\\&70275244851438515265106048595338339402871505719094\\&41798207282164471551373680419703964191743046496589\\&27425623934102086438320211037295872576235850964311\\&05640735015081875106765946292055636855294752135008\\&52879416377328533906109750544334999811150056977236\\&890927563.\end{aligned}$$

Playing with the products of numbers, you notice that by multiplying two 1-digit numbers you can fix the last decimal digit of RSA1024. Multiplying two 2-digit numbers, you can fix the last two decimal digits of RSA1024. Therefore, it seems that RSA1024=pq can easily be factored iteratively "from the right." Denoting by $p^{(m)}$ and $q^{(m)}$ the m-digit "tails" of p and q, show that instead of factoring RSA1024 all you obtain is yet another demonstration of the difficulty of factoring large integers: The number of possibilities to get the tail of RSA1024 right to m digits by multiplying $p^{(m)}$ with $q^{(m)}$ explodes exponentially in m. Illustrate this phenomenon by listing all possibilities for the pairs $\left(p^{(1)}, q^{(1)}\right)$ and $\left(p^{(2)}, q^{(2)}\right)$.

9.6 Summary

Secure communication is an enabling technology for all of modern military, government, and commercial information exchange. Internet commerce, and especially Internet banking, are unthinkable without convenient cryptosystems for the secure transmission of business transactions. There are essentially two different ways to set up a cryptosystem for the secure exchange of messages: symmetric, private-key cryptosystems, and asymmetric, public-key cryptosystems. While private-key cryptosystems display the highest degree of security, they are awkward and inconvenient in practice and suffer from the key-distribution problem. Public-key cryptosystems, in particular the RSA cryptosystem, are the dominant technology in today's Internet applications. RSA, however, is only as safe as integer factorization is difficult to do. Advances in number theory, resulting in a fast factoring algorithm, may render RSA obsolete. But even in the absence of number-theoretic breakthroughs quantum computers have the potential to crack RSA.

A quantum algorithm specifically developed for cracking RSA is Shor's algorithm. We will discuss it in Chapter 10.

Chapter Review Exercises:

1. Reveal the plaintext message hidden in the following Caesar cipher: qpjyxly zynp dlto esle yzmzoj fyopcdelyod bflyefx xpnslytnd. estd td ecfp. hp nly rpe fdpo ez te, mfe hp oz yze fyopcdelyo te.

2. In the case of Caesar ciphers, leaving out the spaces between words and the punctuation marks between sentences, does little to increase the security of messages. This method, however, greatly increases the security of more elaborate substitution ciphers. Reveal the contents of the following message, encrypted with a simple substitution cipher, where the spaces between words and the punctuation marks between sentences have been omitted:

   fljldbcfnftbstxnpnrncbysofelkynsfymsbwjbstsqfelbclm tfwbmlxtsensorpelslulcfelotclwfliwensql badncftwjlxtxotaatwyjfbctmdbxxtgjl

3. The RSA cryptosystem is based on the concept of trap-door functions, i.e., functions that are easy to compute in one direction, but very difficult to compute in the reverse direction. In the case of RSA: Multiplication of integers is easy; factoring a large integer is hard. Another example of a trap-door function is squaring an integer. It is easy to square an integer, but it is hard, without the benefit of an algebraic software package, to take the square root of a large integer. To illustrate, a plaintext message was converted into an integer N using Bob's substitution code of Table 9.3. Following this step, the integer N was squared, which yielded the cipher

$$C = N^2 = 17562415270674216931930772024329344342779\allowbreak 9733\allowbreak 771354979108245154223240154423093691291202788\allowbreak 70695654888473600.$$

   Reveal the plaintext message without using the square-root function of an algebraic software package.

chapter 10

Quantum Factoring

In this chapter:

- ◆ Introduction
- ◆ Miller's Algorithm
- ◆ Quantum Fourier Transform
- ◆ Shor's Algorithm
- ◆ Summary

10.1 Introduction

In Chapter 8 we learned how quantum computing works in principle and studied a few examples of quantum software including Grover's practically important quantum search algorithm. The central focus of this chapter is another important quantum algorithm that may be implemented on a quantum computer: Shor's algorithm. This algorithm allows us to factor large integers in polynomial time. Thus it presents a threat for RSA encryption schemes. In essence, Shor's algorithm is an implementation of Miller's "classical" factoring algorithm on a quantum computer.

Miller's factoring algorithm is described in Section 10.2. The quantum Fourier transform, an important component of Shor's algorithm, is described in Section 10.3. The quantum implementation of Miller's algorithm, Shor's algorithm, is described in Section 10.4.

10.2 Miller's Algorithm

In Section 9.5 we discussed several simple algorithms for factoring large semi-primes n. But these algorithms are not the only ones in existence. Mathematicians constantly work on improving speed and performance of factoring algorithms. The following algorithm, due to Miller, is the basis of Shor's quantum factoring algorithm.

Suppose we have to factor a large semi-prime $n = pq$, i.e., we have to reveal the two prime factors p and q of n. Instead of dividing n by all integers smaller than $\sqrt{n}$ and finding the two prime factors of n (the method we used in Section 9.5), we may use the following, alternative method. Select, at random, an integer x with $1 < x < n$, co-prime with n. The integer x is called the *seed*. Now determine the smallest positive integer ω such that

$$x^\omega \mod n = 1. \tag{10.1}$$

The number ω is called the *order* of the seed x. If it turns out that ω is odd for our choice of x, we have to try another x. We keep trying until we find an x with an even order. This is important, since in the next step we compute

$$M = x^{\omega/2}, \tag{10.2}$$

and we want M to be an integer. Then, compute the two numbers $A = M - 1$ and $B = M + 1$. If $B \mod n = 0$, we have to find yet another x that satisfies the two conditions (1) ω is even and (2) $B \mod n \neq 0$. Finally, determine the greatest common divisor (GCD) of A and n and of B and n. We will then have

$$p = GCD(A, n), \quad q = GCD(B, n). \tag{10.3}$$

As an example, let us compute the two prime factors p and q of $n = pq = 15$. The numbers $x_1 = 1$, $x_2 = 2$, $x_3 = 4$, $x_4 = 7$, $x_5 = 8$, $x_6 = 11$, $x_7 = 13$, and $x_8 = 14$ are co-prime to $n = 15$, since $GCD(x_i, 15) = 1$, $i = 1, \ldots, 8$. Let us choose $x = x_2 = 2$. Then, $2 \cdot 2 \mod 15 = 4$, $4 \cdot 2 \mod 15 = 8$, $8 \cdot 2 \mod 15 = 1$. Therefore, $2^4 \mod 15 = 1$. The exponent $\omega = 4$ is also the smallest exponent for which $2^\omega \mod 15 = 1$. This means that the order ω of $x = 2$ is $\omega(2) = 4$. In addition, $\omega = 4$ fulfills the condition that the order has to be an even number. We compute $A = 2^{\omega/2} - 1 = 2^2 - 1 = 3$ and $B = 2^{\omega/2} + 1 = 2^2 + 1 = 5$. We do not have to search for a different seed x, since $B \mod 15 \neq 0$. We now obtain $p = GCD(3, 15) = 3$ and $q = GCD(5, 15) = 5$. These are indeed the two prime factors of 15.

We mention that Miller's algorithm for the factorization of semi-primes is not a deterministic algorithm. It is probabilistic, since there is always a chance that we select a seed x whose order is odd or yields B mod $n = 0$. This, however, is not a problem, since it can be determined immediately whether a seed x "works" or not. In case x does not work, another seed x' can be chosen at random until we find a seed that works. This is not an onerous task, since not too many attempts are usually necessary to find a suitable seed x. In our case, for instance, it is straightforward to check that all x_j, $j = 2, \ldots, 7$, work as seeds for factoring $n = 15$. Apart from $x_1 = 1$, which cannot be used as a seed because of the condition $x > 1$, the only seed that does not work is $x_8 = 14$. Therefore, in the case of factoring the semi-prime $n = 15$, choosing randomly from the seven possibilities x_j, $j = 2, \ldots, 8$, our chances of finding a suitable seed $x > 1$ are 6/7, i.e., larger than 85%.

Miller's algorithm factors integers n as if by magic. However, the mathematical background of Miller's algorithm is not difficult to understand. Suppose we choose a seed x and obtain its order ω modulo n. Then, ω is the smallest positive integer with

$$x^\omega \mod n = 1. \tag{10.4}$$

This implies that

$$(x^\omega - 1) \mod n = 0, \tag{10.5}$$

i.e., n divides $(x^\omega - 1)$. In symbols:

$$n \mid (x^\omega - 1). \tag{10.6}$$

Since we constructed our algorithm such that ω is even, we may now write:

$$n \mid \left[\left(x^{\omega/2} - 1\right)\left(x^{\omega/2} + 1\right)\right]. \tag{10.7}$$

But since $A = x^{\omega/2} - 1$ and $B = x^{\omega/2} + 1$, we also have

$$n \mid AB, \tag{10.8}$$

i.e., n divides the *product* AB. But n does not divide B. We excluded this case by requiring that $B \mod n \neq 0$ for a valid seed x. If we assume that n divides A, we obtain immediately $x^{\omega/2} \mod n = 1$. But since $\omega \geq 2$, we have $\omega/2 < \omega$. This contradicts the fact that ω is the *smallest* integer that satisfies $x^\omega \mod n = 1$. Therefore, n does not divide A either. But this means that n must have nontrivial factors in common with both A and B. Since n has only two factors, p and q, one of them must be contained in A and the other in B. We may label p and q such that p is a factor of A and q is a factor of B. Since q is not in A, p is the greatest common divisor of n and A. This accounts for the first identity of Equation 10.3. Since p is not in B, q is the greatest common divisor of n and B. This accounts for the second identity of Equation 10.3.

One might think that Miller's algorithm neatly solves the problem of factoring large semi-primes n. However, the moduli used in modern RSA schemes have 200 decimal digits and more. While seeds x exist that have tractably small orders ω, the overwhelming majority of any randomly picked seeds have orders ω that are so large that they cannot be determined on a classical computer in a reasonable time. The security of RSA cryptosystems is based on just this difficulty. For a quantum computer, however, finding the order of a seed x is easy. Quantum computers, therefore, pose a real challenge for RSA cryptosystems.

Exercises:

10.2.1 Compute the order ω of the seeds x_j, $j = 3, \ldots, 8$. Show that $x = x_1 = 1$ and $x = x_8 = 14$ cannot be used for factoring $n = 15$.

10.2.2 Show that the order of $x = 1644$ modulo $n = 2491$ is $\omega = 2$. Use this information to factor $n = pq$ into its two prime factors p and q.

10.2.3 Explain why in Miller's algorithm x and n have to be co-prime.

10.2.4 Show that

$$1 < \omega \leq n - 1. \tag{10.9}$$

10.3 Quantum Fourier Transform

The crucial element of Shor's quantum algorithm for the factorization of large semi-primes is the application of a quantum Fourier transform. The quantum Fourier transform is itself one of the most important quantum algorithms and is frequently called by other quantum algorithms as a *subroutine*. The quantum Fourier transform is based on the discrete Fast Fourier Transform of Cooley and Tukey. The quantum Fourier transform can be executed very rapidly on a quantum computer since it can be broken down into a product of simpler unitary operations. Thus the quantum Fourier transform can be represented and executed as a quantum circuit.

The "classical" discrete Fourier transform is defined in the following way: given a set of N (complex) numbers $x_0, x_1, \ldots, x_{N-1}$, their discrete Fourier transform is defined as

$$y_k = \frac{1}{\sqrt{N}} \sum_{l=0}^{N-1} \exp\left(\frac{2\pi i k l}{N}\right) x_l, \quad k = 0, 1, \ldots, N-1. \tag{10.10}$$

10.3 • QUANTUM FOURIER TRANSFORM

Defining the vectors $\boldsymbol{x}^T = (x_0, x_1, \ldots, x_{N-1})$ and $\boldsymbol{y}^T = (y_0, y_1, \ldots, y_{N-1})$, we may write Equation 10.10 as the following linear, unitary mapping,

$$\boldsymbol{y} = U^{(\text{FT})} \boldsymbol{x}, \tag{10.11}$$

where $U^{(\text{FT})}$ is the transformation matrix with matrix elements

$$U^{(\text{FT})}_{kl} = \frac{1}{\sqrt{N}} \exp\left(\frac{2\pi i k l}{N}\right), \quad k, l \in \{0, 1, \ldots, N-1\}. \tag{10.12}$$

From Equation 10.12 we obtain

$$U^{(\text{FT})}_{kl} = U^{(\text{FT})}_{lk}, \tag{10.13}$$

i.e., $U^{(\text{FT})}$ is symmetric. To prove the unitarity of $U^{(\text{FT})}$, we need two formulas. The first one is the well-known summation formula for geometric sums:

$$S = \sum_{l=0}^{N-1} q^l = \frac{1 - q^N}{1 - q}. \tag{10.14}$$

We prove it in the following way:

$$S = 1 + q + q^2 + \ldots + q^{N-1}, \tag{10.15}$$
$$qS = q + q^2 + \ldots + q^{N-1} + q^N. \tag{10.16}$$

Subtracting Equation 10.16 from Equation 10.15, we obtain $S - qS = 1 - q^N$. Solving this for S, we obtain the summation formula stated in Equation 10.14. The second formula we need is:

$$T = \frac{1}{N} \sum_{l=0}^{N-1} \exp(2\pi i M l / N) = \begin{cases} 1, & \text{for } M = 0, \\ 0, & \text{for } 0 < |M| < N. \end{cases} \tag{10.17}$$

We prove it in two steps.
(a) $M = 0$:

$$T = \frac{1}{N} \sum_{l=0}^{N-1} 1 = \frac{1}{N} N = 1. \tag{10.18}$$

(b) $0 < |M| < N$: With the geometric sum formula in Equation 10.14 we obtain:

$$T = \frac{1}{N} \sum_{l=0}^{N-1} [\exp(2\pi i M/N)]^l = \left(\frac{1}{N}\right)\left[\frac{1 - \exp(2\pi i M)}{1 - \exp(2\pi i M/N)}\right] = 0, \tag{10.19}$$

since $\exp(2\pi i M) = 1$ and $1 - \exp(2\pi i M/N) \neq 0$ for $0 < |M| < N$.

We are now ready for the proof of the unitarity of $U^{(\text{FT})}$. Using the geometric sum formula in Equation 10.14 together with the formula in Equation 10.17, we obtain:

$$\left[U^{(\text{FT})\dagger} U^{(\text{FT})}\right]_{mn} = \sum_{l=0}^{N-1} U^{(\text{FT})*}_{lm} U^{(\text{FT})}_{ln}$$

$$= \sum_{l=0}^{N-1} \frac{1}{\sqrt{N}} \exp(-2\pi i l m/N) \frac{1}{\sqrt{N}} \exp(2\pi i l n/N)$$

$$= \frac{1}{N} \sum_{l=0}^{N-1} \exp[2\pi i (n-m) l/N] = \delta_{nm}. \quad (10.20)$$

This shows that $U^{(\text{FT})}$ is unitary.

Now let $|m\rangle$, $m = 0, \ldots, N-1$, be a set of orthonormal states, i.e., $\langle m|n\rangle = \delta_{mn}$, $m,n = 0, \ldots, N-1$. The action of the quantum Fourier transform, represented by the linear, unitary operator $\hat{U}^{(\text{QFT})}$, maps the state $|k\rangle$ into the superposition state

$$\hat{U}^{(\text{QFT})} |k\rangle = \frac{1}{\sqrt{N}} \sum_{l=0}^{N-1} \exp(2\pi i k l/N) |l\rangle. \quad (10.21)$$

From this we see that the matrix elements of $\hat{U}^{(\text{QFT})}$ are:

$$U^{(\text{QFT})}_{lk} = \langle l|\hat{U}^{(\text{QFT})}|k\rangle = \frac{1}{\sqrt{N}} \exp(2\pi i k l/N) = U^{(\text{FT})}_{lk}$$

$$= U^{(\text{FT})}_{kl} = U^{(\text{QFT})}_{kl}. \quad (10.22)$$

This also shows that $\hat{U}^{(\text{QFT})}$ is symmetric. The operator $\hat{U}^{(\text{QFT})}$ is obviously unitary, since its matrix elements are the same as the matrix elements $U^{(\text{FT})}_{kl}$ of the "classical" Fourier transform, which we already proved to be unitary and symmetric.

Let us look at a few examples. In the single-qubit case, $n = 1$, we have $N = 2^n = 2$ basis states. Their quantum Fourier transforms are:

$$\hat{U}^{(\text{QFT})} |0\rangle = \frac{1}{\sqrt{N}} \sum_{l=0}^{N-1} \exp(2\pi i 0 l/N) |l\rangle = \frac{1}{\sqrt{2}} \{|0\rangle + |1\rangle\},$$

$$\hat{U}^{(\text{QFT})} |1\rangle = \frac{1}{\sqrt{N}} \sum_{l=0}^{N-1} \exp(2\pi i 1 l/N) |l\rangle = \frac{1}{\sqrt{2}} \{|0\rangle - |1\rangle\}. \quad (10.23)$$

Notice that in this case $\hat{U}^{(\text{QFT})}$ produces the same result as the single-qubit Hadamard operator H. Therefore, for $n = 1$, $\hat{U}^{(\text{QFT})} = $ H.

In the two-qubit case, $n = 2$, we have $N = 2^n = 4$ basis states. We make the identification: $|00\rangle \equiv |0\rangle$, $|01\rangle \equiv |1\rangle$, $|10\rangle \equiv |2\rangle$, $|11\rangle \equiv |3\rangle$. Then,

$$\begin{aligned}
\hat{U}^{(\mathrm{QFT})} |00\rangle &= \frac{1}{\sqrt{4}} \sum_{l=0}^{3} \exp(2\pi i 0 l/4) |l\rangle = \frac{1}{2}(|0\rangle + |1\rangle + |2\rangle + |3\rangle) \\
&= \frac{1}{2}(|00\rangle + |01\rangle + |10\rangle + |11\rangle) = \frac{1}{2}(|0\rangle + |1\rangle)(|0\rangle + |1\rangle), \\
\hat{U}^{(\mathrm{QFT})} |01\rangle &= \frac{1}{\sqrt{4}} \sum_{l=0}^{3} \exp(2\pi i 1 l/4) |l\rangle = \frac{1}{2}(|0\rangle + i|1\rangle - |2\rangle - i|3\rangle) \\
&= \frac{1}{2}(|00\rangle + i|01\rangle - |10\rangle - i|11\rangle) = \frac{1}{2}(|0\rangle - |1\rangle)(|0\rangle + i|1\rangle), \\
\hat{U}^{(\mathrm{QFT})} |10\rangle &= \frac{1}{\sqrt{4}} \sum_{l=0}^{3} \exp(2\pi i 2 l/4) |l\rangle = \frac{1}{2}(|0\rangle - |1\rangle + |2\rangle - |3\rangle) \\
&= \frac{1}{2}(|00\rangle - |01\rangle + |10\rangle - |11\rangle) = \frac{1}{2}(|0\rangle + |1\rangle)(|0\rangle - |1\rangle), \\
\hat{U}^{(\mathrm{QFT})} |11\rangle &= \frac{1}{\sqrt{4}} \sum_{l=0}^{3} \exp(2\pi i 3 l/4) |l\rangle = \frac{1}{2}(|0\rangle - i|1\rangle - |2\rangle + i|3\rangle) \\
&= \frac{1}{2}(|00\rangle - i|01\rangle - |10\rangle + i|11\rangle) = \frac{1}{2}(|0\rangle - |1\rangle)(|0\rangle - i|1\rangle).
\end{aligned}$$
(10.24)

We notice that according to Equation 10.24 (for $n = 2$) the Fourier transforms of all four basis states can be written as product states. This holds generally for any number n of qubits. This is both surprising and powerful. It is this feature that makes the quantum Fourier transform practical for actual quantum computations. While in the original definition of the quantum Fourier transform we have to perform a sum over $N = 2^n$ terms, i.e., exponentially many terms, we have to do only n products in the product representation of the quantum Fourier transform. The following is an explicit formula that provides the quantum Fourier transform of an n-qubit state in product form:

$$\hat{U}^{(\mathrm{QFT})} |b_{n-1} b_{n-2} \ldots b_0\rangle = \frac{1}{2^{n/2}} \{|0\rangle + \exp(2\pi i [.b_0])|1\rangle\}$$
$$\{|0\rangle + \exp(2\pi i [.b_1 b_0])|1\rangle\} \ldots \{|0\rangle + \exp(2\pi i [.b_{n-1} b_{n-2} \ldots b_0])|1\rangle\}.$$
(10.25)

Here, $|b_{n-1}b_{n-2}\ldots b_0\rangle$ is the initial n-qubit state in binary representation where $b_0, b_1, \ldots, b_{n-1} \in \{0, 1\}$, and $[.b_0] = b_0 \times 2^{-1}$, $[.b_1 b_0] = b_1 \times 2^{-1} + b_0 \times 2^{-2}$, etc., are numbers smaller than 1, constructed from the binary digits of the initial n-qubit state.

When it comes to a practical implementation of the quantum Fourier transform, we have to implement the relative phases between the states $|0\rangle$ and $|1\rangle$. In order to do this we use the phase-rotation gate R(θ) defined in Equation 7.13.

Exercises:

10.3.1 Compute the quantum Fourier transforms of the states $|0\rangle$ and $|1\rangle$ of a single qubit. Then show that the matrix $\langle a'|\hat{U}^{(\text{QFT})}|a\rangle$, $a, a' \in \{0, 1\}$, is an orthogonal matrix.

10.3.2 With the help of Equation 10.25 compute the quantum Fourier transform of $|101\rangle$.

10.3.3 Use the definition, Equation 10.21, of the quantum Fourier transform to show that for any state $|a\rangle$,

$$\left[\hat{U}^{(\text{QFT})}\right]^\dagger \hat{U}^{(\text{QFT})} |a\rangle = |a\rangle,$$

i.e., $\hat{U}^{(\text{QFT})}$ is a unitary operator.

10.4 Shor's Algorithm

One way of cracking RSA (see Chapter 9) is to factor the semi-prime RSA modulus n. Miller's algorithm (see Section 10.2) does just that, but we learned in Section 10.2 that classical computers are not powerful enough to factor large RSA moduli n using Miller's algorithm. Quantum computers, however, exponentially more powerful than classical computers, may be up to the task. In order to run Miller's algorithm on a quantum computer, we recast the problem of finding the order ω of a seed x modulo n,

$$x^\omega \mod n = 1, \tag{10.26}$$

into the equivalent problem of finding the period of the function

$$f(r) = x^r \mod n. \tag{10.27}$$

The period of $f(r)$ is the smallest integer ω for which

$$f(r + \omega) = f(r). \tag{10.28}$$

The two problems are equivalent because of the following reasoning.

(a) Assume that Equation 10.26 holds. Then, with Equation 9.4:

$$\begin{aligned} f(r+\omega) &= x^{r+\omega} \mod n \\ &= (x^r \mod n)(x^\omega \mod n) \mod n \\ &= x^r \mod n = f(r), \end{aligned} \quad (10.29)$$

i.e., assuming the validity of Equation 10.26, Equation 10.28 holds.

(b) Assume that Equation 10.28 holds. Then, with Equation 9.4:

$$\begin{aligned} x^{r+\omega} \mod n &= x^r \mod n, \\ (x^r \mod n)(x^\omega \mod n) \mod n &= x^r \mod n, \\ (x^r \mod n)[(x^\omega \mod n) - 1] \mod n &= 0. \end{aligned} \quad (10.30)$$

Now, either $x^r \mod n = 0$, which is impossible, since x and n are relatively prime (see Section 10.2), or $x^\omega \mod n = 1$, i.e., assuming the validity of Equation 10.28, Equation 10.26 holds. Thus, in conjunction with (a), we have now shown that Equations 10.26 and 10.28 are equivalent.

We find ω with the following quantum scheme. We work with a quantum computer that has two quantum registers. We call them Register I and Register II. We assume that Register I is capable of representing M quantum states $|a\rangle_\mathrm{I}$, $a = 0, 1, \ldots, M-1$, where $M > n$. We assume further that Register II can represent at least ω quantum states $|b\rangle_\mathrm{II}$, where ω is the order of a suitable seed x. Therefore, according to Equation 10.9, Register II should be able to represent at least $n-1$ quantum states. We denote by $|\mathrm{I}\rangle$ the quantum state of Register I, and by $|\mathrm{II}\rangle$ the quantum state of Register II. The complete quantum state of our quantum computer is labeled $|COMP\rangle$. Factoring a given semi-prime $n = pq$ on our quantum computer into its two prime factors p and q then proceeds according to the following scheme.

First we initialize the computer in the state

$$|COMP\rangle = \frac{1}{\sqrt{M}} \sum_{a=0}^{M-1} |a\rangle_\mathrm{I} |0\rangle_\mathrm{II}. \quad (10.31)$$

This state is not difficult to construct. It is a simple product state of Registers I and II, where Register I is initialized in a linear combination of all possible states $|a\rangle_\mathrm{I}$, each occurring with equal weight $1/\sqrt{M}$, and Register II is set to $|0\rangle_\mathrm{II}$. Since M may be very large, the question arises how we initialize this many states. But this is not difficult to achieve. If we realize Register I as a binary qubit array, then on the order of $\log_2(M)$ applications of Hadamard gates to the qubits of Register I achieves the linear superposition with equal weights necessary for initializing Register I (see the quantum register loading operation in Section 7.5).

Following the initialization step we run our quantum computer and populate Register II with states that correspond to the application of f to the labels of the states of Register I. This results in:

$$|COMP\rangle = \frac{1}{\sqrt{M}} \sum_{a=0}^{M-1} |a\rangle_\mathrm{I} |f(a)\rangle_\mathrm{II}. \qquad (10.32)$$

Since $f(a)$ has period ω, there are precisely $\omega < n$ different possibilities for the values of $f(a)$. Let us label them $k_0 = 1, k_1, \ldots, k_{\omega-1}$.

Suppose we measure Register II. Register $|II\rangle$ would then collapse to one of the ω different states $|k_j\rangle$, $j = 0, \ldots, \omega - 1$. Let us label this state $|\tilde{k}\rangle$. In this case Register I collapses to a superposition state $|S\rangle_\mathrm{I}$ consisting of all basis states $|a\rangle$ with $f(a) = \tilde{k}$. Let $|\tilde{a}\rangle$ be one of these states. Then, because of the periodicity of the function f, the superposition state $|S\rangle_\mathrm{I}$ consists of all basis states $|a\rangle = |\tilde{a}+l\omega\rangle$, where l is an integer. Thus $|S\rangle_\mathrm{I}$ has a periodic structure. This structure is most clearly brought out by applying a Fourier transform to $|S\rangle_\mathrm{I}$, which would show peaks with a spacing that reveals the period ω. This motivates the central step of Shor's algorithm, the application of a quantum Fourier transform to the content of Register I. But instead of measuring $|II\rangle$ and then applying a Fourier transform to $|S\rangle_\mathrm{I}$, Shor's algorithm applies a quantum Fourier transform directly to the still coherent, unmeasured state in Equation 10.32. This results in the state

$$\begin{aligned}|COMP\rangle &= \frac{1}{\sqrt{M}} \sum_{a=0}^{M-1} \left[\hat{U}_\mathrm{QFT} |a\rangle_\mathrm{I}\right] |f(a)\rangle_\mathrm{II} \\ &= \frac{1}{M} \sum_{a=0}^{M-1} \sum_{y=0}^{M-1} \exp\left(\frac{2\pi i a y}{M}\right) |y\rangle_\mathrm{I} |f(a)\rangle_\mathrm{II}. \qquad (10.33)\end{aligned}$$

At this point we perform a measurement of $|COMP\rangle$, i.e., we measure both Registers $|I\rangle$ and $|II\rangle$. The complicated, entangled state $|COMP\rangle$ collapses to the state $|\tilde{y}\rangle_\mathrm{I}|f(\tilde{a})\rangle_\mathrm{II}$, where, without restriction of generality, we may assume $0 \leq \tilde{a} < \omega$. According to Equation 10.33, this state occurs with amplitude

$$A(\tilde{y},\omega,\tilde{a}) = \frac{1}{M} \sum_a \exp\left(\frac{2\pi i a \tilde{y}}{M}\right), \qquad (10.34)$$

where the summation is over all a, ranging from 0 to $M-1$, under the condition that $f(a) = f(\tilde{a})$. Notice that, although not explicitly obvious on the right-hand side of Equation 10.34, because of the ω-periodicity of $f(r)$, the amplitude $A(\tilde{y}, \omega, \tilde{a})$ depends on ω via the summation condition $f(a) = f(\tilde{a})$, which, as we know already, includes all a values of the form $a = \tilde{a} + l\omega$, $l = 0, \ldots, L$, where L is the largest integer with $\tilde{a} + L\omega \leq M - 1$. Therefore, we have

$$A(\tilde{y}, \omega, \tilde{a}) = \frac{1}{M} \sum_{l=0}^{L} \exp\left[\frac{2\pi i(\tilde{a} + l\omega)\tilde{y}}{M}\right]$$

$$= \frac{1}{M} \exp\left(\frac{2\pi i \tilde{a} \tilde{y}}{M}\right) \sum_{l=0}^{L} \exp\left(\frac{2\pi i l \omega \tilde{y}}{M}\right). \quad (10.35)$$

Since the exponential factor in front of the l-sum in Equation 10.35 is uni-modular, the probability of collapse into the state $|\tilde{y}\rangle_{\mathrm{I}} |f(\tilde{a})\rangle_{\mathrm{II}}$ is:

$$P(\tilde{y}, \omega) = |A(\tilde{y}, \omega, \tilde{a})|^2 = \frac{1}{M^2} \left| \sum_{l=0}^{L} \exp\left(\frac{2\pi i l \omega \tilde{y}}{M}\right) \right|^2. \quad (10.36)$$

Notice that $P(\tilde{y}, \omega)$ in Equation 10.36 no longer depends on $\tilde{a}$, but still depends on ω, the order of the seed x. If $\omega \tilde{y}/M$ is an integer, we can sum Equation 10.36 trivially. In this case the arguments of the exponential functions in Equation 10.36 are integer multiples of $2\pi i$. Therefore, all the exponential terms in Equation 10.36 are equal to 1 and we obtain immediately:

$$P(\tilde{y}, \omega) = \left(\frac{L+1}{M}\right)^2 \quad \text{for } \omega\tilde{y}/M \text{ integer.} \quad (10.37)$$

Even if $\omega \tilde{y}/M$ is not an integer, we can sum Equation 10.36 analytically. Using the summation formula for geometric sums (see Equation 10.14), we obtain:

$$P(\tilde{y}, \omega) = \frac{1}{M^2} \left| \frac{\exp[2\pi i(L+1)\omega\tilde{y}/M] - 1}{\exp(2\pi i \omega \tilde{y}/M) - 1} \right|^2$$

$$= \left\{ \frac{\sin[(L+1)\pi\omega\tilde{y}/M]}{M \sin[\pi\omega\tilde{y}/M]} \right\}^2, \quad \text{for } \omega\tilde{y}/M \text{ not integer.} \quad (10.38)$$

Whenever $\omega\tilde{y}/M$ is close to an integer, the denominator of Equation 10.38 is close to 0 resulting in a large peak in $P(\tilde{y}, \omega)$. This effect is illustrated in Figure 10.1 for the example of factoring $n = 15$, choosing the seed $x = 4$ and $M = 16$. In this case we have $\omega = 2$ and $L = 7$, since (1)

$\omega = 2$ is the smallest integer that yields $4^\omega \mod 15 = 1$ and (2) since $\omega = 2$, there are only two representatives $\tilde{a}$, i.e., $\tilde{a} = 0$ and $\tilde{a} = 1$, and $a + L\omega \leq M - 1 = 15$ is solved for $L = 7$ for both representatives, $\tilde{a} = 0$ and $\tilde{a} = 1$. Figure 10.1 shows $P(\tilde{y}, \omega)$ (plot symbols) as a function of $\tilde{y}$ for $\tilde{y} = 0, \ldots, 15$. In this case the quantity $\omega \tilde{y}/M = \tilde{y}/8$ is an integer for precisely two $\tilde{y}$ values, $\tilde{y} = 0$ and $\tilde{y} = 8$. This is clearly reflected in Figure 10.1. To guide the eye, a smooth interpolation of $P(\tilde{y}, \omega)$, treating $\tilde{y}$ as a continuous, real variable in Equation 10.36, is also shown (smooth line in Figure 10.1). The maximal height of $P(\tilde{y}, \omega)$, $P_{\max} = 1/4$, is consistent with Equation 10.37, which predicts $P_{\max} = [(L+1)/M]^2 = (8/16)^2 = 1/4$.

Because of the peak structure of $P(\tilde{y}, \omega)$, a measurement of $|COMP\rangle$ will most likely result in a $\tilde{y}$ value corresponding to one of the peaks of $P(\tilde{y}, \omega)$. This means that with a large probability our quantum computer will collapse into a state $|\tilde{y}\rangle$ for which $\omega \tilde{y}/M$ is very close to an integer. What we measure is $\tilde{y}$ and what we know is M. Therefore, we know the ratio $\tilde{y}/M$. Converting $\tilde{y}/M$ into a proper fraction $\tilde{y}/M \approx P/Q$, where P and Q are relatively prime natural numbers, we know that $\omega P/Q$ is close to an integer. This happens only if Q divides ω, i.e., ω is a multiple of Q. Thus, our quantum computer does not give us ω directly; it gives

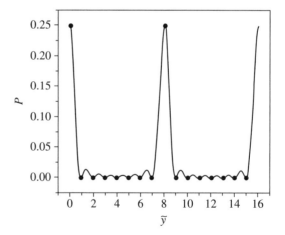

Figure 10.1 Probability $P(\tilde{y}, \omega)$ as a function of $\tilde{y}$ for $M = 16$, $\omega = 2$, and $L = 7$. Plot symbols: $P(\tilde{y}, \omega)$ evaluated at the $M = 16$ discrete values $\tilde{y} = 0, \ldots, 15$. Smooth line: Interpolation function $P(\tilde{y}, \omega)$, interpreting $\tilde{y}$ as a continuous variable. The smooth curve is drawn to guide the eye. The maxima in $P(\tilde{y}, \omega)$ correspond to integer values of the quantity $\omega \tilde{y}/M = \tilde{y}/8$, in this case $\tilde{y} = 0$ and $\tilde{y} = 8$. These peaks are important for the quantum factoring of semi-primes n.

us only a divisor Q of ω. But systematically trying multiples of Q will usually produce ω with high probability. This step of Shor's algorithm is called *classical post processing*, since it is most efficiently done on a classical computer.

Let us illustrate the measurement procedure together with classical post processing with our example of factoring $n = 15$ with seed $x = 4$. As shown in Figure 10.1, the probability for obtaining a certain $\tilde{y}$ value is zero, except for $\tilde{y} = 0$ and $\tilde{y} = 8$. If a measurement of our quantum computer results in the state $|\tilde{y}\rangle = |0\rangle$, no information on ω is gained. But it is equally probable that our quantum computer collapses into the state $|\tilde{y}\rangle = 8$. From this measurement we obtain $\tilde{y}/M = 8/16 = 1/2$. The denominator of this proper fraction is $Q = 2$, which is a divisor of ω. We now know that ω is a multiple of 2. Trying the first multiple, namely $\omega = 2$ already works, since $4^2 \bmod 15 = 1$. Thus we found $\omega = 2$ as the order of $x = 4$ modulo 15. Having determined ω on our quantum computer, we may now compute the factors $p = \text{GCD}(x^{\omega/2} - 1, n) = \text{GCD}(3, 15) = 3$ and $q = \text{GCD}(x^{\omega/2} + 1, n) = \text{GCD}(5, 15) = 5$ of $n = 15$.

Exercises:

10.4.1 A four-qubit register $|a_1 a_2 a_3 a_4\rangle$, $a_j \in \{0, 1\}$, $j = 1, \ldots, 4$, is capable of realizing $M = 16$ basis states $|0000\rangle, |0001\rangle, \ldots, |1111\rangle$. The initialization step of Shor's algorithm requires us to bring these states into a coherent superposition with equal weights. Show that if we can bring each of the qubits of the register into the superposition state $(|0\rangle + |1\rangle)/\sqrt{2}$, the initialization state of the register may be constructed as a simple product state of the four individual superposition states.

10.4.2 Use Shor's algorithm to factor the semi-prime $n = 35$.

(a) What is the minimum number m of qubits in Registers I and II necessary to factor n?

(b) Assume $M = 2^m$. What is the starting state of your quantum computer?

(c) Choose the seed $x = 6$. Determine ω, L, and the $\omega - 1$ different values $k_1, k_2, \ldots, k_{\omega-1}$ of $f(a)$.

(d) Compute $P(\tilde{y}, \omega)$.

(e) How many times, on average, do you have to run your quantum computer to be able to determine ω?

10.5 Summary

According to present scientific consensus, classical computers cannot crack RSA cryptosystems with RSA moduli that have 300 decimal digits or more. The reason is that classical factoring algorithms require an execution time that grows exponentially with the number of digits of the RSA modulus. Quantum computers, however, have the capability of executing exponentially many instructions in parallel, thus outclassing any classical computer. However, a quantum computer without quantum software is powerless. In this chapter we presented the basis for a possible future quantum factorization software, Shor's algorithm (see Section 10.4), that has the ability to crack the RSA cryptosystem. It is based on an ingenious combination of the "classical" Miller algorithm for factoring semi-primes (see Section 10.2) and the quantum Fourier transform (see Section 10.3).

Chapter Review Exercises:

1. Use Miller's algorithm with seed $x = 64$ to factor $n = 221$.

2. If $|x\rangle$ and $|y\rangle$ are two orthogonal states, i.e., $\langle x|y\rangle = 0$, show that their Fourier transforms $|x'\rangle = \hat{U}^{(\mathrm{QFT})}|x\rangle$ and $|y'\rangle = \hat{U}^{(\mathrm{QFT})}|y\rangle$ are orthogonal, too.

3. Since the two states $|x\rangle = |0b_{n-2}b_{n-3}\ldots b_0\rangle$ and $|y\rangle = |1b_{n-2}b_{n-3}\ldots b_0\rangle$ differ in their first bit, they are orthogonal. Since the quantum Fourier transform is a unitary operation, we know that the Fourier transforms $|x'\rangle$ and $|y'\rangle$ of $|x\rangle$ and $|y\rangle$, respectively, are orthogonal, too. Test the correctness of the formula defined in Equation 10.25 by first computing $|x'\rangle$ and $|y'\rangle$ according to Equation 10.25, and then showing, by explicit computation, that $\langle x'|y'\rangle = 0$.

chapter 11

Ion-Trap Quantum Computers

In this chapter:

- Introduction
- Linear Radio-Frequency Ion Trap
- Laser Cooling
- Cirac-Zoller Scheme
- Ca^+ Quantum Computer
- Summary

11.1 Introduction

Conceptually, the world of classical computing is divided into software and hardware. Quantum computing is no different: Its two components are quantum software (quantum algorithms) and quantum hardware (quantum computers). We studied examples of quantum software in Chapters 8 and 10. But what about quantum hardware? What do the machines look like that actually execute the quantum algorithms?

One of the most promising quantum computer architectures is the ion-trap quantum computer. This design was suggested theoretically in 1995 by Cirac and Zoller. It is shown schematically in Figure 11.1. The qubits are realized by the internal electronic states of ions that are held essentially motionless on the axis of a linear radio-frequency ion trap consisting of four conducting rods. A schematic sketch of such an ion-trap quantum computer is shown in Figure 11.1. The quantum states of

282   CHAPTER 11 • ION-TRAP QUANTUM COMPUTERS

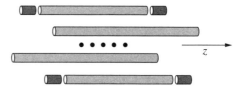

Figure 11.1   Schematic sketch of an ion-trap quantum computer.

the qubits are manipulated with lasers that are tightly focused on the individual, trapped ions. Quantum communication between the ions is established by phonons (quantized vibrational modes) due to the repulsive Coulomb force that acts between the trapped ions. Ion-trap quantum computers satisfy the five DiVincenzo criteria that need to be fulfilled by any viable quantum computer: (1) Scalable system of well-defined qubits, (2) method to reliably initialize the quantum computer, (3) long coherence times, (4) existence of universal gates, and (5) efficient measurement scheme. Because of their conceptual simplicity, we will study Ca^+ ion-trap quantum computers in this chapter. Alternative computer architectures are briefly discussed in Chapter 12.

We start our discussion of ion-trap quantum computers in Section 11.2 with a discussion of the physics of the linear radio-frequency trap that holds the ions, i.e., the qubits, in place. Trapping of the ions, however, is not enough for an ion-trap quantum computer to work. The kinetic energy of the ions, i.e., their temperature, has to be low enough so that their vibrational motions are in the quantum regime. This is accomplished by using laser cooling discussed in Section 11.3. With the basic hardware in place, how do we implement quantum gates between the ions stored in the radio-frequency trap? This question is answered in Section 11.4 where we present the scheme of Cirac and Zoller that implements the fundamental CNOT gate with the help of laser pulses applied to individual trapped ions. Finally, in Section 11.5, we put all the pieces together and discuss the Ca^+ ion-trap quantum computer. This specific implementation of the ion-trap quantum computer architecture has already been demonstrated in the lab.

11.2   Linear Radio-Frequency Ion Trap

The centerpiece of an ion-trap quantum computer is a linear radio-frequency (rf) quadrupole trap. As shown in Figure 11.1 it consists of four rods parallel to the z-axis of the trap. An x-y section of the trap is sketched in Figure 11.2. As shown in Figure 11.2, the two pairs of diametrically opposite rods are electrically connected. An rf voltage of amplitude V_0 and frequency ω is applied to the rods. Close to the axis

11.2 • LINEAR RADIO-FREQUENCY ION TRAP

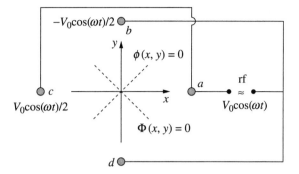

Figure 11.2 x-y section through the electrodes shown in Figure 11.1.

of the rods (the origin of the x-y coordinate system in Figure 11.2), the electric potential is

$$\Phi(x, y; t) = \frac{1}{2}\Phi_0 \left(x^2 - y^2\right) \cos(\omega t), \tag{11.1}$$

where Φ_0 is a constant. The electric field, derived from Equation 11.1, is

$$\boldsymbol{E}(x, y; t) = -\boldsymbol{\nabla}\,\Phi(x, y; t) = -\Phi_0 \begin{pmatrix} x \\ -y \end{pmatrix} \cos(\omega t). \tag{11.2}$$

An ion of charge q placed in this electric field, close to the axis of the trap, experiences a force

$$\boldsymbol{F}(x, y; t) = q\,\boldsymbol{E}(x, y; t). \tag{11.3}$$

If we denote by m the ion's mass, the force $\boldsymbol{F}$ in Equation 11.3 gives rise to two decoupled Newtonian equations of motion:

$$\begin{aligned} m\ddot{x} &= -q\Phi_0 x \cos(\omega t), \\ m\ddot{y} &= q\Phi_0 y \cos(\omega t), \end{aligned} \tag{11.4}$$

which describe the dynamics of the ion in the x-y plane.

We solve the x equation first. Technically, this is a known, second-order differential equation, the Mathieu equation, which is solved by a type of higher-order mathematical functions, the Mathieu functions. These functions are known analytically, but only as series expansions. Since these functions provide little direct insight into the physics of the ion's motion, we resort to an approximate solution of Equations 11.4.

From experiment it is known that charged particles in the force field in Equation 11.3 execute a motion that consists of a rapid, small-amplitude oscillation around a slow, large-amplitude drift motion. The rapid oscillation is known as the micromotion; the slower drift motion is known as the secular motion. Reflecting this physical insight, we write the x component of the ion's motion as a sum of the slow, secular motion $X(t)$ and the fast micromotion $\xi(t)$ according to:

$$x(t) = X(t) + \xi(t). \tag{11.5}$$

Inserting this into the x component of the Newtonian Equations 11.4, we obtain:

$$m\left[\ddot{X}(t) + \ddot{\xi}(t)\right] = -q\Phi_0\left[X(t) + \xi(t)\right]\cos(\omega t). \tag{11.6}$$

Since the secular motion is assumed to be slow, its acceleration $\ddot{X}(t)$ is small. The acceleration of the micromotion, however, is large, since it executes fast oscillations around the secular trajectory $X(t)$. Therefore, in the first step of our solution method, we neglect $\ddot{X}(t)$ with respect to $\ddot{\xi}(t)$ on the left-hand side of Equation 11.6. Since the amplitude of the micromotion $\xi(t)$ is small compared with the amplitude of the secular motion $X(t)$, we neglect $\xi(t)$ on the right-hand side of Equation 11.6 and obtain the differential equation:

$$\ddot{\xi}(t) = -\frac{q\Phi_0}{m}X(t)\cos(\omega t). \tag{11.7}$$

Since $X(t)$ is slow compared with $\xi(t)$, we treat $X(t)$ in Equation 11.7 as a constant. Integrating twice and neglecting the integration constants, we obtain:

$$\xi(t) = \frac{q\Phi_0}{m\omega^2}X(t)\cos(\omega t). \tag{11.8}$$

With $\xi(t)$ known, we return to Equation 11.6 and solve for $X(t)$. Since $X(t)$ is slow and $\xi(t)$ is fast, we extract an equation for $X(t)$ by averaging the differential Equation 11.6 over one period of the rf field. Indicating the operation of averaging with the angle brackets $\langle\ldots\rangle$, we have:

$$\langle\ddot{X}(t) + \ddot{\xi}(t)\rangle \approx \langle\ddot{X}(t)\rangle \approx \ddot{X}(t),$$

$$\langle X(t)\cos(\omega t)\rangle \approx 0,$$

$$\langle\xi(t)\cos(\omega t)\rangle \approx \frac{q\Phi_0}{m\omega^2}\langle X(t)\cos^2(\omega t)\rangle \approx \frac{q\Phi_0}{2m\omega^2}X(t), \tag{11.9}$$

where we used $\langle \cos^2(\omega t)\rangle = 1/2$, and again treated $X(t)$ as a constant on the scale of one oscillation period of the micromotion. With the results summarized in Equation 11.9, averaging Equation 11.6 over one cycle of the micromotion now yields:

$$m\ddot{X}(t) = -\frac{q^2\Phi_0^2}{2m\omega^2}X(t) = -\frac{\partial}{\partial X}V_{\text{eff}}(X(t)), \qquad (11.10)$$

where we introduced the effective potential

$$V_{\text{eff}}(X) = \frac{q^2\Phi_0^2}{4m\omega^2}X^2. \qquad (11.11)$$

The effective potential can be written as an oscillator potential

$$V_{\text{eff}}(X) = \frac{1}{2}m\Omega^2 X^2, \qquad (11.12)$$

where

$$\Omega = \frac{q\Phi_0}{m\omega\sqrt{2}} \qquad (11.13)$$

is the oscillator frequency of the secular motion. What this reveals physically is the following: On average the ions in the linear ion trap are confined in an oscillator potential. Consequently, they experience a strong focusing force, the oscillator's restoring force, toward the axis of the trap. This explains why, as shown in Figure 11.1, ions confined in a linear quadrupole trap tend to line up on the axis of the trap. Defining

$$y(t) = Y(t) + \eta(t), \qquad (11.14)$$

where $Y(t)$ describes the ion's secular motion in the y-direction and $\eta(t)$ describes the micromotion in the y-direction, replacing X with Y and ξ with η, and following the steps outlined in Equations 11.6–11.10, we obtain the ion's effective potential in the y-direction:

$$V_{\text{eff}}(Y) = \frac{q^2\Phi_0^2}{4m\omega^2}Y^2, \qquad (11.15)$$

which leads to the same frequency Ω defined in Equation 11.13 for the ion's secular oscillations in the y-direction. The combined, effective potential of the ion in the x- and y-directions is the sum of the effective potentials in Equations 11.11 and 11.15 in the x- and y-directions, respectively. It can be written as

$$V_{\text{eff}}(X,Y) = \frac{1}{2}m\Omega^2(X^2 + Y^2), \qquad (11.16)$$

where Ω is defined in Equation 11.13. The potential defined in Equation 11.16 is a completely confining, two-dimensional harmonic oscillator potential, which confines the ion to the vicinity of the trap's axis. In the presence of damping, kinetic energy is removed from the oscillatory secular motion of the ion. As a consequence, the ion's oscillation amplitude decreases in time, the ion is driven toward the trap's axis, and eventually, neglecting the quantum mechanical zero-point motion of the secular oscillator, comes to rest on the trap's axis, which implies $X(t) = Y(t) = 0$. Since, according to Equation 11.8, the micromotion $\xi(t)$ is proportional to $X(t)$ (and $\eta(t)$ is proportional to $Y(t)$), the micromotion ceases if the ion is located on the axis of the trap. Thus, when located on the trap's axis, the ion is perfectly "still" (up to quantum zero-point motion). If many ions are stored in the linear trap, they will all tend to line up on the trap's axis, with only the repulsive Coulomb force acting between them. If no additional forces are acting along the axis of the trap, the ions, under the influence of their mutual repulsive Coulomb forces, will be pushed out of the trap. In the linear ion trap this is counteracted by positively charged electrodes that generate an additional oscillator potential acting along the z-axis of the trap, keeping the stored ions in place. In equilibrium, the repulsive Coulomb forces in conjunction with the confining restoring force due to the z oscillator, produce a one-dimensional "lattice" of ions with approximately equal spacings as shown in Figure 11.1. This ion arrangement is ideally suited for targeted manipulation of the quantum states of the stored ions with tightly focused lasers.

When the ions are first loaded into the linear trap, they have a large kinetic energy. Confined to the the trap by the x, y, and z oscillator potentials, they form a cloud-like state called a *nonneutral plasma*. Because of the irregular, gas-like motion of the ions in the plasma state we may assign a temperature to the cloud. Then, in order to bring the ions into the crystalline state in which they align on the axis of the trap, the ions have to be cooled, i.e., their kinetic energy has to be reduced. In practice this is accomplished by using the method of *laser cooling* described in the next section.

Exercises:

11.2.1 A strong focusing potential in the x- and y-directions restricts the motion of two ^{24}Mg$^+$ ions to the z-axis of a linear rf trap. In addition to their mutual repulsive Coulomb force the ions experience a confining harmonic oscillator potential $V(z) = m\Omega_z^2 z^2/2$ along the z axis of the trap, where $\Omega_z = 2 \times 10^6 \, \text{s}^{-1}$ and m is the mass of a ^{24}Mg$^+$ ion.

(a) Calculate the equilibrium spacing d of the two ions.

(b) Calculate the frequency ω_c of the center-of-mass motion of the two ions. In this mode the two ions move synchronously, i.e., in phase, inside of the harmonic oscillator potential.

(c) Show that in the limit of small vibrations in the z-direction around the equilibrium positions of the two ions, the frequency ω_s of the symmetric stretch mode of the two ions is $\omega_s = \sqrt{3}\,\omega_c$. In the symmetric stretch mode the two ions oscillate $180°$ out of phase with respect to each other.

11.2.2 A five-qubit quantum computer consists of five $^{40}\text{Ca}^+$ ions stored on the z-axis of a linear rf trap. The end caps of the trap produce the on-axis harmonic oscillator potential $V(z) = m\Omega_z^2 z^2/2$, where $\Omega_z = 3 \times 10^6\,\text{s}^{-1}$ and m is the mass of the $^{40}\text{Ca}^+$ ions.

(a) Calculate the spacings between the ions. This is hard to do exactly, so you may want to introduce appropriate approximations.

(b) Although not essential, it is desirable that the ions in an ion-trap quantum computer are approximately equally spaced along the axis of the trap. How well is this hope fulfilled in the five-ion case?

11.3 Laser Cooling

We have all seen industrial-grade lasers that cut through metal like a knife cuts through butter. Thus, we tend to associate lasers with heating up solid objects to their melting points and beyond. But according to an ingenious scheme due to the two Nobel Prize winners Hänsch and Schawlow, lasers can also be used to cool objects, in particular atoms and ions. This is ideally suited for cooling the ions in our linear trap, since we have already decided to use lasers to manipulate the trapped ions and their quantum states.

Suppose we have an ion with a ground state $|g\rangle$ and an excited state $|e\rangle$ with corresponding energy levels E_g and E_e, respectively, as shown in Figure 11.3. We define the energy spacing

$$\Delta E = E_e - E_g \qquad (11.17)$$

and the *resonance frequency*

$$\omega_0 = \Delta E/\hbar. \qquad (11.18)$$

Let us assume that we can use laser light to promote an electron from the ground state $|g\rangle$ to the excited state $|e\rangle$. Then, if we irradiate the ion

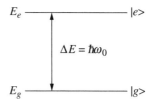

Figure 11.3 Trapped ion as a two-level system.

with laser light of frequency ω in the vicinity of ω_0, we observe that some of the incident laser light is scattered by the ions into random directions different from the direction of the incident laser beam. This process is called *fluorescence*. The number $N(\omega)$ of fluorescence photons emitted by the ion per unit time is a function of ω. For many ion species $N(\omega)$ is well described by a Lorentzian function

$$N(\omega) \; = \; N_0 \, \frac{(\Gamma/2)^2}{(\omega - \omega_0)^2 + (\Gamma/2)^2}, \qquad (11.19)$$

where $N_0 = N(\omega_0)$ and Γ is a parameter that describes the width of $N(\omega)$. A sketch of the Lorentzian function defined in Equation 11.19 is shown in Figure 11.4. For $\omega = \omega_0 \pm \Gamma/2$ the Lorentzian $N(\omega)$ drops to $N_0/2$, i.e., half its maximum height of N_0 at $\omega = \omega_0$. For this reason Γ is called the full width at half maximum (FWHM) of the Lorentzian function in Equation 11.19. The fluorescence rate $N(\omega)$ is peaked at $\omega = \omega_0$. This is the reason why we call ω_0 the resonance frequency.

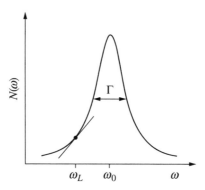

Figure 11.4 Lorentzian line shape of the rate $N(\omega)$ of fluorescence photons in the vicinity of the resonance frequency ω_0. The straight line approximates the function $N(\omega)$ in the vicinity of a red-detuned laser frequency ω_L.

The mechanism of fluorescence is as follows. Let us assume that we irradiate a stationary ion in its ground state $|g\rangle$ with laser light of frequency ω. A laser photon may be absorbed and the ion's electron will make a transition from $|g\rangle$ to $|e\rangle$. Once in the excited state $|e\rangle$, the electron may jump back to $|g\rangle$ under simultaneous emission of a photon. This photon is a *spontaneous photon*, which is emitted in a random direction. Whenever the ion absorbs a laser photon, it receives a momentum kick of magnitude $\hbar k \hat{\boldsymbol{k}}_L$, where $k = \omega/c$ and $\hat{\boldsymbol{k}}_L$ is a unit vector pointing into the direction of the laser beam. Following absorption of a laser photon, a spontaneous photon may be emitted with wave number $\boldsymbol{k}' = k' \hat{\boldsymbol{k}}_r$, where $\hat{\boldsymbol{k}}_r$ is a random unit vector. The wave number k' is different from k since in the emission process both energy and momentum have to be conserved. However, in the limit of large ion mass m, i.e., $\hbar\omega/(mc^2) \ll 1$, which, for laser photons, is an excellent approximation, we can safely replace k' with k. In this limit, emission of a spontaneous photon imparts a momentum kick $-k\hat{\boldsymbol{k}}_r$ onto the ion. This momentum kick is the *recoil momentum*.

Since the spontaneous photons are emitted in random directions, the net recoil momentum imparted on the ion is zero on average. But since all of the initially absorbed laser photons have the same direction of momentum, $\hat{\boldsymbol{k}}_L$, the net momentum transferred to the ion by the absorbed photons per unit time is nonzero and results in a *radiation pressure force*. To compute this force, we use the fact that at frequency ω, $N(\omega)$ photons are absorbed and emitted per second and each photon transfers an amount $\hbar k$ of momentum. Therefore, the radiation-pressure force acting on the ion is

$$\boldsymbol{F}_{\text{rad}}(\omega) = N(\omega)\hbar k \hat{\boldsymbol{k}}_L = \frac{\hbar\omega}{c} N(\omega) \hat{\boldsymbol{k}}_L. \quad (11.20)$$

The "trick" of laser cooling is to make ingenious use of the nonlinear frequency dependence (via $N(\omega)$) of the radiation pressure force $\boldsymbol{F}_{\text{rad}}(\omega)$ in Equation 11.20. In order to illustrate the physical principle of laser cooling, it is enough to study laser cooling in one dimension. Consider the situation shown in Figure 11.5. A two-level ion with resonance transition frequency ω_0 is trapped in the oscillator potential

$$V_{\text{osc}}(x) = \frac{1}{2} m\Omega^2 x^2 \quad (11.21)$$

and executes harmonic motion in the x-direction. A laser, situated to the left of the ion (see Figure 11.5), irradiates the ion with *red-detuned* laser light of frequency $\omega_L < \omega_0$ as shown in Figure 11.4. The orientation of the laser, as shown in Figure 11.5, corresponds to $\hat{\boldsymbol{k}}_L = \boldsymbol{e}_x$. Since the ion

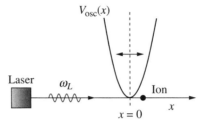

Figure 11.5 Red-detuned laser light ($\omega_L < \omega_0$) is incident from the left on an ion executing harmonic oscillations in an oscillator potential. Because of the Doppler shift, the radiation pressure on the ion is larger when it approaches the laser (it is shifted into resonance) than when it moves away from the laser (it is shifted out of resonance). This results in the damping of the ion's motion, i.e., laser cooling.

is in motion, the frequency ω', as seen in the ion's rest frame, is Doppler shifted:

$$\omega' = \omega_L \sqrt{\frac{1-v/c}{1+v/c}}, \qquad (11.22)$$

where v is the speed of the ion and c is the speed of light. Since $v \ll c$, we expand Equation 11.22 to lowest order in v/c. In this approximation Equation 11.22 becomes:

$$\omega' = (1 - v/c)\omega_L. \qquad (11.23)$$

Via this equation the radiation pressure force is:

$$\begin{aligned} F_{\rm rad} &= \frac{\hbar}{c}\omega' N(\omega') = \frac{\hbar\omega_L}{c}\left(1-\frac{v}{c}\right) N\left(\omega_L - \omega_L \frac{v}{c}\right) \\ &\approx \frac{\hbar\omega_L}{c}\left(1-\frac{v}{c}\right)\left[N(\omega_L) - \frac{v\omega_L}{c} N'(\omega_L)\right] \\ &= \frac{\hbar\omega_L}{c} N(\omega_L) \left\{1 - \left[1 + \frac{\omega_L N'(\omega_L)}{N(\omega_L)}\right]\frac{v}{c}\right\}, \qquad (11.24) \end{aligned}$$

where, again, we kept only first-order terms in v/c, and $N'(\omega_L)$ denotes the first derivative of $N(\omega)$ with respect to ω at $\omega = \omega_L$. As shown in Figure 11.4, in actual laser-cooling applications, the detuning $\Delta\omega = \omega_0 - \omega_L$ of the laser light is typically chosen to be on the order of $\Gamma/2$. Therefore, with $N(\omega)$ defined in Equation 11.19,

$$\frac{\omega_L N'(\omega_L)}{N(\omega_L)} = \frac{\omega_L(-2)(\omega_L - \omega_0)}{(\omega_L - \omega_0)^2 + (\Gamma/2)^2} = \frac{2\omega_L(\Delta\omega)}{(\Delta\omega)^2 + (\Gamma/2)^2}$$

$$\approx \frac{2\omega_L(\Gamma/2)}{\Gamma^2/2} = \frac{2\omega_L}{\Gamma} \gg 1, \qquad (11.25)$$

since Γ is typically on the order of 100 MHz, whereas ω_L, an optical frequency, is on the order of 10^{15} Hz. Therefore, in Equation 11.24, we may neglect 1 with respect to $\omega_L N'(\omega_L)/N(\omega_L)$ and obtain

$$F_{\text{rad}} = \frac{\hbar \omega_L}{c} N(\omega_L) - \frac{\hbar \omega_L^2}{c^2} N'(\omega_L) v. \quad (11.26)$$

Thus, it turns out that the radiation pressure force is velocity dependent. Since, as shown in Figure 11.4, for red-detuned laser light ($\omega_L < \omega_0$) the first derivative of $N(\omega)$ in ω_L, i.e., the slope of $N(\omega)$ in ω_L, is positive, we may write the radiation pressure force in Equation 11.26 in the form

$$F_{\text{rad}} = F_0 - m\gamma v, \quad (11.27)$$

where

$$F_0 = \frac{\hbar \omega_L}{c} N(\omega_L) \quad (11.28)$$

and

$$\gamma = \frac{\hbar \omega_L^2}{mc^2} N'(\omega_L) > 0. \quad (11.29)$$

Under the combined influence of the restoring force (see Equation 11.21)

$$F_{\text{rest}}(x) = -\frac{\partial V_{\text{osc}}(x)}{\partial x} = -m\Omega^2 x \quad (11.30)$$

and the radiation-pressure force defined in Equation 11.27, the equation of motion of the trapped ion is

$$\ddot{x} + \gamma \dot{x} + \Omega^2 x = F_0/m, \quad (11.31)$$

where we used $v = \dot{x}$.

This is the familiar equation of motion of a damped, driven harmonic oscillator. The solution of Equation 11.31 is

$$x(t) = x_s + x_0 e^{-\gamma t/2} \cos(\tilde{\Omega} t) + \frac{p_0}{m\tilde{\Omega}} e^{-\gamma t/2} \sin(\tilde{\Omega} t), \quad (11.32)$$

where

$$x_s = \frac{F_0}{m\Omega^2}, \quad \tilde{\Omega} = \sqrt{\Omega^2 - (\gamma/2)^2}, \quad (11.33)$$

and x_0, p_0 are constants. The validity of the Solution 11.32 may immediately be verified by inserting $x(t)$ defined in Equation 11.32 into Equation 11.31. Physically, the Solution 11.32 represents the motion of a damped, harmonic oscillator around the equilibrium point x_s. The origin of x_s is easily explained. The cooling laser exerts the radiation pressure force $\boldsymbol{F}_{\text{rad}}$ onto the ion, which shifts the ion from $x = 0$ (the minimum of the harmonic oscillator potential defined in Equation 11.21) to x_s defined in Equation 11.33. Then, x_0 is revealed as the oscillation amplitude with respect to x_s at time $t = 0$, and p_0 is the initial momentum of the ion at $t = 0$. Since $\gamma > 0$, the ion's oscillation amplitude decreases exponentially in time such that eventually the ion will settle down at $x = x_s$. This is the principle of laser cooling. Since the radiation pressure force, especially in the final stages of laser cooling, can be made very small, the ion, as a result of laser cooling, will settle down nearly motionless, i.e., with very low temperature, close to the axis of the trap. Since laser cooling works by emitting spontaneous photons, we cannot cool the ions to kinetic energies lower than the recoil energy defined as:

$$E_{\text{recoil}} = \frac{\hbar^2 k_L^2}{2m}. \tag{11.34}$$

Since the spontaneous photons are emitted in random directions, we may convert E_{recoil} into a temperature. We define:

$$T_{\text{recoil}} = \frac{E_{\text{recoil}}}{k_B} = \frac{\hbar^2 k_L^2}{2m k_B}, \tag{11.35}$$

where $k_B = 1.38 \times 10^{-23}$ J/K is Boltzmann's constant. The lowest temperature that can be achieved with our laser-cooling scheme is on the order of T_{recoil} defined in Equation 11.35. It is known as the *recoil limit* of laser cooling.

Exercises:

11.3.1 A ^{24}Mg$^+$ ion is cooled with laser light of wavelength $\lambda = 280$ nm. Estimate the recoil temperature T_{recoil}.

11.3.2 A ^{24}Mg$^+$ ion, confined in a harmonic-oscillator potential with $\Omega = 3 \times 10^6$ s^{-1}, is laser-cooled with laser light of wavelength $\lambda = 280$ nm. It emits, on average, 10^8 spontaneous photons per second. Compute x_s.

11.3.3 Confirm that $x(t)$ defined in Equation 11.32 solves the Equation of Motion 11.31.

11.3.4 An ionic two-level system consists of a ground state $|g\rangle$ and an excited state $|e\rangle$. The lifetime of the excited state $|e\rangle$ is 1 ns. The excited state $|e\rangle$ is populated with a powerful pump laser such that whenever $|e\rangle$ decays into $|g\rangle$, the laser, essentially immediately, repopulates $|e\rangle$. What is the spontaneous rate N of fluorescence photons?

11.4 Cirac-Zoller Scheme

Laser-cooled and aligned on the z-axis of the linear rf trap, the ions of the ion-trap quantum computer are ready to receive instructions. We assume that each of the ions is well approximated by a quantum mechanical two-level system consisting of a ground state $|g\rangle$ and an excited state $|e\rangle$ as shown in Figure 11.3. The ions are assumed to be separated far enough from each other such that, without disturbing the other ions, each of them may be manipulated individually with tightly focused laser light. This requirement, by its very nature, seems to preclude any communication between ions for the purpose of realizing quantum gates. It was the idea of Cirac and Zoller to use the center-of-mass vibration of the trapped ion chain (see Figure 11.1 and Section 11.2) to act as a "quantum bus" that allows exchange of quantum information between the ions. If we allow for vibrations, the energy level scheme of an individual ion consists of more than two levels. As shown in Figure 11.6, both the ground-state level and the excited-state level appear in many replicas, separated by the various vibrational energies $\epsilon_1, \epsilon_2, \ldots$ of the ion chain.

Although, in principle, one might envision using several of the acoustic modes of the trapped ion chain as channels for information transmission, the Cirac-Zoller scheme makes use of only the center-of-mass mode. Thus, suppressing all the other levels, we represent each ion in the rf trap as a four-level system as shown in Figure 11.7. The next step is to find a mechanism by which the vibrational mode, the "bus mode," of the quantum computer can be excited. Suppose one of the ions of the ion chain is in its excited state $|e\rangle$. Then, irradiating this ion with laser light of frequency $\omega_0 - \omega_c$, where ω_c is the frequency of the center-of-mass mode, induces a transition to the state $|g\rangle|1\rangle$ as shown in Figure 11.7.

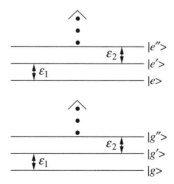

Figure 11.6 Level scheme of a trapped ion including phonon excitations.

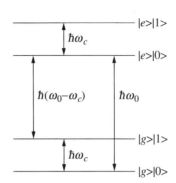

Figure 11.7 Level scheme of a trapped ion including only zero- and one-phonon excitations.

In this state the bus mode contains one quantum of vibration, i.e., one *phonon* of frequency ω_c. The phonon can be eliminated again by a second laser pulse with frequency $\omega_0 - \omega_c$, which returns the ion into the state $|e\rangle$. Since all ions in the trapped ion chain share the center-of-mass motion, we can use this mechanism to communicate between any two ions of the ion chain.

Suppose we want to realize a CNOT gate between ions number m and n of the ion chain. The computational basis states of the (m, n) subsystem of the ion-trap quantum computer are

$$|a\rangle_m |b\rangle_n |c\rangle, \tag{11.36}$$

where $a \in \{e, g\}$ refers to the state of ion number m, $b \in \{e, g\}$ refers to the state of ion number n, and $c \in \{0, 1\}$ refers to the state of the phonon with $c = 0$ indicating that no phonon is in the bus, and $c = 1$ indicating that the bus contains one phonon. Thus, there are a total of eight computational basis states.

We start the realization of the CNOT gate with application of a laser pulse on ion number m. We assume that initially no phonon is in the bus. Thus, initially, including the phonon state, there are four two-qubit computational states to be considered:

$$|g\rangle_m |g\rangle_n |0\rangle, \quad |g\rangle_m |e\rangle_n |0\rangle, \quad |e\rangle_m |g\rangle_n |0\rangle, \quad |e\rangle_m |e\rangle_n |0\rangle. \tag{11.37}$$

The laser pulse applied to ion number m is red-detuned to the frequency $\omega = \omega_0 - \omega_c$ such that, according to Figure 11.7, only a transition from $|e\rangle_m |0\rangle$ to $|g\rangle_m |1\rangle$ is possible. All other states are invariant under the action of the laser pulse. In addition, the duration, polarization, intensity, and phase of the laser pulse are chosen such that the pulse performs the following transformations on the four computational states defined in Equation 11.37:

$$\begin{aligned}\hat{U}_m: \quad |g\rangle_m|g\rangle_n|0\rangle &\to |g\rangle_m|g\rangle_n|0\rangle, \\ |g\rangle_m|e\rangle_n|0\rangle &\to |g\rangle_m|e\rangle_n|0\rangle, \\ |e\rangle_m|g\rangle_n|0\rangle &\to -i|g\rangle_m|g\rangle_n|1\rangle, \\ |e\rangle_m|e\rangle_n|0\rangle &\to -i|g\rangle_m|e\rangle_n|1\rangle. \end{aligned} \quad (11.38)$$

Next, we apply a laser pulse to ion number n. This laser pulse is designed in such a way that it changes the sign of the state $|g\rangle_n|1\rangle$ leaving all other states invariant. Application of this laser pulse transforms the states on the right-hand side of Equation 11.38 in the following way:

$$\begin{aligned}\hat{V}_n: \quad |g\rangle_m|g\rangle_n|0\rangle &\to |g\rangle_m|g\rangle_n|0\rangle, \\ |g\rangle_m|e\rangle_n|0\rangle &\to |g\rangle_m|e\rangle_n|0\rangle, \\ -i|g\rangle_m|g\rangle_n|1\rangle &\to i|g\rangle_m|g\rangle_n|1\rangle, \\ -i|g\rangle_m|e\rangle_n|1\rangle &\to -i|g\rangle_m|e\rangle_n|1\rangle. \end{aligned} \quad (11.39)$$

Finally, we apply another red-shifted laser pulse to ion number m. This pulse has the same characteristics as the pulse we had already applied on ion number m before. This time, however, the bus contains one phonon. Thus, according to Figure 11.7 a transition may only occur from $|g\rangle_m|1\rangle$ to $|e\rangle_m|0\rangle$. Therefore, the laser pulse transforms the states on the right-hand side of Equation 11.39 according to:

$$\begin{aligned}\hat{U}_m: \quad |g\rangle_m|g\rangle_n|0\rangle &\to |g\rangle_m|g\rangle_n|0\rangle, \\ |g\rangle_m|e\rangle_n|0\rangle &\to |g\rangle_m|e\rangle_n|0\rangle, \\ i|g\rangle_m|g\rangle_n|1\rangle &\to |e\rangle_m|g\rangle_n|0\rangle, \\ -i|g\rangle_m|e\rangle_n|1\rangle &\to -|e\rangle_m|e\rangle_n|0\rangle. \end{aligned} \quad (11.40)$$

Thus, according to Equations 11.38–11.40, application of the three laser pulses has the following net effect:

$$\begin{aligned}\hat{W} = \hat{U}_m \hat{V}_n \hat{U}_m: \quad |g\rangle_m|g\rangle_n|0\rangle &\to |g\rangle_m|g\rangle_n|0\rangle, \\ |g\rangle_m|e\rangle_n|0\rangle &\to |g\rangle_m|e\rangle_n|0\rangle, \\ |e\rangle_m|g\rangle_n|0\rangle &\to |e\rangle_m|g\rangle_n|0\rangle, \\ |e\rangle_m|e\rangle_n|0\rangle &\to -|e\rangle_m|e\rangle_n|0\rangle. \end{aligned} \quad (11.41)$$

Inspection of Equation 11.41 shows that the phonon that appeared as an intermediate result of the application of the first laser pulse to ion number m has vanished after application of the third laser pulse. Therefore, after application of the three-pulse sequence the bus is "clean" and ready for other gate operations. We also notice that the net effect of the three-pulse sequence $\hat{W}$ is to produce a minus sign in front of the computational state $|e\rangle_m|e\rangle_n|0\rangle$, while leaving all other states invariant. This result can be used to construct an actual CNOT gate in the following way. First, we define the states:

$$|\pm\rangle_n = \frac{1}{\sqrt{2}}(|g\rangle_n \pm |e\rangle_n). \qquad (11.42)$$

Then, suppressing the phonon states, which are $|0\rangle$ anyway before and after application of the three-pulse laser sequence, we have:

$$\hat{W}|g\rangle_m|\pm\rangle_n = |g\rangle_m|\pm\rangle_n,$$
$$\hat{W}|e\rangle_m|\pm\rangle_n = |e\rangle_m|\mp\rangle_n. \qquad (11.43)$$

Apparently, the qubit associated with ion number m acts as a control qubit: the state of ion n gets flipped only if ion number m is in the state $|e\rangle_m$. If ion number m is in $|g\rangle_m$, the state of ion number n remains invariant. We can produce the state defined in Equation 11.42 with the help of an H gate applied to ion number n. This way we obtain:

$$\mathrm{H}_n\hat{W}\mathrm{H}_n|g\rangle_m|g\rangle_n = |g\rangle_m|g\rangle_n,$$
$$\mathrm{H}_n\hat{W}\mathrm{H}_n|g\rangle_m|e\rangle_n = |g\rangle_m|e\rangle_n,$$
$$\mathrm{H}_n\hat{W}\mathrm{H}_n|e\rangle_m|g\rangle_n = |e\rangle_m|e\rangle_n,$$
$$\mathrm{H}_n\hat{W}\mathrm{H}_n|e\rangle_m|e\rangle_n = |e\rangle_m|g\rangle_n. \qquad (11.44)$$

Clearly,

$$\mathrm{CNOT} = \mathrm{H}_n\hat{W}\mathrm{H}_n. \qquad (11.45)$$

This concludes our demonstration of the Cirac-Zoller scheme. Using laser pulses applied to individual ions trapped in a linear rf ion trap, it is possible to implement the universal CNOT gate. Since all other quantum gates can be constructed from CNOT gates, the Cirac-Zoller scheme defines a possible architecture of a quantum computer. That this scheme is more than theory is demonstrated in the following section, which discusses an actual implementation of the Cirac-Zoller scheme using trapped $^{40}\mathrm{Ca}^+$ ions.

Exercises:

11.4.1 The Cirac-Zoller scheme works with eight computational states,

$$|a\rangle_m |b\rangle_n |c\rangle, \quad a,b \in \{g,e\}, \quad c \in \{0,1\}. \qquad (11.46)$$

Equations 11.38–11.40 specify the actions of $\hat{U}_m$ and $\hat{V}_n$, on a subset of the eight computational states defined in Equation 11.46. Assume that $\hat{U}_m$ and $\hat{V}_n$ leave the unspecified states invariant. Show that both $\hat{U}_m$ and $\hat{V}_n$ are unitary operators in the space of the eight computational states in Equation 11.46.

11.4.2 Suppose a different set of laser pulses produces the following net effect $\hat{W}'$ on the computational states $|a\rangle_m|b\rangle_n|0\rangle$, $a,b \in \{g,e\}$:

$$\hat{W}' : \quad |g\rangle_m|g\rangle_n|0\rangle \rightarrow |g\rangle_m|g\rangle_n|0\rangle,$$
$$|g\rangle_m|e\rangle_n|0\rangle \rightarrow |g\rangle_m|e\rangle_n|0\rangle,$$
$$|e\rangle_m|g\rangle_n|0\rangle \rightarrow |e\rangle_m|g\rangle_n|0\rangle,$$
$$|e\rangle_m|e\rangle_n|0\rangle \rightarrow i|e\rangle_m|e\rangle_n|0\rangle.$$

How do you realize the CNOT gate in this case?

11.5 Ca$^+$ Quantum Computer

Not long after publication of Cirac and Zoller's proposal, several research groups implemented and demonstrated linear ion-trap quantum computers based on the Cirac-Zoller design. This proved (1) that the Cirac-Zoller scheme works in practice and (2) that quantum computers can be constructed and operated in the lab. In this section we briefly discuss an implementation of the Cirac-Zoller scheme by a research team working at the University of Innsbruck. The Innsbruck quantum computer is based on ^{40}Ca$^+$ ions confined in a linear rf trap with $\Omega_z \approx 2\pi \times 1.7\,\text{MHz}$. The electronic energy levels most important for the implementation of the ^{40}Ca$^+$ ion-trap quantum computer are shown in Figure 11.8. The Innsbruck team chose ^{40}Ca$^+$ because the structure of its energy levels readily lends itself to the implementation of the two basic requirements

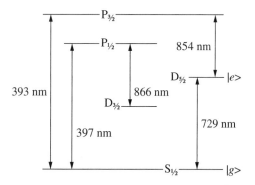

Figure 11.8 Level scheme of the valence electron of the ^{40}Ca$^+$ ion.

for a working linear ion-trap quantum computer: laser cooling and implementation of qubits.

Referring to Figure 11.8, the ^{40}Ca$^+$ ion-trap quantum computer works in the following way. Shortly after creating and trapping the ^{40}Ca$^+$ ions, they are cooled to very low temperatures using laser cooling (see Section 11.3) on the $S_{1/2}$–$P_{1/2}$ transition. This transition is driven by UV laser light of $\lambda = 397$ nm. The qubit is implemented with the help of the $S_{1/2}$ and $D_{5/2}$ energy levels. The computational states are $|S_{1/2}\rangle \equiv |0\rangle$ and $|D_{5/2}\rangle \equiv |1\rangle$. The $S_{1/2}$–$D_{5/2}$ two-level system was chosen for the implementation of a qubit since the lifetime of the $D_{5/2}$ state is about 1 second. This is long compared to the time it takes to perform gate operations on the qubit. Therefore, it is unlikely that the two-level $S_{1/2}$–$D_{5/2}$ qubit decays during a quantum computation. As shown in Figure 11.8, the $S_{1/2}$–$D_{5/2}$ transition is driven by laser light of $\lambda = 729$ nm.

The ^{40}Ca$^+$ level scheme shown in Figure 11.8 also illustrates the experimental difficulties that arise when one attempts to implement a quantum computer in the lab: the ideal two-level systems postulated in theoretical quantum computing schemes never exist this pure in practice. In the case of ^{40}Ca$^+$, for example, Figure 11.8 shows that apart from the "working levels" $S_{1/2}$, $P_{1/2}$, and $D_{5/2}$, there are at least two "parasitic" levels, $P_{3/2}$ and $D_{3/2}$. The presence of these two levels is undesirable, but also unavoidable, since they are an integral part of the electronic structure of the ^{40}Ca$^+$ ion. For instance, when laser cooling on the $S_{1/2}$–$P_{1/2}$ transition, the electron has a small, but nonzero chance to jump from the $P_{1/2}$ level to the $D_{3/2}$ level, instead of jumping back to the $S_{1/2}$ level. Once in the $D_{3/2}$ level, the electron will spend a long time in it before jumping back to the $S_{1/2}$ level. During all this time the electron is unavailable for laser cooling. This unwanted process, if it occurs, is taken care of in the following way. In case a transition from the $P_{1/2}$ level to the $D_{3/2}$ level actually occurs, a diode laser tuned to $\lambda = 866$ nm will ensure that the electron is pumped back to the $S_{1/2}$–$P_{1/2}$ system. Another diode laser, tuned to $\lambda = 854$ nm, serves a similar purpose in case the electron gets trapped in the $D_{5/2}$ level. Once laser cooling is accomplished and the quantum computing on the $S_{1/2}$–$D_{5/2}$ transition begins, the two diode lasers are switched off since they would seriously disrupt the quantum computation. The 854 nm laser, for example, would transport the electron from the computational state $|1\rangle \equiv D_{5/2}$ into the parasitic state $P_{3/2}$ and thus take the electron out of the $\{|0\rangle, |1\rangle\}$ computational space.

Once a quantum computation is completed, a measurement has to be performed on the $S_{1/2}$–$D_{5/2}$ two-level system. This is done by switching on the 397 nm laser. If the electron is in the $S_{1/2}$ state, i.e., in the computational state $|0\rangle$, the 397 nm laser will make the ion light up by inducing rapid $S_{1/2}$–$P_{1/2}$ transitions that produce strong laser fluorescence. If the electron, instead, is in the state $D_{5/2}$, i.e., in the computational state $|1\rangle$, then no laser fluorescence occurs and the ion is dark. This way the quantum state of the $S_{1/2}$–$D_{5/2}$ two-level system can be measured with near 100% efficiency.

Using their ^{40}Ca$^+$ ion-trap quantum computer, the Innsbruck group has demonstrated all basic gate operations of a quantum computer. Thus, the Innsbruck computer as well as other quantum computers operated in many laboratories around the world have demonstrated that quantum computers are no longer fiction but firmly grounded in reality.

Exercises:

11.5.1 The Innsbruck group chose ^{40}Ca$^+$ to realize their ion-trap quantum computer. Ions with a qualitatively similar electronic level structure are ^{24}Mg$^+$ and ^{138}Ba$^+$, where ^{24}Mg$^+$ is lighter than ^{40}Ca$^+$ and ^{138}Ba$^+$ is heavier than ^{40}Ca$^+$.

(a) Draw a qualitative level scheme similar to the one shown in Figure 11.8 for (i) ^{24}Mg$^+$ and (ii) ^{138}Ba$^+$. You may compile the data for the energy levels of ^{24}Mg$^+$ and ^{138}Ba$^+$ by either using spectroscopic data published in books or scientific papers, or by searching the Internet, for instance the NIST (National Institute of Standards and Technology) website. When searching the Internet, be careful to use only trusted sources of information (NIST is one of them).

(b) What is the main qualitative difference between the level scheme of ^{24}Mg$^+$ and the level schemes of ^{40}Ca$^+$ and ^{138}Ba$^+$? Would this provide an argument in favor or against using ^{24}Mg$^+$ to realize qubits in an ion-trap quantum computer?

(c) The level schemes of ^{40}Ca$^+$ and ^{138}Ba$^+$ are qualitatively similar. However, based on the wavelengths for radiative transitions between levels, what do you think was the main reason that led the Innsbruck group to choose ^{40}Ca$^+$ instead of ^{138}Ba$^+$ as their favorite ion?

11.5.2 As mentioned in Section 3.7, electrons in a magnetic field are an ideal two-level system and one might be tempted to use electrons instead of ions as qubit realizations in a quantum computer. On the downside, since laser cooling does not work on quasi-free electrons (non-neutral electron plasma), it is hard to cool electrons and line them up on the axis of a linear radio-frequency trap. But let us be optimistic and assume that the electron-cooling problem is solved. Changing the polarity of the end-cap electrodes, we trap the electrons on-axis in a copy of the Innsbruck trap in which ^{40}Ca$^+$ ions have an axial vibration frequency of $\Omega_z = 2\pi \times 1.7\,\text{MHz}$. An axial magnetic field $\boldsymbol{B} = B_0\boldsymbol{e}_z$ is switched on, producing an energy gap E_0 (see Equation 3.327) between the ground state $|g\rangle \equiv |\downarrow\rangle$ and the excited state $|e\rangle \equiv |\uparrow\rangle$.

(a) What is the axial vibration frequency ω_c of the electrons in the trap?

(b) Referring to Figure 11.7, we would like to have the energy level corresponding to the state $|g\rangle|1\rangle \equiv |\downarrow\rangle|1\rangle$ well separated from the state $|e\rangle|0\rangle \equiv |\uparrow\rangle|0\rangle$, i.e., we would like to have $\omega_c/\omega_0 \ll 1$, where $\omega_0 = E_0/\hbar$. What is the smallest B_0 such that $\omega_c/\omega_0 < 0.1$?

11.6 Summary

In theory quantum computers are powerful devices. But do they work in practice? In this chapter we showed that the answer to this question is an emphatic "yes"! Nothing is more convincing than demonstrating a working quantum computer in the lab. And several experimental groups throughout the world have done just that: These groups are operating quantum computers that are based on various quantum computer architectures, one of them the ion-trap quantum computer design. Due to its conceptual simplicity, we chose the ^{40}Ca$^+$ ion-trap quantum computer of the Innsbruck group as a representative example (see Section 11.5). The centerpiece of the Innsbruck quantum computer is a linear radio-frequency (rf) ion trap (see Section 11.2). Initially, the trap is loaded with several ^{40}Ca$^+$ ions, which are dynamically confined to the trap, but form a relatively "hot" ($\approx$ room temperature) charged gas, a non-neutral plasma. In this state the ions are useless for quantum computing. However, application of a cooling laser (see Section 11.3), reduces the kinetic energy of the ions, eventually lining them up on the axis of the trap. This state of the ions is sometimes referred to as the crystalline state of the ^{40}Ca$^+$ ions, in which each ^{40}Ca$^+$ ion represents one qubit of the quantum computer. As far as the hardware is concerned (the trap and the ions), the qubits are now ready for a quantum calculation. This is where the *microprogramming*, i.e., the realization of basic quantum gates, comes into play. According to the Cirac-Zoller scheme (see Section 11.4), laser pulses and the ion-crystal's center-of-mass motion, are used to implement the quantum CNOT gate. Since the CNOT gate is universal (see Chapter 7), all quantum computations can be done on a quantum computer once the CNOT gate is implemented. Thus, the trap hardware together with the Cirac-Zoller scheme, implements a universal quantum computer. The details of how this is done in practice are presented in Section 11.5, providing positive proof that quantum computers are now a reality.

Chapter Review Exercises:

1. In order to test your two-qubit $^{40}\text{Ca}^+$ ion-trap quantum computer, you decide to run the Deutsch algorithm for the function f_1 defined in Equation 8.7, using the Cirac-Zoller scheme. Specify the sequence of laser pulses you have to apply to the two $^{40}\text{Ca}^+$ ions of your quantum computer.

2. An $^{24}\text{Mg}^+$ ion is loaded into a linear rf trap with $\Omega_z = 3 \times 10^6 \, \text{s}^{-1}$. It executes oscillations with an amplitude of $200 \, \mu\text{m}$. Estimate the time it takes a cooling laser with $\lambda = 280 \, \text{nm}$ to reduce the ion's oscillation amplitude to $20 \, \mu\text{m}$. Assume that, on average, the ion emits $N(\omega_L) = 10^7$ spontaneous photons per second. Assume further that the laser is red-detuned to $\omega_L = \omega_0 - \Gamma/2$, where $\Gamma = 2\pi \times 150 \, \text{MHz}$.

3. The effectiveness of laser cooling strongly depends on the detuning $\Delta\omega$. How do you have to choose $\Delta\omega$ to optimize the effect of laser cooling, i.e., at what $\Delta\omega$ is γ maximal?

chapter 12

Outlook

In this chapter:

- ◆ Introduction
- ◆ Quantum Internet
- ◆ Quantum Cryptography
- ◆ Quantum Computing
- ◆ Summary

12.1 Introduction

Although not yet noticeable in everyday life, where we still rely exclusively on classical computers and machines based on classical working principles, we have already entered the quantum information age. Its icons, quantum communication devices, and quantum computers are currently researched, implemented, and perfected by scientists and engineers in university, industrial, and government laboratories all around the world.

As we saw in this course, there are two aspects of quantum mechanics, the "machinery" and the "spook." The machinery of quantum mechanics are the technical, mathematical tools we need to solve the Schrödinger equation. These tools are usually taught in introductory courses on quantum mechanics and are briefly reviewed in Chapter 3. There is nothing particularly quantum about the machinery part of quantum mechanics, since the same tools may be used to solve wave equations in any of the classical areas of physics such as acoustics, wave optics, and hydrodynamics, for example. The "spook" is much more interesting. It deals with aspects of quantum mechanics that have no classical analogue, not even on the conceptual level. Examples are irreducible randomness (Chapter 4) and entanglement (Chapter 6).

In this course we focused mainly on the spooky aspects of quantum mechanics, which up until recently, were the domain of philosophers. However, based on significant advances in laboratory technique, physicists were recently able to demonstrate many of the manifestly nonclassical aspects of quantum mechanics, among them interaction-free measurement (Chapter 5), teleportation (Chapter 7), and quantum computing (Chapter 8). This proves that the "spook" is real, and no longer just the domain of philosophers. It also opens the possibility of creating quantum machines, i.e., machines that use quantum effects, such as EPR correlations (Chapter 6), as one of their main working principles. Since they rely on quantum effects with no classical analogues, quantum machines are not just better, but fundamentally different from classical machines. The creation of quantum machines, in turn, requires a whole new field of physics and technology, *quantum engineering*.

How far along are we on our way into the quantum information age? The topics presented in the following three sections show that most of the basic technology for implementing the quantum revolution already exist, and some have already been developed into marketable products. Even rudimentary local quantum networks are already operating in several countries and form the basis of a future, global quantum Internet (Section 12.2). Commercially available quantum encryption devices, already in active use by banks and governments (Section 12.3), provide unprecedented security, and the first quantum computers are already working (Sections 11.5 and 12.4). In the way of an outlook on where the road into the quantum information age may take us, the following three sections briefly discuss three areas of interest in quantum information technology: the quantum Internet, quantum cryptography, and quantum computing. In addition to the development of technological applications of quantum mechanics, work continues on the foundations of quantum mechanics in particular in the area of quantum measurement (Chapter 4).

12.2 Quantum Internet

The currently existing Internet exchanges classical messages encoded in the form of classical bits. A quantum Internet exchanges quantum information encoded in the form of qubits. Basic to this technology is the technique of quantum teleportation introduced and discussed in Section 7.6. Following the theoretical proposal of a teleportation scheme in 1993, the first laboratory demonstration of the teleportation of photon states was accomplished in 1997. In 2004 the teleportation of material particles was demonstrated, followed, in 2006, by the demonstration of teleportation of a photon state directly into the quantum state of a material particle. This type of "heterogeneous teleportation" demonstrates the possibility of exchange of quantum information between different species of particles. This is of immense practical importance for the construction of quantum computer random access memory (QuRAM): While photons are optimal in quantum communication for transmitting quantum information over long distances, in a finite-size quantum circuit, such as in a quantum computer, confined photons are too short-lived to provide long-term quantum storage. Material particles are better suited for this purpose, since they can store quantum information over much longer times, for instance in their spin states or their meta-stable electronic states.

While the first laboratory demonstrations involved teleportation over distances of a few inches, successful teleportation of photons is nowadays routinely achieved over distances of up to 100 km and beyond. Each successful teleportation event confirms that EPR correlations, Einstein's "spooky action at a distance" is real. Although teleportation ranges of 100 km are impressive, international quantum communication requires quantum information exchange over much longer distances. Extending the teleportation range is currently the subject of active laboratory and technological research and development.

One of the first long-range point-to-point quantum teleportations was accomplished in 2004 when an Austrian team of researchers successfully transmitted quantum information over a distance of 600 m across the river Danube from one of its banks to the other. This experiment involved the teleportation of quantum information carried by photons through 800 m of optical fiber that was strung through a tunnel under the river. Although the data transmission rate was very small, this experiment demonstrated that quantum information can be transmitted under real-world conditions over commercially interesting distances. Moreover, by using optical fiber for their experiment this group of researchers demonstrated that quantum networks are possible using the already existing commercial optical telecommunications network.

Indeed, it did not take long after the demonstration of long-distance teleportation for quantum networks to appear. The first quantum network, a mini quantum Internet, started in 2006 and connected several sites in Boston. In 2008 a quantum network started to operate in Vienna connecting several industrial sites separated by several kilometers each.

The fundamental technical limitation of quantum networks is to guarantee quantum entanglement over a distance of many kilometers. However, from about 15 km on, photon loss rates in optical fibers become an increasingly serious problem. In normal classical communications this problem is solved by using devices called *repeaters*. A repeater receives a weakened signal at one end of an optical transmission line, amplifies it, and sends it out again on the other end. In classical communication networks this simple method works because there is no problem cloning, or copying, classical data. But as we saw in Section 4.4, the quantum no-cloning theorem forbids the construction of *quantum repeaters* that work on this simple amplification principle. One qubit of quantum information in an optical fiber is carried by a single photon, and this photon is either present or not; it cannot be "amplified." However, this does not mean that quantum repeaters are impossible. As quantum machines, as previously mentioned, they are just going to be different. Indeed, scientists have already defined schemes (quantum protocols) that allow the construction of quantum repeaters. This technology is currently perfected and will allow the transmission of quantum entanglement over very long distances. Once this is accomplished, nothing will stand in the way of constructing a global quantum Internet. A possible use for a quantum Internet is to connect quantum computers worldwide to achieve a network of quantum computers for massive quantum grid computing.

12.3 Quantum Cryptography

Current Internet traffic relies almost exclusively on public-key encryption schemes that rest on the perceived difficulty of solving certain mathematical problems such as integer factorization (Chapter 9). However, in Chapter 10 we saw that classical, public-key encryption methods, such as RSA, are vulnerable to attacks by quantum computers. We also saw that symmetric private-key encryption schemes are safer than public-key methods, but suffer from the key-distribution problem we illustrated in Section 9.2. Obviously, the best encryption scheme is one that combines the safety of symmetric private-key cryptosystems with the ease and convenience of currently employed public-key schemes. Solving the key-distribution problem would do the trick. And indeed, again, quantum mechanics provides the solution. In 1984, Bennett and Brassard realized that a secret key may be exchanged safely using entangled EPR pairs. Turning a "weakness" of quantum mechanics into a strength, these researchers noticed that eavesdropping on the entangled pairs would mess up their quantum correlations making any such illicit intrusion attempt immediately noticeable to the two parties engaged in creating a secret key. The technical implementation of this idea is known as the BB84 protocol. Another protocol, E91, was described by Ekert in 1991. These two quantum key-distribution (QKD) systems have recently been developed into products that are sold by commercial companies to banks and government institutions worldwide. In 2007, for instance, for added security, the Swiss government transmitted election results from local Geneva polling stations to a central data processing and storing center using a QKD quantum encryption device. This was a historic event in several respects. (1) It showed that governments are trusting the new quantum machines based on technology that is unlike any classical technology, (2) it proved that quantum transmissions are reliable to the point that they meet government standards, and (3) it demonstrated once and for all that quantum information can be transmitted over considerable distances without corruption. The next obvious step is to standardize QKD protocols for quantum Internet applications.

12.4 Quantum Computing

While quantum networks and quantum cryptography have already found their way into commercial and government use, quantum computers are still a bit further away from realizing their full commercial and scientific potential. But what a potential it is! Completely outclassing classical computers, quantum computers have the potential to solve problems in code breaking, drug synthesis, and industrial scheduling at lightning speed.

To construct a working quantum computer we need quantum software and quantum hardware. For some special purpose applications, such as code breaking and the sorting of databases, quantum software is already available. However, as we saw in Chapters 8 and 10, quantum algorithms are still formulated using mainly the language of physics, making their description cumbersome to anyone who is not a trained physicist. This is similar to the situation at the beginning of the computer age in the 1940s and 1950s when computers were programmed on the machine level in the way of early telephone switch boards, using plug-in wires for making electrical connections between various electronic circuits. Only the subsequent development of assembly languages and higher-level computer languages made classical computers into the conveniently programmable tools we have nowadays in the form of personal computers on nearly everybody's desk.

The same development of higher-level quantum languages has not yet happened in quantum computing and currently prevents programmers who are not trained physicists to develop quantum software and quantum applications. Obviously, the development of a higher-level, user-friendly quantum programming language is high on the list of priorities in quantum computing.

On the hardware side various quantum computer architectures are currently under investigation. For universal quantum computing applications ion-trap quantum computers, as discussed in Chapter 11, look particularly promising. Prototype ion-trap quantum computers with up to three qubits were already demonstrated in the lab. The modular design of ion-trap quantum computers make this computer architecture conceptually straightforward and, at least in principle, easily scalable. It is also possible, and advantageous, to combine several ion-trap quantum computers into a cluster of ion-trap quantum computers by shuttling ions or photons between the individual computers. This quantum cluster design has certain advantages, since loading too many ions into a single linear ion trap may cause technical problems with the acoustic quantum bus. Because of their transparent design, ion-trap quantum computers are prime candidates for the eventual construction of universal, large-scale quantum computers for scientific and commercial use.

In addition to the ion-trap quantum computers discussed in Chapter 11, there are several other promising quantum computer designs. Of particular importance are quantum computers based on nuclear magnetic resonance (NMR). Instead of using electron spins or the electronic states of ions and atoms, NMR quantum computers use nuclear spins to implement qubits. Nuclear spins are particularly well shielded from environmental noise and at present offer the fastest progress in increasing the number of qubits. Indeed, Shor's algorithm has already been implemented on a seven-qubit NMR quantum computer and has factored the number 15. While this seems like a small feat today, the working principle is demonstrated and all that is needed now is to scale NMR quantum computers up to work with more qubits.

A third type of quantum computer works quite differently from ion-trap and NMR quantum computers. While ion-trap quantum computers and NMR quantum computers may be considered the quantum versions of classical digital computers, *adiabatic quantum computers* may be considered the quantum versions of classical analogue computers. Their working principle is the following. Suppose we are given a Hamiltonian $\hat{\mathcal{H}}(\lambda)$ that depends on a parameter λ. Suppose also that for $\lambda = 0$ the Hamiltonian $\hat{\mathcal{H}}(\lambda)$ is simple and its ground state is known. Suppose, finally, that the ground state of $\hat{\mathcal{H}}(\lambda = 1)$ represents the solution of a complicated classical or quantum problem. Then, starting in the known ground state of $\hat{\mathcal{H}}(\lambda = 0)$ and slowly changing the parameter λ from 0 to 1, we end up in the ground state of $\hat{\mathcal{H}}(\lambda = 1)$. This is guaranteed by a theorem called the *adiabatic theorem of quantum mechanics*. Since over the entire time evolution from $\lambda = 0$ to $\lambda = 1$ the quantum computer, represented by the Hamiltonian $\hat{\mathcal{H}}(\lambda)$ remains in its ground state, adiabatic quantum computers are not too sensitive to ambient noise and thus maintain quantum coherence over relatively large times. This is a particular strength of adiabatic quantum computers. The complexity of the problems that can be solved with an adiabatic quantum computer depends on the complexity of the Hamiltonian $\hat{\mathcal{H}}(\lambda)$ and the number of qubits governed by $\hat{\mathcal{H}}(\lambda)$. Realizing the qubits as superconducting patches on a chip, adiabatic quantum computers have already been demonstrated in university and industrial labs.

As with all new technologies we ask ourselves: Will the quantum computer ever replace the now existing powerful PC technology? The answer is: "No." Classical and quantum computing are complementary technologies. Just because quantum computers are more powerful in certain applications, the PC will not go away. People are "classical." They need to interface with quantum computers "on their terms." The interface is provided by classical computers and classical computing technology. Therefore, what we will most likely see in the future is a harmonic side-by-side of the classical and the quantum computer. Both are made for vastly different purposes and fulfill vastly different functions.

12.5 Summary

Not so long ago it was inconceivable how pure research in the foundations of quantum mechanics could possibly lead to practical applications. Yet, here we are, with Einstein's "spooky action-at-a-distance" providing the foundation for a completely new industry, quantum information technology. The relatively short time it took to turn quantum paradoxes into practical quantum devices provides one of the success stories of how curiosity-driven research results in a major scientific and industrial revolution.

Appendix

- **Physical Constants**
  (rounded to four significant figures)

Boltzmann's constant	k_B	$= 1.381 \times 10^{-23}$ J/K
Planck's constant	h	$= 6.626 \times 10^{-34}$ Js
	$\hbar$	$= 1.055 \times 10^{-34}$ Js
speed of light in vacuum	c	$= 2.998 \times 10^{8}$ m/s
elementary charge	e	$= 1.602 \times 10^{-19}$ C
Bohr magneton	μ_B	$= 9.274 \times 10^{-24}$ J/T
gravitational constant	G	$= 6.674 \times 10^{-11}$ m^3/kg s^2
electron mass	m_e	$= 9.109 \times 10^{-31}$ kg
proton mass	m_p	$= 1.673 \times 10^{-27}$ kg
neutron mass	m_n	$= 1.675 \times 10^{-27}$ kg
permittivity of the vacuum	ϵ_0	$= 8.854 \times 10^{-12}$ C^2 s^2 / kg m^3

Index

absorption
    of electromagnetic waves, 46
    of a laser photon, 289
    of quantum particles, 46
acoustic
    modes, 293
    quantum bus, 308
action at a distance, 184
addition modulo 2, defined, 207
adiabatic
    quantum computer, 308-309
    theorem, 308
d'Alembert, Jean le Rond
    equation, 4
algorithm
    deterministic, 239, 242
    probabilistic, 239, 242, 268
amplitude
    defined, 34, 92
    Dirac notation of, 34
    of an event, 34
    partial, 35
    of a state, 92
AND gate, defined, 204
angular frequency
    of matter waves, 33
    of light, 4
angular momentum
    and Bohm's EPR system, 183
    commutator identities, 58-59
    defined, 54
    intrinsic, 20, 21
    operator, defined, 54
    spin, 21
anticoincidence
    defined, 14
anti-commutator, defined, 119

barrier penetration
    amplitude, 85
    defined, 85
beam
    reflected, 14
    transmitted, 14
    weak, 13
beam splitter
    for light waves, 13-15
    for matter waves, 87
    polarizing, defined, 163
Bell, John Stewart
    argument, 185
    inequality, 188, 189
    and local hidden variable theories, 185-190
    and nonlocality of quantum mechanics, 184
    proof of nonlocality, 184-190
    state measurement, 215-217
    states, defined, 215

Bennett, Charles H. and
    Brassard, Gilles
    BB84, 306
    quantum key distribution,
        306
binary
    argument, 227, 233
    representation, 235
    valued function, 227, 232-
        233
    variables, 232
bit
    defined, 199
    implementation, 199
    values, 199
Bloch, Felix
    sphere, defined, 201
Bohm, David Joseph
    EPR system, 182-184
Bohr, Niels Henrik David
    magneton, defined,
        109
Boltzmann, Ludwig Eduard
    constant, 292
Born, Max
    and commutators, 57
    Nobel Prize, 35
    rule, 35, 36, 39, 45
boundary condition, 47, 65
bound states
    delta-function potential,
        86-87
    infinite square well, 66-
        67
bound-state energies
    delta-function potential,
        87
    finite square well, 79
    harmonic oscillator, 70
    infinite square well, 66
box and marble system, 182-
    183, 184, 186

branch point, defined, 209
bra state
    defined, 91
    Hermitian conjugate of,
        94
bus mode, 293-294

Caesar, Gaius Julius
    cipher, 252
    encryption scheme, 252
    shift, 252-253
    substitution, 252-253
calcium
    atom, 14, 16
    atomic vapor, 16
    energy levels, 297
    ion, 282, 296, 297-300
    ions, crystalline state, 300
    quantum computer, 282,
        297-299
canonical quantization, 54, 55,
    109
Cauchy, Augustin Louis and
    Schwarz, Hermann Amandus
    inequality, 18
center-of-mass
    mode, 293
    motion, 294, 300
    vibration, 293
certification yield, defined, 157
Chinese Remainder Theorem,
    260-261
chosen plain-text attack, 258
cipher, 251, 252-253
Cirac, Juan Ignacio and Zoller,
    Peter
    and measurement, 298-299
    quantum computer, 297-
        299
    scheme, 293-296, 297, 300-
        301

INDEX 315

classical
    algorithm, 224, 226, 230, 239, 242, 245
    areas of physics, 303
    bit, 199-200
    circuits, 209-212
    communication channel, 216, 219
    computation, 197-199, 210
    computer, 220, 247, 279, 303
    discrete Fourier transform, 270-272
    front-end computer, 212
    gates, 202-204, 207
    information, 197-210, 217
    parallel computer, 232
    reasoning, 224, 228
    search algorithm, 224, 239
    working principles, 303
classically forbidden regions
    defined, 81
    and tunneling, 82
cloning, defined, 138
CNOT gate
    computation, 224
    construction, 296
    defined, 206-208
    f-controlled, defined, 229
    as a quantum computer, 224-226
    realization, 294
    universality, 210
coincidence
    detection, 14
    false, 16
    signal, 14
collector electrode, 10, 11
column vector, defined, 92
communication
    secret, 251, 254
    secure, 251-252, 265
    sensitive, 252

commutator
    of angular momentum, 57-59
    defined, 56
    of position and momentum, 57
    relation, 135
commuting diagram, 173
Composition Rule
    defined, 36
    example of, 36, 38
Compton, Arthur Holly
    effect, 22
    scattering, 23
    wavelength, 23
computation, defined, 201
computational
    space, 298
    state, 294, 295, 298, 299
concatenation, 255
continuity equation
    in differential form, 47
    in integral form, 47
continuum
    defined, 72
    wave functions, 71-74
control qubit, 206, 296
controlled-NOT gate, defined, 206
convergence generating factor
    defined, 72
conversion function, 255
Cooley, James William and Tukey, John Wilder
    Fast Fourier Transform, 270
Copenhagen
    interpretation, 169
    probabilistic world view, 170
co-prime, 268
copy operator, 138-139
corpuscles, 1-2, 8

cryptanalysis, defined, 251
cryptography, defined, 251
cryptology, defined, 251
cryptosystem
    algorithm, 252, 253
    broken, 253
    key, 252, 253
    symmetric, 253
current density, 46

damping, 286, 290
database
    access, 239, 241-242
    quantum search algorithm, 224
    search, 241
    states, 239-240
    unsorted, 224, 238-239, 245

de Broglie, Louis-Victor-Pierre-Raymond
    matter waves, 32-33, 44, 51
    wavelength, 32
    and wave nature of electrons, 26
    and wave-particle duality, 32
de l'Hospital, Guillaume Francois Antoine
    rule, 144
delta potential
    boundary conditions, 86
    bound-state energy, 87
    defined, 86
    matching condition, 86-87
    reflection amplitude, 87
    Schrödinger equation, 86
    transmission amplitude, 87
    wave functions, 86
    wave number, 86

detector
    bright, 155-156
    dark, 155-156
    polarization sensitive, 163
detuning
    blue, 110
    defined, 110, 290
    red, 110, 288, 291, 294, 295
Deutsch, David Elieser
    algorithm, circuit diagram, 230
    algorithm, defined, 228-230
    problem, defined, 228
Deutsch, David Elieser and Jozsa, Richard
    algorithm, defined, 234-237
    problem, defined, 232-233
diffraction
    defined, 2-3
    pattern, 28, 29, 30-31
diode laser, 298
Dirac, Paul Adrien Maurice
    delta function, 73, 74-75
    notation, 34
discrete Fourier transform, 270-272
dispersion relation, 4
DiVincenzo, David P.
    criteria, 282
Doppler, Christian Andreas
    shift, 290
dot
    notation, defined, 93
    product, defined, 93
double-slit experiment
    with classical light, 2-10, 22
    with matter, 32-33
    with photons, 29-32

INDEX 317

and seeing in the dark, 152

eavesdropping, 253, 258
eigenfunction
   defined, 52-53
eigenstates
   of momentum, 100-101
   of position, 99-100
eigenvalue
   defined, 52-53
   equation, 52-53
Einstein, Albert
   cosmic speed limit, 199
   energy-momentum formula, 19
   formula, photoelectric effect, 11
   and foundations of quantum mechanics, 170
   grainy light, 10
   and interpretation of quantum mechanics, 169
   and light quantum, 169
   mass-energy equivalence, 32
   and Newtonian tradition, 195
   Nobel Prize, 12
   and objective reality, 195
   photoelectric effect, 1, 10-12
   photon hypothesis, 12, 13
   and quantum mechanics, 169
   and "spooky action at a distance", 179, 182, 305, 309
   and stimulated emission, 169
Einstein, Albert, Podolsky, Boris, and Rosen, Nathan (EPR)
   argument, 180-183, 195
   and collapse of the wave function, 178-179
   and completeness of quantum mechanics, 175-181
   and correctness of quantum mechanics, 174
   correlation, 218, 221, 304
   incompleteness of quantum mechanics, 182, 195
   and local realism, 170, 183, 184
   pair, 217-218, 306
   paper, 170, 174, 195
   paradox, 170, 177, 180-181, 195
   particle, 216, 218
   and reality, 170, 182, 184
   singlet state, 214
   and space-time diagram, 177
   system, 177-181, 182, 195
Ekert, Artur
   E91, 306
   quantum key distribution, 306
electric field
   amplitude, 4
   of gap, 5
   of light, 6
   secondary, 3
   strength, 2, 3, 5
   transmitted, 3, 4
   vector, 2
electron
   charge, 11, 105
   de Broglie wavelength, 32
   diffraction, 26
   ejected, 10, 11
   g-factor, 109
   interference pattern, 32
   magnetic moment, 109
   microscope, 26, 33, 41
   as point particle, 37-38

properties, 105
rest mass, 19, 105
spin, 105
electron-positron pair, 19, 20
element of reality, defined, 175
Elitzur, Avshalom Cyrus and Vaidman, Lev
grain certification scheme, 155-159
interaction-free measurement, 152
and Mach-Zehnder interferometer, 155
scheme, 152
seeing in the dark, 152
emission
of a spontaneous photon, 289, 292
enabling
circuit, 212
technology, 120-121, 265
encryption scheme
algorithm, 252, 253
key, 252, 253
energy
barrier, 11
kinetic, maximal, 10, 11
levels, 66, 287
spacing, defined, 287
splitting, defined, 110
ensemble average, 16
entanglement, 118, 180, 304
event
amplitude of, defined, 34
probability of occurrence, 35
excited state
of two-level system, 110, 287, 293
expectation value
defined, 50, 53
of energy, 55
of momentum, 52
of observables, defined, 53
of an operator, defined, 97
of position, 53
as scalar product, 59
experiment of
Aspect and collaborators, 2, 12, 13-18
Hanbury-Brown and Twiss, 2, 14
Kwiat and collaborators, 159-165
Pound and Rebka, 20, 23
exponential
proliferation, 233
speed-up, 224

Fast Fourier Transform, 270
f-CNOT gate
defined, 229
generalization, 234
Fermat, Pierre
Little Theorem, 260, 261
Feynman, Richard Phillips
double-slit experiment, 41
and mystery of quantum mechanics, 31
routes, 35
rule, 35, 36, 38, 154
field strength
complex, 6
physical, 6
finite square well
energy levels, 79
graphical solution, 79
potential, defined, 71
Schrödinger equation, 76, 80
spectral equation, 78
wave functions, 76-77, 80-81, 82-83

fluorescence
    defined, 288
    mechanism, 289
Fourier, Jean-Baptiste-Joseph
    discrete transform, 270-272
    Theorem, 74
    transform, 270, 272, 273-274
four-level system, 293
Fraunhofer, Joseph von
    limit 4-5, 8
free motion
    Schrödinger equation, 71
    wave functions, 71-74
full width at half maximum,
    defined, 288
function
    balanced, 228, 230, 233-234, 237
    constant, 228, 230, 233-234, 236, 237
    integrable, 47
    nonintegrable, 47
    of operators, 102
    period of, 274
    square-integrable, 47

gate
    irreversible, 204
    reversible, 204

Gauss, Carl Friedrich
    integrals, 88
    theorem, 46
GOOGLE, 238
gravitational
    acceleration, 19
    constant, defined, 20
    energy, 19, 20
    field, 19, 20
    red shift, 20, 23

greatest common divisor, 268, 270
ground state
    of two-level system, 109, 287, 293
group velocity, defined, 42
Grover, Lov Kumar
    algorithm, defined, 239-245
    iteration, 241-242, 244-245
    operator, 240-241
    ratio, 244
    rotation, 244

Hadamard, Jacques Salomon
    formula, applied, 236
    formula, defined, 234
    gate, defined, 206
Hamilton, William Rowan
    function, 54
    operator, 54, 61, 63
Hänsch, Theodor Wolfgang and
        Schawlow, Arthur Leonard
    and laser cooling, 287
    Nobel Prize, 287
harmonic oscillator
    damped, driven, 291-292
    energy levels, 70
    equation, 65
    frequency, defined, 68
    potential, defined, 68
    Schrödinger equation of, 68
    spectral equation, 70
    spectrum, 70
    wave functions, 70
Heisenberg, Werner Karl
    operator, defined, 102, 104
    equation of motion, 104
    formulation of quantum
        mechanics, 102-104
    Uncertainty Principle, 127, 132, 136-137

Hermite, Charles
    conjugate, 59
    differential equation, 69
    polynomials, 69-70
Hertz, Heinrich Rudolf
    electromagnetic waves, 2
    photoelectric effect, 10
H gate, defined, 206
hidden variables
    defined, 184
    theory, 184-190
Hilbert, David
    space, 93, 105
Huygens, Christiaan
    Principle, 4, 5
    wave theory of light, 1, 22
hyperexponential
    growth, 233
    proliferation, 233

infinite square-well potential
    boundary condition, 65
    defined, 64
    energy levels, 66
    energy quantization, 66
    Schrödinger equation of, 65
    spectral equation, 65
    spectrum, 66
    wave functions, 67
    wave number quantization, 65
information
    age, 197
    classical, 199
    processing, 200, 207
Innsbruck quantum computer, 297-299, 300
inserting a one, defined, 99
integer factorization
    basic algorithm, 262
    difficulty of, 258, 262-264, 306
    with MATHEMATICA, 263-264
    on a PowerBook G4 computer, 263-264
    via prime-number division, 263
    with Shor's algorithm, 267
intensity
    of light, 6, 7
    maximum, 8
    minimum, 8
    time-averaged, 6, 9, 16
intensity distribution
    function, 28
    of light, 6
    normalized, 7
    observed, 9
    of single-slit experiment, 27
interaction
    energy, 109
    strength, 114
    time, 114
interaction-free imaging, 151, 165
interaction-free measurement
    certification yield, defined, 157
    conceptual scheme, 152-153
    defined, 149
    efficiency, 157
    efficiency barrier, 161
    and experimental verification, 151
    optimal, 159-165
    and reflectivity of beam splitters, 158-159
    Renninger's thought experiment, 149-151
interference, 1-2, 31
    pattern, 31, 32, 39-40
    term, 37

interferometer
    for matter waves, 88, 122
ion trap
    centerpiece of quantum computer, 282, 300
    electric field, 283
    equations of motion, 283, 291
    physics and design, 282-286
ion-trap quantum computer
    with calcium ions, 282, 297-299
    design, 281-282, 300
    and DiVincenzo criteria, 282
    cluster, 308
    promise, 308
    sketch, 282
irreducible randomness, 134, 304

ket state
    defined, 91
    Hermitian conjugate of, 94
key-distribution problem, 253, 265, 306
Kronecker, Leopold
    symbol, 67, 73

Landauer, Rolf William
    Principle, 205
Laplace, Pierre-Simon
    operator, 44
laser
    beam, 289
    cooling, 282, 287-292, 298
    fluorescence, 299
    light, 288, 293, 298
    photon, 289
    pulse, 282, 294-296, 300
    red-detuned, 289-291, 294, 295

leads, defined, 203
Lenard, Philipp Anton Eduard
    counterfield method, 10
lifetime, 298
light
    angular frequency, 4
    clustered, 14
    corpuscular theory, 1, 2, 8, 22
    diffraction, 3
    field, 8, 31
    frequency, 10
    grainy, 1, 12
    intensity, 10
    intensity distribution, 6, 7
    intensity, time averaged, 6, 7
    nature of, 22
    particle theory, 10
    photon theory, 14, 16
    quanta, 10
    quantum theory, 10
    speed, 4, 7, 18, 290
    transmitted, 3
    wave length, 7
    wave theory, 1, 2, 9, 15-17, 22
    wave vector, 4
linear
    algebra, 90-94
    combination, 92
    independence, 91
    operator, 95
local
    hidden variable theory, 185, 186, 188-190
    realism, 183-185, 190
logic
    circuit, 197, 209-212
    gates, 203-204
    network, 197
    operations, 203

Lorentz, Hendrik Antoon
    function, 73, 288

Mach, Ernst and Zehnder,
    Ludwig Louis Albert
    interferometer, 155-156
Mach-Zehnder interferometer
    classical mode, 156
    single-photon mode, 156
magnetic
    field, 109
    moment, 109
mapping
    bijective, 172
    complete, 172
    incomplete, 172, 174
    injective, 172
    redundant, 172
    surjective, 172
matching condition
    defined, 48
Mathieu, Emile Leonard
    equation, 283
    functions, 283
matter wave
    angular frequency, 33
    equation, 44
    momentum, 32
    wave number, 33
Maxwell, James Clerk
    electromagnetic theory of
        light, 2, 6, 9
mean value, defined, 50
measurement
    active role in quantum mechanics, 133, 179
    chain, 133
    classical, 132, 146
    continuous, 143
    destructive, 131, 132
    and information, 134
    and instantaneous response, 179
    as interface, 126
    and "more is different", 126
    nondestructive, 130, 132
    as nondeterministic collapse, 146
    passive role in classical mechanics, 133
    of position, 50
    position-sensitive, 154
    and quantum computing, 179
    and random number generator, 134
    and random results, 131, 133
    and reality, 133
    repeated, 130, 131
    and sharp value, 131
    and simultaneously sharp values, 132
    and "spooky action at a distance", 179
    three steps of, 129
Mermin, Norman David
    genes, 134, 192-194
    reality machine, 191-194
message
    encrypted, 253
    secret, 251
    secure, 251
    superluminal, 180
Michelson, Albert Abraham
    interferometer, 159-160, 161-165
micromotion
    amplitude, 284
    and damping, 286
    defined, 284
    oscillation period, 285

Miller, Gary L.
    algorithm, 268-270
    algorithm and cracking RSA, 274
    mathematical background, 269-270
Millikan, Robert Andrews
    Nobel Prize, 11
    and photoelectric effect, 11
miniaturization limit, 198
mirror
    half-silvered, 13
modular algebra
    formula, 257
molecular
    dipoles, 3
    oscillators, 3
momentum
    conservation, 183
    eigenfunctions, 72
    and element of reality, 176
    expectation value, 52
    kick, 289
    linear, 183
    local, 52
    operator, 52
    representation, defined, 94
    variance of, 52
Moore, Gordon Earle
    law, 198

NAND gate
    defined, 204
    universality, 210
Newton, Isaac
    corpuscular theory, 1-2, 10, 22
    equations of motion, 283
    point particle, 32
NMR quantum computer, 308
no-broadcasting theorem, 217

no-cloning theorem, 128, 138-140
nonlocality, 180, 184, 190, 194
nonneutral plasma, 286, 300
NOR gate, defined, 204
normalization
    constant, 66
    of continuum wave functions, 73-74
    integral, 63, 69
    of square-normalizable wave functions, 73
    of the wave function, 45
NOT gate, defined, 203

observables
    and assignment of operators, 51
    compatible, 132
    eigenvalue equation, 53
    as Hermitian operators, 62
    incompatible, 132
    as measurable quantity, 128
    and sharp eigenvalues, 131
    and simultaneously sharp values, 132
observation
    angle, 5
    point, 4, 5, 7, 8, 9
observer
    defined, 125
    as part of the system, 127
one-bit gates, 203
one-qubit gates, 205
operator
    analogue, 54
    of angular momentum, 54, 59, 102
    commuting, 56
    defined, 51

of energy, 55
equivalence, defined, 55
expectation value, 97
functions, 102
Hermitian, 60-62
identity, 56, 57
linear, 95
matrix representation of, 95
of momentum, 51-52
non-commuting, 56
of position, 53
of potential energy, 102
in quantum mechanics, 95
of spin, defined, 106-108
time dependent, defined, 102
of time evolution, defined, 103
time independent, defined, 102
in vector algebra, 95
optical
    path length, 155-156
    Zeno effect, 161-162
order
    defined, 268
    even, 268
    odd, 268, 269
OR gate, defined, 204
orthonormal
    defined, 68
oscillation
    fast, 9
    slow, 9
oscillator
    frequency, 68, 285
    length, defined, 68
    potential, 68, 285
    restoring force, 285, 291

pair
    annihilation, 19
    production, 19
pangram, 254
parallel processing, 210, 226
parity
    even, 122
    odd, 122
path
    information, 39-40
    integral, 35
    and stages, 36
Pauli, Wolfgang Ernst
    matrices, 108
permittivity, 7
phase
    convention, 108
    rotation gate, 206, 274
phenomenon
    collective, 26
    emergent, 26
phonon, 282, 294-296
photocurrent, 10
photoelectric effect, 1, 9, 10-13,
photoelectron, 10-12
photographic emulsion
    density of dots, 27-28
photon
    charge, 19
    coincidence experiment, 13
    confined, 305
    defined, 1
    diffraction experiment, 26-28
    diffraction pattern, 28, 29
    existence of, 1, 2, 13-18
    gravitational energy, 20
    gravitational red shift, 20
    as point particle, 10, 14, 15, 18, 20-22, 27-29
    probability density, 28
    properties of, 18-21
    rest mass, 19
    speed, 18, 22

INDEX 325

spin, 21
splitting, 15
spontaneous, 289, 292
as wave, 28
and wave-particle duality, 28
photonic nanograins
    certification, 154, 156-159
    defined, 154
    dud, 154, 156, 157
    live, 154, 156-158
π-pulse, 114
$\pi/2$-pulse, 114
plaintext, 251-253, 261, 262
Planck, Max
    constant, 2
    constant, defined, 10, 18
plane wave, 32-33
point particle, defined, 25
polarization
    angle, 21
    defined, 21
    degree-of-freedom, 21
    direction, 21, 161
    filter, 21, 161
    horizontal, 21, 161-162
    linear, 21
    rotator, defined, 161
    vertical, 21, 161-162
polarizing beam splitter, 163
position
    operator, defined, 53
    representation, defined, 94
positron, 19
potential
    complex, 46
    delta-function, 86
    discontinuous, 48, 49
    effective, 285
    electric, 10, 283
    energy, 44
    finite square well, 71
    harmonic-oscillator, 68

    infinite-square-well, 64
    smooth, 48
    time-dependent, 55, 62
    time-independent, 62
prime factor, 258, 268, 275
prime number
    counting function, 263
    theorem, 263
private-key cryptosystems, 252-254, 265, 306
probability
    current, 46, 48
    density, 28, 45
    distribution, 30, 31, 32, 39
product form
    of the quantum Fourier transform, 273-274
product state, 225, 273, 275
public-key cryptosystems, 252, 254-259, 265, 306

QNOT gate, defined, 205
quantum
    algorithm, 221, 223, 226, 227, 239, 270
    barrier, 198
    bus, 293, 294, 295
    circuit, 270, 305
    communication, 282, 303, 305
    computing, 223-250, 304, 307-309
    copying machine, 138
    cryptography, 304
    effect, 105
    electrodynamics, 109
    encryption device, 304
    engineering, 304
    entanglement, 180, 217, 305, 306
    experiment, stages, 125

Fourier transform, 270-274, 276
gate, 204, 205-208
grid computing, 306
hardware, 281, 307
information technology, 304, 309
interference, 226
Internet, 304-306
machines, 304, 306
network, 304, 305
parallel processing, 221, 226
probability distribution, 36
protocol, 306
register, 201, 275
register loading, 210-212, 221, 235
repeater, 306
revolution, 304
rules, 34-37
search algorithm, 239
software, 224, 280, 281, 307
speed-up, 226-227
states, 91
subroutine, 270
superposition, 226
teleportation, 213-221, 304
transition, defined, 110
word, 201
world, 40-41
Zeno effect, 128, 141-145
quantum computer
adiabatic, 308-309
architectures, 281, 300, 307-308
vs. classical computer, 247
classical mode, 224
CNOT implementation, 294-296
and collapse, 276-278
and control, 114
and cryptosystems, 252
and measurement, 230, 236-237, 276, 304
and microprogramming, 300
network, 306
as a new computing paradigm, 223
and nuclear magnetic resonance (NMR), 308
outclassing classical computers, 307
vs. PC technology, 309
programming language, 307
random access memory, 305
and RSA, 252
and RSA code breaking, 259, 270, 274-279
and Shor's algorithm, 275
sketch of, 282
starting state, 240, 275
and two-level systems, 114
quantum information
age, 200, 303, 304
processing, 211
storage, 305
technology, 304, 309
transmission, 217, 305
swapping, 217
quantum key distribution (QKD)
BB84, 306
E91, 306
and endorsement by the Swiss government, 307
and long-distance transmission, 307
and reliability, 307
quantum mechanics
as complete theory of nature, 181
foundations, 304

machinery, 43-123, 303
nonclassical aspects, 304
three rules of, 34-37, 154
spook, 303-304
qubit
    and Bloch sphere, 201
    defined, 199-201
    implementation with calcium ions, 298
    implementation in two-level systems, 200
    possible implementations, 201

Rabi, Isidor Isaac
    flopping, 113
    frequency, defined, 112
radiation
    field, 4
    pressure, 22, 289-292
    source, 3, 4
radio-frequency
    field, 284
    quadrupole trap, 282
ray, defined, 25
reality
    discovering vs. creating, 133
    elements of, 172, 175
    local, 186
    model of, 174
    objective, 132, 170, 172, 195
    prediction of, 174
    preexisting, 182, 183
    simultaneous, 176-177
recoil
    energy, 292
    limit, 292
    momentum, 289
    temperature, 292
reflection
    amplitude, 82-84
    defined, 82
    probability, defined, 83
relatively prime, defined, 255
Renninger, Mauritius
    and interaction-free measurement, 151
    negative-result measurement, 150-151, 165
    thought experiment, 149-151
repeater, 306
representation
    of a state, 92
    of a vector, 92
resonance frequency
    defined, 110
    in two-level system, 287, 288
Rivest, Ronald Linn, Shamir, Adi, and Adleman, Leonard Max (RSA)
    cipher, 256, 260-261
    code breaking with quantum computers, 264, 265-266
    cryptosystem, 252, 254-259, 265
    decryption, 256, 260-261, 262
    encryption, 256, 260-261
    example, 256-258
    mathematical underpinning, 260-262
    modulus, 254
    preparation steps, 254-255
    private-key exponent, 255
    public key, 255
    public-key exponent, 255
    and Shor's algorithm, 274, 280
    vulnerability to quantum computer attacks, 306

328 INDEX

rotating wave approximation, 112
rotation
    angle, 244
    matrix, 244
row vector, defined, 93
RSA-1024, 265
rules of quantum mechanics
    Born, 35
    Composition, 36
    Feynman, 35

scalar product
    anti-linearity, 90-91
    of continuum wave functions, 72
    of data base states, 239
    definiteness, 90-91
    and dot notation, 93
    linearity, 90-91
    modulo 2, 234
    of ordinary vectors, defined, 4, 90
    and orthogonality, 67
    and orthonormality, 68
    properties of, 90-91
    of states, 91
    symmetry, 90-91
    of wave functions, defined, 59
scattering
    defined, 83
    potential, 122
Schrödinger, Erwin Rudolf Josef Alexander
    Nobel Prize, 45
    operator, defined, 102
Schrödinger equation
    defined, 44
    deterministic, 127, 180
    in Dirac notation, 96
    of the finite square well, 76

    one-dimensional, 63
    solution technique, 63-64
    standard form, 54
    stationary, 64
Schwarz, Hermann Amandus
    inequality, 101
screen
    opaque, 2
secular motion
    amplitude, 284
    and damping, 286
    trajectory, 284
seed, 268, 269, 277
seeing
    in the dark, 151, 152
    five steps of, 151
semi-prime,
    defined, 254
    factoring, 268, 274-275
Shor, Peter Williston
    algorithm, 274-279
    algorithm, example, 279
    classical post processing, 279
    and cracking RSA, 274
    and NMR quantum computer, 308
singlet state,
    defined, 183
    and Mermin's reality machine, 191
Sirius, 179, 180
slit
    double, 8-9
    left, 8, 34, 36
    parallel, 8
    region, 4, 5
    right, 8, 34, 37
    single, 3-8
spectral equation
    defined, 66
    in Dirac notation, 96

of the finite square well, 78
graphical solution, 79
of the harmonic oscillator, 70
of the infinite square well, 65, 66
spectrum
   continuous, 73, 82, 87
   defined, 66
   discrete, 70
   and measurement, 128
   non-degenerate, defined, 128
spin
   commutator identities, 106
   down, defined, 105
   of the electron, 105
   flip, 113
   non-classical property, 21
   operators, defined, 106-108
   of the photon, 21
   projection, 105
   up, defined, 105
square integrable, defined, 47
square hump
   and barrier penetration, 85
   potential, defined, 71
   reflection amplitude, 85
   transmission amplitude, 85
   and tunneling, 84
square normalizability
   absence thereof, 73
   and continuum wave functions, 72-73
state
   anti-symmetric, 122
   defined, 90
   empty, 91
   entangled, 118, 180, 207

   as fundamental concept, 175
   irreducibly random, 134
   random string, 133
   separable, 117, 207, 208
   symmetric, 122
substitution cipher, 253
summation formula
   for geometric sums, 6, 271
   proof, 271
superluminal
   information transmission, 180
   "message", 180
superposition
   principle, 35, 211, 226
   state, 132, 214, 235, 272, 275
   of waves, 4
switchable mirror, 162-163

target qubit, defined, 206
Taylor, Brook
   series expansion, 61, 102
teleportation
   across Danube river, 305
   and EPR correlations, 218
   heterogeneous, 305
   long distance, 305
   of material particles, 305
   of photons, 304-305
   protocol, 213-218
   real-world conditions, 305
   and the Uncertainty Principle, 218
   and unitary transformation, 216
test function, 55-56
theory
   bad, 173
   building, 173
   complete, 171, 173-175, 181
   and concepts, 172

good, 173-174
incomplete, 171, 174, 175
overcomplete, 171
physical, 172
quality, 171, 173
redundant, 171
time-evolution operator, defined, 103
totient, defined, 254
transmission
    amplitude, 38, 39, 82-84
    defined, 82
    probability, defined, 83
    resonance, 122
trap-door function, 266
truth table, defined, 204
tunneling
    defined, 82
    regime, 85
two-bit gates, 203-204
two-level system
    with continuous drive, 111-114
    control, 113, 114, 117
    defined, 110
    as enabling technology, 120-121
    with impulsive drive, 114-117
    and quantum computing, 117
    and qubits, 200
    Schrödinger equation, 111
    time-evolution operator, 116
two-particle
    state, 117
    system, 117
two-qubit gates, 206

uncertainty
    defined, 50
    principle, derived, 136-137
    product, 147
    relationship, 136
unit
    matrix, defined, 108
    operator, defined, 98-99
    vector, 48, 57, 114, 191, 289
unitarity
    defined, 103

variance
    defined, 50, 53
    of energy, 55
    of momentum, 52
    of observables, defined, 53
    of position, 53
vector
    addition, 90
    and basis, 91
    column, 92
    coordinates, 91
    defined, 90
    linearly independent, 91
    representation of, 92
    row, 93
    scalar multiplication, 90
    space, 93
vibrational
    energies, 293
    mode, 293
voltage
    critical, 10
    negative, 10
von Neumann, John
    axiom of measurement, 127, 146
    collapse of the wave function, 127, 129
    theory of measurement, 127, 128-134
    three steps of measurement, 129

watched pot effect, 145

wave
    defined, 25
    electromagnetic, 2
    equation, 44
    reflected, 82, 87
    transmitted, 82, 87
    vector, 4
wave function
    as amplitude, 44
    defined, 44
    entangled, 118, 180
    extraction of physical information, 45
    instantaneous collapse, 180
    normalization, 45
    spherical, 150
    square-normalizable, 73
    not square-normalizable, 73
wave number
    of finite square well, 76
    of infinite square well, 65
    of matter waves, 33
    quantization, 65

wave-particle duality
    defined, 26, 28, 42
    discussed, 40-41
    of electrons, 26, 37
    and interaction-free measurement, 165
    of objects in general, 2, 34
    of photons, 22, 37
    take-home message, 41, 42
which-way experiment
    defined, 38
    example of, 38-40

Young, Thomas
    double-slit experiment, 2-10, 22, 29-33
    interference fringes, 1, 2
    single slit, 8

Zeno, of Elea
    paradox, 144